CAMBRIDGE LIBRARY COLLECTION

Books of enduring scholarly value

Darwin

Two hundred years after his birth and 150 years after the publication of 'On the Origin of Species', Charles Darwin and his theories are still the focus of worldwide attention. This series offers not only works by Darwin, but also the writings of his mentors in Cambridge and elsewhere, and a survey of the impassioned scientific, philosophical and theological debates sparked by his 'dangerous idea'.

Darwinism

Alfred Russel Wallace (1823–1913) is regarded as the co-discoverer with Darwin of the theory of evolution. It was an essay which Wallace sent in 1858 to Darwin (whom he greatly admired and to whom he dedicated his most famous book, The Malay Archipelago) which impelled Darwin to publish an article on his own long-pondered theory simultaneously with that of Wallace. As a travelling naturalist and collector in the Far East and South America, Wallace already inclined towards the Lamarckian theory of transmutation of species, and his own researches convinced him of the reality of evolution. On the publication of On the Origin of Species, Wallace became one of its most prominent advocates, and Darwinism, published in 1889, supports the theory and counters many of the arguments put forward by scientists and others who opposed it.

Cambridge University Press has long been a pioneer in the reissuing of out-of-print titles from its own backlist, producing digital reprints of books that are still sought after by scholars and students but could not be reprinted economically using traditional technology. The Cambridge Library Collection extends this activity to a wider range of books which are still of importance to researchers and professionals, either for the source material they contain, or as landmarks in the history of their academic discipline.

Drawing from the world-renowned collections in the Cambridge University Library, and guided by the advice of experts in each subject area, Cambridge University Press is using state-of-the-art scanning machines in its own Printing House to capture the content of each book selected for inclusion. The files are processed to give a consistently clear, crisp image, and the books finished to the high quality standard for which the Press is recognised around the world. The latest print-on-demand technology ensures that the books will remain available indefinitely, and that orders for single or multiple copies can quickly be supplied.

The Cambridge Library Collection will bring back to life books of enduring scholarly value (including out-of-copyright works originally issued by other publishers) across a wide range of disciplines in the humanities and social sciences and in science and technology.

Darwinism

An Exposition of the Theory of Natural Selection, with Some of its Applications

ALFRED RUSSEL WALLACE

CAMBRIDGE UNIVERSITY PRESS

Cambridge, New York, Melbourne, Madrid, Cape Town, Singapore,
São Paolo, Delhi, Dubai, Tokyo

Published in the United States of America by Cambridge University Press, New York

www.cambridge.org
Information on this title: www.cambridge.org/9781108001328

This edition first published 1889
This digitally printed version 2009

ISBN 978-1-108-00132-8 Paperback

DARWINISM

DARWINISM

AN EXPOSITION OF THE

THEORY OF NATURAL SELECTION

WITH SOME OF ITS APPLICATIONS

BY

ALFRED RUSSEL WALLACE

LL.D., F.L.S., ETC.

WITH MAP AND ILLUSTRATIONS

London

MACMILLAN AND CO.

AND NEW YORK

1889

PREFACE

THE present work treats the problem of the Origin of Species on the same general lines as were adopted by Darwin; but from the standpoint reached after nearly thirty years of discussion, with an abundance of new facts and the advocacy of many new or old theories.

While not attempting to deal, even in outline, with the vast subject of evolution in general, an endeavour has been made to give such an account of the theory of Natural Selection as may enable any intelligent reader to obtain a clear conception of Darwin's work, and to understand something of the power and range of his great principle.

Darwin wrote for a generation which had not accepted evolution, and which poured contempt on those who upheld the derivation of species from species by any natural law of descent. He did his work so well that "descent with modification" is now universally accepted as the order of nature in the organic world; and the rising generation of naturalists can hardly realise the novelty of this idea, or that their fathers considered it a scientific heresy to be condemned rather than seriously discussed.

The objections now made to Darwin's theory apply, solely, to the particular means by which the change of species has been brought about, not to the fact of that change. The objectors seek to minimise the agency of natural selection and to subordinate it to laws of variation, of use and disuse, of intelligence, and of heredity. These views and objections are urged with much force and more confidence, and for the most part by the modern school of laboratory naturalists, to whom the peculiarities and distinctions of species, as such, their distribution and their affinities, have little interest as compared with the problems of histology and embryology, of physiology and morphology. Their work in these departments is of the greatest interest and of the highest importance, but it is not the kind of work which, by itself, enables one to form a sound judgment on the questions involved in the action of the law of natural selection. These rest mainly on the external and vital relations of species to species in a state of nature—on what has been well termed by Semper the "physiology of organisms," rather than on the anatomy or physiology of organs.

It has always been considered a weakness in Darwin's work that he based his theory, primarily, on the evidence of variation in domesticated animals and cultivated plants. I have endeavoured to secure a firm foundation for the theory in the variations of organisms in a state of nature; and as the exact amount and precise character of these variations is of paramount importance in the numerous problems that arise when we apply the theory to explain the facts of nature, I have endeavoured, by means of a series of diagrams, to exhibit to the eye the actual variations as they are found to exist in a sufficient number of species. By doing this, not only does the reader obtain a better and more precise idea of variation than can be given by any number of tabular statements or cases of extreme individual variation, but we obtain a basis of fact by which to test the statements and objections usually put forth on the subject of specific variability; and it will be found that, throughout the work, I have frequently to appeal to these diagrams and the facts they illustrate, just as Darwin was accustomed to appeal to the facts of variation among dogs and pigeons.

I have also made what appears to me an important change in the arrangement of the subject. Instead of treating first the comparatively difficult and unfamiliar details of variation, I commence with the Struggle for Existence, which is really the fundamental phenomenon on which natural selection depends, while the particular facts which illustrate it are comparatively familiar and very interesting. It has the further advantage that, after discussing variation and the effects of artificial selection, we proceed at once to explain how natural selection acts.

Among the subjects of novelty or interest discussed in this volume, and which have important bearings on the theory of natural selection, are : (1) A proof that all *specific* characters are (or once have been) either useful in themselves or correlated with useful characters (Chap. VI); (2) a proof that natural selection can, in certain cases, increase the sterility of crosses (Chap. VII); (3) a fuller discussion of the colour relations of animals, with additional facts and arguments on the origin of sexual differences of colour (Chaps. VIII–X); (4) an attempted solution of the difficulty presented by the occurrence of both very simple and very complex modes of securing the cross-fertilisation of plants (Chap. XI); (5) some fresh facts and arguments on the wind-carriage of seeds, and its bearing on the wide dispersal of many arctic and alpine plants (Chap. XII); (6) some new illustrations of the non-heredity of acquired characters, and a proof that the effects of use and disuse, even if inherited, must be overpowered by natural selection (Chap. XIV); and (7) a new argument as to the nature and origin of the moral and intellectual faculties of man (Chap. XV).

Although I maintain, and even enforce, my differences from some of Darwin's views, my whole work tends forcibly to illustrate the overwhelming importance of Natural Selection over all other agencies in the production of new species.

I thus take up Darwin's earlier position, from which he somewhat receded in the later editions of his works, on account of criticisms and objections which I have endeavoured to show are unsound. Even in rejecting that phase of sexual selection depending on female choice, I insist on the greater efficacy of natural selection. This is pre-eminently the Darwinian doctrine, and I therefore claim for my book the position of being the advocate of pure Darwinism.

I wish to express my obligation to Mr. Francis Darwin for lending me some of his father's unused notes, and to many other friends for facts or information, which have, I believe, been acknowledged either in the text or footnotes. Mr. James Sime has kindly read over the proofs and given me many useful suggestions; and I have to thank Professor Meldola, Mr. Hemsley, and Mr. E. B. Poulton for valuable notes or corrections in the later chapters in which their special subjects are touched upon.

GODALMING, *March* 1889.

CONTENTS

CHAPTER I

WHAT ARE "SPECIES" AND WHAT IS MEANT BY THEIR "ORIGIN"

CHAPTER II

THE STRUGGLE FOR EXISTENCE

CHAPTER III

THE VARIABILITY OF SPECIES IN A STATE OF NATURE

CHAPTER IV

VARIATION OF DOMESTICATED ANIMALS AND CULTIVATED PLANTS

CHAPTER V

NATURAL SELECTION BY VARIATION AND SURVIVAL OF THE FITTEST

CHAPTER VI

DIFFICULTIES AND OBJECTIONS

CHAPTER VII

ON THE INFERTILITY OF CROSSES BETWEEN DISTINCT SPECIES AND THE USUAL STERILITY OF THEIR HYBRID OFFSPRING

CHAPTER VIII

THE ORIGIN AND USES OF COLOUR IN ANIMALS

CHAPTER IX

WARNING COLORATION AND MIMICRY

CHAPTER X

COLOURS AND ORNAMENTS CHARACTERISTIC OF SEX

CHAPTER XI

THE SPECIAL COLOURS OF PLANTS: THEIR ORIGIN AND PURPOSE

CHAPTER XII

THE GEOGRAPHICAL DISTRIBUTION OF ORGANISMS

CHAPTER XIII

THE GEOLOGICAL EVIDENCES OF EVOLUTION

CHAPTER XIV

FUNDAMENTAL PROBLEMS IN RELATION TO VARIATION AND HEREDITY

CHAPTER XV

DARWINISM APPLIED TO MAN

LIST OF ILLUSTRATIONS

CHAPTER I

WHAT ARE "SPECIES," AND WHAT IS MEANT BY THEIR "ORIGIN"

Definition of species—Special creation—The early Transmutationists—Scientific opinion before Darwin—The problem before Darwin—The change of opinion effected by Darwin—The Darwinian theory—Proposed mode of treatment of the subject.

The title of Mr. Darwin's great work is—*On the Origin of Species by means of Natural Selection and the Preservation of Favoured Races in the Struggle for Life.* In order to appreciate fully the aim and object of this work, and the change which it has effected not only in natural history but in many other sciences, it is necessary to form a clear conception of the meaning of the term "species," to know what was the general belief regarding them at the time when Mr. Darwin's book first appeared, and to understand what he meant, and what was generally meant, by discovering their "origin." It is for want of this preliminary knowledge that the majority of educated persons who are not naturalists are so ready to accept the innumerable objections, criticisms, and difficulties of its opponents as proofs that the Darwinian theory is unsound, while it also renders them unable to appreciate, or even to comprehend, the vast change which that theory has effected in the whole mass of thought and opinion on the great question of evolution.

The term "species" was thus defined by the celebrated botanist De Candolle: "A species is a collection of all the individuals which resemble each other more than they resemble anything else, which can by mutual fecundation produce fertile individuals, and which reproduce themselves by generation, in such a manner that we may from analogy suppose them all to have sprung from one single individual." And the zoologist Swainson gives a somewhat similar definition: "A species, in the usual acceptation of the term, is an animal which, in a state of nature, is distinguished by certain peculiarities of form, size, colour, or other circumstances, from another animal. It propagates, 'after its kind,' individuals perfectly resembling the parent; its peculiarities, therefore, are permanent." [1]

To illustrate these definitions we will take two common English birds, the rook (Corvus frugilegus) and the crow (Corvus corone). These are distinct *species*, because, in the first place, they always differ from each other in certain slight peculiarities of structure, form, and habits, and, in the second place, because rooks always produce rooks, and crows produce crows, and they do not interbreed. It was therefore concluded that all the rooks in the world had descended from a single pair of rooks, and the crows in like manner from a single pair of crows, while it was considered impossible that crows could have descended from rooks or *vice versâ*. The "origin" of the first pair of each kind was a mystery. Similar remarks may be applied to our two common plants, the sweet violet (Viola odorata) and the dog violet (Viola canina). These also produce their like and never produce each other or intermingle, and they were therefore each supposed to have sprung from a single individual whose "origin" was unknown. But besides the crow and the rook there are about thirty other kinds of birds in various parts of the world, all so much like our species that they receive the common name of crows; and some of them differ less from each other than does our crow from our rook. These are all *species* of the genus Corvus, and were therefore believed to have been always as distinct as they are now, neither more nor less, and to have each descended from one pair of ancestral crows of the same identical species, which themselves had an unknown "origin." Of violets there are more than a hundred different kinds in various parts of the world, all differing very slightly from each other and forming distinct

[1] *Geography and Classification of Animals*, p. 350.

species of the genus Viola. But, as these also each produce their like and do not intermingle, it was believed that every one of them had always been as distinct from all the others as it is now, that all the individuals of each kind had descended from one ancestor, but that the "origin" of these hundred slightly differing ancestors was unknown. In the words of Sir John Herschel, quoted by Mr. Darwin, the origin of such species was "the mystery of mysteries."

The Early Transmutationists.

A few great naturalists, struck by the very slight difference between many of these species, and the numerous links that exist between the most different forms of animals and plants, and also observing that a great many species do vary considerably in their forms, colours, and habits, conceived the idea that they might be all produced one from the other. The most eminent of these writers was a great French naturalist, Lamarck, who published an elaborate work, the *Philosophie Zoologique*, in which he endeavoured to prove that all animals whatever are descended from other species of animals. He attributed the change of species chiefly to the effect of changes in the conditions of life—such as climate, food, etc.—and especially to the desires and efforts of the animals themselves to improve their condition, leading to a modification of form or size in certain parts, owing to the well-known physiological law that all organs are strengthened by constant use, while they are weakened or even completely lost by disuse. The arguments of Lamarck did not, however, satisfy naturalists, and though a few adopted the view that closely allied species had descended from each other, the general belief of the educated public was, that each species was a "special creation" quite independent of all others; while the great body of naturalists equally held, that the change from one species to another by any known law or cause was impossible, and that the "origin of species" was an unsolved and probably insoluble problem. The only other important work dealing with the question was the celebrated *Vestiges of Creation*, published anonymously, but now acknowledged to have been written by the late Robert Chambers. In this work the action of general laws was traced throughout the universe as a system of growth and development, and it was argued that the various species of animals and plants had been produced in orderly succession from each other by the action of unknown laws of development aided by the action of external conditions. Although this work had a considerable effect in influencing public opinion as to the extreme improbability of the doctrine of the independent "special creation" of each species, it had little effect upon naturalists, because it made no attempt to grapple with the problem in detail, or to show in any single case how the allied species of a genus could have arisen, and have preserved their numerous slight and apparently purposeless differences from each other. No clue whatever was afforded to a law which should produce from any one species one or more slightly differing but yet permanently distinct species, nor was any reason given why such slight yet constant differences should exist at all.

Scientific Opinion before Darwin.

In order to show how little effect these writers had upon the public mind, I will quote a few passages from the writings of Sir Charles Lyell, as representing the opinions of the most advanced thinkers in the period immediately preceding that of Darwin's work. When recapitulating the facts and arguments in favour of the invariability and permanence of species, he says: "The entire variation from the original type which any given kind of change can produce may usually be effected in a brief period of time, after which no further deviation can be obtained by continuing to alter the circumstances, though ever so gradually, indefinite divergence either in the way of improvement or deterioration being prevented, and the least possible excess beyond the defined limits being fatal to the existence of the individual." In another place he maintains that "varieties of some species may differ more than other species do from each other without shaking our confidence in the reality of species." He further adduces certain facts in geology as being, in his opinion, "fatal to the theory of progressive development," and he explains the fact that there are so often distinct species in countries of similar climate and vegetation by "special creations" in each country; and these conclusions were arrived at after a careful study of Lamarck's work, a full abstract of which is given in the earlier editions of the *Principles of Geology*.[1]

Professor Agassiz, one of the greatest naturalists of the last generation, went even further, and maintained not only that each species was specially created, but that it was created in the proportions and in the localities in which we now find it to exist. The following extract from his very instructive book on Lake Superior explains this view: "There are in animals peculiar adaptations which are characteristic of their species, and which cannot be supposed to have arisen from subordinate influences. Those which live in shoals cannot be supposed to have been created in single pairs. Those which are made to be the food of others cannot have been created in the same proportions as those which live upon them. Those which are everywhere found in innumerable specimens must have been introduced in numbers capable of maintaining their normal proportions to those which live isolated and are comparatively and constantly fewer. For we know that this harmony in the numerical proportions between animals is one of the great laws of nature. The circumstance that species occur within definite limits where no obstacles prevent their wider distribution leads to the further inference that these limits were assigned to them from the beginning, and so we should come to the final conclusion that the order which prevails throughout nature is intentional, that it is regulated by the limits marked out on the first day of creation, and that it has been maintained unchanged through ages with no other modifications than those which the higher intellectual powers of man enable him to impose on some few animals more closely connected with him."[2]

These opinions of some of the most eminent and influential writers of the pre-Darwinian age seem to us, now, either altogether obsolete or positively absurd; but they nevertheless exhibit the mental condition of even the most advanced section of scientific men on the problem of the

[1] These expressions occur in Chapter IX. of the earlier editions (to the ninth) of the *Principles of Geology*.

[2] L. Agassiz, *Lake Superior*, p. 377.

nature and origin of species. They render it clear that, notwithstanding the vast knowledge and ingenious reasoning of Lamarck, and the more general exposition of the subject by the author of the *Vestiges of Creation*, the first step had not been taken towards a satisfactory explanation of the derivation of any one species from any other. Such eminent naturalists as Geoffroy Saint Hilaire, Dean Herbert, Professor Grant, Von Buch, and some others, had expressed their belief that species arose as simple varieties, and that the species of each genus were all descended from a common ancestor; but none of them gave a clue as to the law or the method by which the change had been effected. This was still "the great mystery." As to the further question—how far this common descent could be carried; whether distinct families, such as crows and thrushes, could possibly have descended from each other; or, whether all birds, including such widely distinct types as wrens, eagles, ostriches, and ducks, could all be the modified descendants of a common ancestor; or, still further, whether mammalia, birds, reptiles, and fishes, could all have had a common origin;—these questions had hardly come up for discussion at all, for it was felt that, while the very first step along the road of "transmutation of species" (as it was then called) had not been made, it was quite useless to speculate as to how far it might be possible to travel in the same direction, or where the road would ultimately lead to.

The Problem before Darwin.

It is clear, then, that what was understood by the "origin" or the "transmutation" of species before Darwin's work appeared, was the comparatively simple question whether the allied species of each genus had or had not been derived from one another and, remotely, from some common ancestor, by the ordinary method of reproduction and by means of laws and conditions still in action and capable of being thoroughly investigated. If any naturalist had been asked at that day whether, supposing it to be clearly shown that all the different species of each genus had been derived from some one ancestral species, and that a full and complete explanation were to be given of how each minute difference in form, colour, or structure might have originated, and how the several peculiarities of habit and of geographical distribution might have been brought about—whether, if this were done, the "origin of species" would be discovered, the great mystery solved, he would undoubtedly have replied in the affirmative. He would probably have added that he never expected any such marvellous discovery to be made in his lifetime. But so much as this assuredly Mr. Darwin has done, not only in the opinion of his disciples and admirers, but by the admissions of those who doubt the completeness of his explanations. For almost all their objections and difficulties apply to those larger differences which separate genera, families, and orders from each other, not to those which separate one species from the species to which it is most nearly allied, and from the remaining species of the same genus. They adduce such difficulties as the first development of the eye, or of the milk-producing glands of the mammalia; the wonderful instincts of bees and of ants; the complex arrangements for the fertilisation of orchids, and numerous other points of structure or habit, as not being satisfactorily explained. But it is evident that these peculiarities had their origin at a very remote period of the earth's history, and no theory, however complete, can do more than afford a probable conjecture as to how they were produced. Our ignorance of the state of the earth's surface and of the conditions of life at those remote periods is very great; thousands of animals and plants must have existed of which we have no record; while we are usually without any information as to the habits and general life-history even of those of which we possess some fragmentary remains; so that the truest and most complete theory would not enable us to solve *all* the difficult problems which the whole course of the development of life upon our globe presents to us.

What we may expect a true theory to do is to enable us to comprehend and follow out in some detail those changes in the form, structure, and relations of animals and plants which are effected in short periods of time, geologically speaking, and which are now going on around us. We may expect it to explain satisfactorily most of the lesser and superficial differences which distinguish one species from another. We may expect it to throw light on the mutual relations of the animals and plants which live together in any one country, and to give some rational account of the phenomena presented by their distribution in different parts of the world. And, lastly, we may expect it to explain many difficulties and to harmonise many incongruities in the excessively complex affinities and relations of living things. All this the Darwinian theory undoubtedly does. It shows us how, by means of some of the most universal and ever-acting laws in nature, new species are necessarily produced, while the old species become extinct; and it enables us to understand how the continuous action of these laws during the long periods with which geology makes us acquainted is calculated to bring about those greater differences presented by the distinct genera, families, and orders into which all living things are classified by naturalists. The differences which these present are all of the same *nature* as those presented by the species of many large genera, but much greater in *amount;* and they can all be explained by the action of the same general laws and by the extinction of a larger or smaller number of intermediate species. Whether the distinctions between the higher groups termed Classes and Sub-kingdoms may be accounted for in the same way is a much more difficult question. The differences which separate the mammals, birds, reptiles, and fishes from each other, though vast, yet seem of the same nature as those which distinguish a mouse from an elephant or a swallow from a goose. But the vertebrate animals, the mollusca, and the insects, are so radically distinct in their whole organisation and in the very plan of their structure, that objectors may not unreasonably doubt whether they can all have been derived from a common ancestor by means of the very same laws as have sufficed for the differentiation of the various species of birds or of reptiles.

The Change of Opinion effected by Darwin.

The point I wish especially to urge is this. Before Darwin's work appeared, the great majority of naturalists, and almost without exception the whole literary and scientific world, held firmly to the belief that *species* were realities, and had not been derived from other species by any process accessible to us; the different species of crow and of violet were believed to have been always as distinct and separate as they are now, and to have originated by some totally unknown process so far removed from ordinary reproduction that it was usually spoken of as "special creation." There was, then, no question of the origin of families, orders, and classes, because the very first step of all, the "origin of species," was believed to be an insoluble problem. But now this is all changed. The whole scientific and literary world, even the whole educated public, accepts, as a matter of common knowledge, the origin of species from other allied species by the ordinary process of natural birth. The idea of special creation or any altogether exceptional mode of production is absolutely extinct! Yet more: this is held also to apply to many higher groups as well as to the species of a genus, and not even Mr. Darwin's severest critics venture to suggest that the primeval bird, reptile, or fish must have been "specially created." And this vast, this totally unprecedented change in public opinion has been the result of the work of one man, and was brought about in the short space of twenty years! This is the answer to those who continue to maintain that the "origin of species" is not yet discovered; that there are still doubts and difficulties; that there are divergencies of structure so great that we cannot understand how they had their beginning. We may admit all this, just as we may admit that there are enormous difficulties in the way of a complete comprehension of the origin and nature of all the parts of the solar system and of the stellar universe. But we claim for Darwin that he is the Newton of natural history, and that, just so surely as that the discovery and demonstration by Newton of the law of gravitation established order in place of chaos and laid a sure foundation for all future study of the starry heavens, so surely has Darwin, by his discovery of the law of natural selection and his demonstration of the great principle of the preservation of useful variations in the struggle for life, not only thrown a flood of light on the process of development of the whole organic world, but also established a firm foundation for all future study of nature.

In order to show the view Darwin took of his own work, and what it was that he alone claimed to have done, the concluding passage of the introduction to the *Origin of*

Species should be carefully considered. It is as follows: "Although much remains obscure, and will long remain obscure, I can entertain no doubt, after the most deliberate and dispassionate judgment of which I am capable, that the view which most naturalists until recently entertained and which I formerly entertained—namely, that each species has been independently created—is erroneous. I am fully convinced that species are not immutable; but that those belonging to what are called the same genera are lineal descendants of some other and generally extinct species, in the same manner as the acknowledged varieties of any one species are the descendants of that species. Furthermore, I am convinced that Natural Selection has been the most important, but not the exclusive, means of modification."

It should be especially noted that all which is here claimed is now almost universally admitted, while the criticisms of Darwin's works refer almost exclusively to those numerous questions which, as he himself says, "will long remain obscure."

The Darwinian Theory.

As it will be necessary, in the following chapters, to set forth a considerable body of facts in almost every department of natural history, in order to establish the fundamental propositions on which the theory of natural selection rests, I propose to give a preliminary statement of what the theory really is, in order that the reader may better appreciate the necessity for discussing so many details, and may thus feel a more enlightened interest in them. Many of the facts to be adduced are so novel and so curious that they are sure to be appreciated by every one who takes an interest in nature, but unless the need of them is clearly seen it may be thought that time is being wasted on mere curious details and strange facts which have little bearing on the question at issue.

The theory of natural selection rests on two main classes of facts which apply to all organised beings without exception, and which thus take rank as fundamental principles or laws. The first is, the power of rapid multiplication in a geometrical progression; the second, that the offspring always vary slightly from the parents, though generally very closely resembling them. From the first fact or law there follows, necessarily, a constant struggle for existence; because, while the offspring always exceed the parents in number, generally to an enormous extent, yet the total number of living organisms in the world does not, and cannot, increase year by year. Consequently every year, on the average, as many die as are born, plants as well as animals; and the majority die premature deaths. They kill each other in a thousand different ways; they starve each other by some consuming the food that others want; they are destroyed largely by the powers of nature—by cold and heat, by rain and storm, by flood and fire. There is thus a perpetual struggle among them which shall live and which shall die; and this struggle is tremendously severe, because so few can possibly remain alive—one in five, one in ten, often only one in a hundred or even one in a thousand.

Then comes the question, Why do some live rather than others? If all the individuals of each species were exactly alike in every respect, we could only say it is a matter of chance. But they are not alike. We find that they vary in many different ways. Some are stronger, some swifter, some hardier in constitution, some more cunning. An obscure colour may render concealment more easy for some, keener sight may enable others to discover prey or escape from an enemy better than their fellows. Among plants the smallest differences may be useful or the reverse. The earliest and strongest shoots may escape the slug; their greater vigour may enable them to flower and seed earlier in a wet autumn; plants best armed with spines or hairs may escape being devoured; those whose flowers are most conspicuous may be soonest fertilised by insects. We cannot doubt that, on the whole, any beneficial variations will give the possessors of it a greater probability of living through the tremendous ordeal they have to undergo. There may be something left to chance, but on the whole *the fittest will survive.*

Then we have another important fact to consider, the principle of heredity or transmission of variations. If we grow plants from seed or breed any kind of animals year after year, consuming or giving away all the increase we do not wish to keep just as they come to hand, our plants or animals will continue much the same; but if every year we carefully save the best seed to sow and the finest or brightest coloured animals to breed from, we shall soon find that an improvement will take place, and that the average quality of our stock will be raised. This is the way in which all our fine garden fruits and vegetables and flowers have been produced, as well as all our splendid breeds of domestic animals; and they have thus become in many cases so different from the wild races from which they originally sprang as to be hardly recognisable as the same. It is therefore proved that if any particular kind of variation is preserved and bred from, the variation itself goes on increasing in amount to an enormous extent; and the bearing of this on the question of the origin of species is most important. For if in each generation of a given animal or plant the fittest survive to continue the breed, then whatever may be the special peculiarity that causes "fitness" in the particular case, that peculiarity will go on increasing and strengthening *so long as it is useful to the species.* But the moment it has reached its maximum of usefulness, and some other quality or modification would help in the struggle, then the individuals which vary in the new direction will survive; and thus a species may be gradually modified, first in one direction, then in another, till it differs from the original parent form as much as the greyhound differs from any wild dog or the cauliflower from any wild plant. But animals or plants which thus differ in a state of nature are always classed as distinct species, and thus we see how, by the continuous survival of the fittest or the preservation of favoured races in the struggle for life, new species may be originated.

This self-acting process which, by means of a few easily demonstrated groups of facts, brings about change in the organic world, and keeps each species in harmony with the conditions of its existence, will appear to some persons so clear and simple as to need no further demonstration. But to the great majority of naturalists and men of science endless difficulties and objections arise, owing to the wonderful variety of animal and vegetable forms, and the intricate relations of the different species and groups of species with each other; and it was to answer as many of these objections as possible, and to show that the more we know of nature the more we find it to harmonise with the development hypothesis, that Darwin devoted the whole of his life to collecting facts and making experiments, the record of a portion of which he has given us in a series of twelve masterly volumes.

Proposed Mode of Treatment of the Subject.

It is evidently of the most vital importance to any theory that its foundations should be absolutely secure. It is therefore necessary to show, by a wide and comprehensive array of facts, that animals and plants *do* perpetually vary in the manner and to the amount requisite; and that this takes place in wild animals as well as in those which are domesticated. It is necessary also to prove that all organisms *do* tend to increase at the great rate alleged, and that this increase actually occurs, under favourable conditions. We have to prove, further, that variations of all kinds can be increased and accumulated by selection; and that the struggle for existence to the extent here indicated actually occurs in nature, and leads to the continued preservation of favourable variations.

These matters will be discussed in the four succeeding chapters, though in a somewhat different order—the struggle for existence and the power of rapid multiplication, which is its cause, occupying the first place, as comprising those facts which are the most fundamental and those which can be perfectly explained without any reference to the less generally understood facts of variation. These chapters will be followed by a discussion of certain difficulties, and of the vexed question of hybridity. Then will come a rather full account of the more important of the complex relations of organisms to each other and to the earth itself, which are either fully explained or greatly elucidated by the theory. The concluding chapter will treat of the origin of man and his relations to the lower animals.

CHAPTER II

THE STRUGGLE FOR EXISTENCE

Its importance—The struggle among plants—Among animals—Illustrative cases—Succession of trees in forests of Denmark—The struggle for existence on the Pampas—Increase of organisms in a geometrical ratio—Examples of great powers of increase of animals—Rapid increase and wide spread of plants—Great fertility not essential to rapid increase—Struggle between closely allied species most severe—The ethical aspect of the struggle for existence.

THERE is perhaps no phenomenon of nature that is at once so important, so universal, and so little understood, as the struggle for existence continually going on among all organised beings. To most persons nature appears calm, orderly, and peaceful. They see the birds singing in the trees, the insects hovering over the flowers, the squirrel climbing among the tree-tops, and all living things in the possession of health and vigour, and in the enjoyment of a sunny existence. But they do not see, and hardly ever think of, the means by which this beauty and harmony and enjoyment is brought about. They do not see the constant and daily search after food, the failure to obtain which means weakness or death; the constant effort to escape enemies; the ever-recurring struggle against the forces of nature. This daily and hourly struggle, this incessant warfare, is nevertheless the very means by which much of the beauty and harmony and enjoyment in nature is produced, and also affords one of the most important elements in bringing about the origin of species. We must, therefore, devote some time to the consideration of its various aspects and of the many curious phenomena to which it gives rise.

It is a matter of common observation that if weeds are allowed to grow unchecked in a garden they will soon destroy a number of the flowers. It is not so commonly known that if a garden is left to become altogether wild, the weeds that first take possession of it, often covering the whole surface of the ground with two or three different kinds, will themselves be supplanted by others, so that in a few years many of the original flowers and of the earliest weeds may alike have disappeared. This is one of the very simplest cases of the struggle for existence, resulting in the successive displacement of one set of species by another; but the exact causes of this displacement are by no means of such a simple nature. All the plants concerned may be perfectly hardy, all may grow freely from seed, yet when left alone for a number of years, each set is in turn driven out by a succeeding set, till at the end of a considerable period—a century or a few centuries perhaps—hardly one of the plants which first monopolised the ground would be found there.

Another phenomenon of an analogous kind is presented by the different behaviour of introduced wild plants or animals into countries apparently quite as well suited to them as those which they naturally inhabit. Agassiz, in his work on Lake Superior, states that the roadside weeds of the north-eastern United States, to the number of 130 species, are all European, the native weeds having disappeared westwards; and in New Zealand there are no less than 250 species of naturalised European plants, more than 100 species of which have spread widely over the country, often displacing the native vegetation. On the other hand, of the many hundreds of hardy plants which produce seed freely in our gardens, very few ever run wild, and hardly any have become common. Even attempts to naturalise suitable plants usually fail; for A. de Candolle states that several botanists of Paris, Geneva, and especially of Montpellier, have sown the seeds of many hundreds of species of hardy exotic plants in what appeared to be the most favourable situations, but that, in hardly a single case, has any one of them become naturalised.[1] Even a plant like the potato—so widely cultivated, so hardy, and so well adapted to spread by means of its many-eyed tubers—has not established itself in a wild state in any part of Europe. It would be thought that Australian plants would easily run

[1] *Géographie Botanique*, p. 798.

wild in New Zealand. But Sir Joseph Hooker informs us that the late Mr. Bidwell habitually scattered Australian seeds during his extensive travels in New Zealand, yet only two or three Australian plants appear to have established themselves in that country, and these only in cultivated or newly moved soil.

These few illustrations sufficiently show that all the plants of a country are, as De Candolle says, at war with each other, each one struggling to occupy ground at the expense of its neighbour. But, besides this direct competition, there is one not less powerful arising from the exposure of almost all plants to destruction by animals. The buds are destroyed by birds, the leaves by caterpillars, the seeds by weevils; some insects bore into the trunk, others burrow in the twigs and leaves; slugs devour the young seedlings and the tender shoots, wire-worms gnaw the roots. Herbivorous mammals devour many species bodily, while some uproot and devour the buried tubers.

In animals, it is the eggs or the very young that suffer most from their various enemies; in plants, the tender seedlings when they first appear above the ground. To illustrate this latter point Mr. Darwin cleared and dug a piece of ground three feet long and two feet wide, and then marked all the seedlings of weeds and other plants which came up, noting what became of them. The total number was 357, and out of these no less than 295 were destroyed by slugs and insects. The direct strife of plant with plant is almost equally fatal when the stronger are allowed to smother the weaker. When turf is mown or closely browsed by animals, a number of strong and weak plants live together, because none are allowed to grow much beyond the rest; but Mr. Darwin found that when the plants which compose such turf are allowed to grow up freely, the stronger kill the weaker. In a plot of turf three feet by four, twenty distinct species of plants were found to be growing, and no less than nine of these perished altogether when the other species were allowed to grow up to their full size.[1]

But besides having to protect themselves against competing plants and against destructive animals, there is a yet deadlier

[1] *The Origin of Species*, p. 53.

enemy in the forces of inorganic nature. Each species can sustain a certain amount of heat and cold, each requires a certain amount of moisture at the right season, each wants a proper amount of light or of direct sunshine, each needs certain elements in the soil; the failure of a due proportion in these inorganic conditions causes weakness, and thus leads to speedy death. The struggle for existence in plants is, therefore, threefold in character and infinite in complexity, and the result is seen in their curiously irregular distribution over the face of the earth. Not only has each country its distinct plants, but every valley, every hillside, almost every hedgerow, has a different set of plants from its adjacent valley, hillside, or hedgerow—if not always different in the actual species yet very different in comparative abundance, some which are rare in the one being common in the other. Hence it happens that slight changes of conditions often produce great changes in the flora of a country. Thus in 1740 and the two following years the larva of a moth (Phalæna graminis) committed such destruction in many of the meadows of Sweden that the grass was greatly diminished in quantity, and many plants which were before choked by the grass sprang up, and the ground became variegated with a multitude of different species of flowers. The introduction of goats into the island of St. Helena led to the entire destruction of the native forests, consisting of about a hundred distinct species of trees and shrubs, the young plants being devoured by the goats as fast as they grew up. The camel is a still greater enemy to woody vegetation than the goat, and Mr. Marsh believes that forests would soon cover considerable tracts of the Arabian and African deserts if the goat and the camel were removed from them.[1] Even in many parts of our own country the existence of trees is dependent on the absence of cattle. Mr. Darwin observed, on some extensive heaths near Farnham, in Surrey, a few clumps of old Scotch firs, but no young trees over hundreds of acres. Some portions of the heath had, however, been enclosed a few years before, and these enclosures were crowded with young fir-trees growing too close together for all to live; and these were not sown or planted, nothing having been done to the ground beyond enclosing it

[1] *The Earth as Modified by Human Action*, p. 51.

so as to keep out cattle. On ascertaining this, Mr. Darwin was so much surprised that he searched among the heather in the unenclosed parts, and there he found multitudes of little trees and seedlings which had been perpetually browsed down by the cattle. In one square yard, at a point about a hundred yards from one of the old clumps of firs, he counted thirty-two little trees, and one of them had twenty-six rings of growth, showing that it had for many years tried to raise its head above the stems of the heather and had failed. Yet this heath was very extensive and very barren, and, as Mr. Darwin remarks, no one would ever have imagined that cattle would have so closely and so effectually searched it for food.

In the case of animals, the competition and struggle are more obvious. The vegetation of a given district can only support a certain number of animals, and the different kinds of plant-eaters will compete together for it. They will also have insects for their competitors, and these insects will be kept down by birds, which will thus assist the mammalia. But there will also be carnivora destroying the herbivora; while small rodents, like the lemming and some of the field-mice, often destroy so much vegetation as materially to affect the food of all the other groups of animals. Droughts, floods, severe winters, storms and hurricanes will injure these in various degrees, but no one species can be diminished in numbers without the effect being felt in various complex ways by all the rest. A few illustrations of this reciprocal action must be given.

Illustrative Cases of the Struggle for Life.

Sir Charles Lyell observes that if, by the attacks of seals or other marine foes, salmon are reduced in numbers, the consequence will be that otters, living far inland, will be deprived of food and will then destroy many young birds or quadrupeds, so that the increase of a marine animal may cause the destruction of many land animals hundreds of miles away. Mr. Darwin carefully observed the effects produced by planting a few hundred acres of Scotch fir, in Staffordshire, on part of a very extensive heath which had never been cultivated. After the planted portion was about twenty-five years old he observed that the change in the native vegetation was greater than is often seen in passing from one quite different soil to another. Besides a great change in the proportional numbers of the native heath-plants, twelve species which could not be found on the heath flourished in the plantations. The effect on the insect life must have been still greater, for six insectivorous birds which were very common in the plantations were not to be seen on the heath, which was, however, frequented by two or three different species of insectivorous birds. It would have required continued study for several years to determine all the differences in the organic life of the two areas, but the facts stated by Mr. Darwin are sufficient to show how great a change may be effected by the introduction of a single kind of tree and the keeping out of cattle.

The next case I will give in Mr. Darwin's own words: "In several parts of the world insects determine the existence of cattle. Perhaps Paraguay offers the most curious instance of this; for here neither cattle nor horses nor dogs have ever run wild, though they swarm southward and northward in a feral state; and Azara and Rengger have shown that this is caused by the greater numbers, in Paraguay, of a certain fly which lays its eggs in the navels of these animals when first born. The increase of these flies, numerous as they are, must be habitually checked by some means, probably by other parasitic insects. Hence, if certain insectivorous birds were to decrease in Paraguay, the parasitic insects would probably increase; and this would lessen the number of the navel-frequenting flies—then cattle and horses would become feral, and this would greatly alter (as indeed I have observed in parts of South America) the vegetation: this again would largely affect the insects, and this, as we have just seen in Staffordshire, the insectivorous birds, and so onward in ever-increasing circles of complexity. Not that under nature the relations will ever be as simple as this. Battle within battle must be continually recurring with varying success; and yet in the long run the forces are so nicely balanced, that the face of nature remains for a long time uniform, though assuredly the merest trifle would give the victory to one organic being over another."[1]

[1] *The Origin of Species*, p. 56.

Such cases as the above may perhaps be thought exceptional, but there is good reason to believe that they are by no means rare, but are illustrations of what is going on in every part of the world, only it is very difficult for us to trace out the complex reactions that are everywhere occurring. The general impression of the ordinary observer seems to be that wild animals and plants live peaceful lives and have few troubles, each being exactly suited to its place and surroundings, and therefore having no difficulty in maintaining itself. Before showing that this view is, everywhere and always, demonstrably untrue, we will consider one other case of the complex relations of distinct organisms adduced by Mr. Darwin, and often quoted for its striking and almost eccentric character. It is now well known that many flowers require to be fertilised by insects in order to produce seed, and this fertilisation can, in some cases, only be effected by one particular species of insect to which the flower has become specially adapted. Two of our common plants, the wild heart's-ease (Viola tricolor) and the red clover (Trifolium pratense), are thus fertilised by humble-bees almost exclusively, and if these insects are prevented from visiting the flowers, they produce either no seed at all or exceedingly few. Now it is known that field-mice destroy the combs and nests of humble-bees, and Colonel Newman, who has paid great attention to these insects, believes that more than two-thirds of all the humble-bees' nests in England are thus destroyed. But the number of mice depends a good deal on the number of cats; and the same observer says that near villages and towns he has found the nests of humble-bees more numerous than elsewhere, which he attributes to the number of cats that destroy the mice. Hence it follows, that the abundance of red clover and wild heart's-ease in a district will depend on a good supply of cats to kill the mice, which would otherwise destroy and keep down the humble-bees and prevent them from fertilising the flowers. A chain of connection has thus been found between such totally distinct organisms as flesh-eating mammalia and sweet-smelling flowers, the abundance or scarcity of the one closely corresponding to that of the other!

The following account of the struggle between trees in the forests of Denmark, from the researches of M. Hansten-

Blangsted, strikingly illustrates our subject.[1] The chief combatants are the beech and the birch, the former being everywhere successful in its invasions. Forests composed wholly of birch are now only found in sterile, sandy tracts; everywhere else the trees are mixed, and wherever the soil is favourable the beech rapidly drives out the birch. The latter loses its branches at the touch of the beech, and devotes all its strength to the upper part where it towers above the beech. It may live long in this way, but it succumbs ultimately in the fight—of old age if of nothing else, for the life of the birch in Denmark is shorter than that of the beech. The writer believes that light (or rather shade) is the cause of the superiority of the latter, for it has a greater development of its branches than the birch, which is more open and thus allows the rays of the sun to pass through to the soil below, while the tufted, bushy top of the beech preserves a deep shade at its base. Hardly any young plants can grow under the beech except its own shoots; and while the beech can flourish under the shade of the birch, the latter dies immediately under the beech. The birch has only been saved from total extermination by the facts that it had possession of the Danish forests long before the beech ever reached the country, and that certain districts are unfavourable to the growth of the latter. But wherever the soil has been enriched by the decomposition of the leaves of the birch the battle begins. The birch still flourishes on the borders of lakes and other marshy places, where its enemy cannot exist. In the same way, in the forests of Zeeland, the fir forests are disappearing before the beech. Left to themselves, the firs are soon displaced by the beech. The struggle between the latter and the oak is longer and more stubborn, for the branches and foliage of the oak are thicker, and offer much resistance to the passage of light. The oak, also, has greater longevity; but, sooner or later, it too succumbs, because it cannot develop in the shadow of the beech. The earliest forests of Denmark were mainly composed of aspens, with which the birch was apparently associated; gradually the soil was raised, and the climate grew milder; then the fir came and formed large forests. This tree ruled for centuries, and then ceded the

[1] See *Nature*, vol. xxxi. p. 63.

first place to the holm-oak, which is now giving way to the beech. Aspen, birch, fir, oak, and beech appear to be the steps in the struggle for the survival of the fittest among the forest-trees of Denmark.

It may be added that in the time of the Romans the beech was the principal forest-tree of Denmark as it is now, while in the much earlier bronze age, represented by the later remains found in the peat bogs, there were no beech-trees, or very few, the oak being the prevailing tree, while in the still earlier stone period the fir was the most abundant. The beech is a tree essentially of the temperate zone, having its northern limit considerably southward of the oak, fir, birch, or aspen, and its entrance into Denmark was no doubt due to the amelioration of the climate after the glacial epoch had entirely passed away. We thus see how changes of climate, which are continually occurring owing either to cosmical or geographical causes, may initiate a struggle among plants which may continue for thousands of years, and which must profoundly modify the relations of the animal world, since the very existence of innumerable insects, and even of many birds and mammals, is dependent more or less completely on certain species of plants.

The Struggle for Existence on the Pampas.

Another illustration of the struggle for existence, in which both plants and animals are implicated, is afforded by the pampas of the southern part of South America. The absence of trees from these vast plains has been imputed by Mr. Darwin to the supposed inability of the tropical and sub-tropical forms of South America to thrive on them, and there being no other source from which they could obtain a supply; and that explanation was adopted by such eminent botanists as Mr. Ball and Professor Asa Gray. This explanation has always seemed to me unsatisfactory, because there are ample forests both in the temperate regions of the Andes and on the whole west coast down to Terra del Fuego; and it is inconsistent with what we know of the rapid variation and adaptation of species to new conditions. What seems a more satisfactory explanation has been given by Mr. Edwin Clark, a civil engineer, who resided nearly two years in the country and paid much attention to its natural history. He says: "The peculiar characteristics of these vast level plains which descend from the Andes to the great river basin in unbroken monotony, are the absence of rivers or water-storage, and the periodical occurrence of droughts, or 'siccos,' in the summer months. These conditions determine the singular character both of its flora and fauna.

"The soil is naturally fertile and favourable for the growth of trees, and they grow luxuriantly wherever they are protected. The eucalyptus is covering large tracts wherever it is enclosed, and willows, poplars, and the fig surround every estancia when fenced in.

"The open plains are covered with droves of horses and cattle, and overrun by numberless wild rodents, the original tenants of the pampas. During the long periods of drought, which are so great a scourge to the country, these animals are starved by thousands, destroying, in their efforts to live, every vestige of vegetation. In one of these 'siccos,' at the time of my visit, no less than 50,000 head of oxen and sheep and horses perished from starvation and thirst, after tearing deep out of the soil every trace of vegetation, including the wiry roots of the pampas-grass. Under such circumstances the existence of an unprotected tree is impossible. The only plants that hold their own, in addition to the indestructible thistles, grasses, and clover, are a little herbaceous oxalis, producing viviparous buds of extraordinary vitality, a few poisonous species, such as the hemlock, and a few tough, thorny dwarf-acacias and wiry rushes, which even a starving rat refuses.

"Although the cattle are a modern introduction, the numberless indigenous rodents must always have effectually prevented the introduction of any other species of plants; large tracts are still honeycombed by the ubiquitous biscacho, a gigantic rabbit; and numerous other rodents still exist, including rats and mice, pampas-hares, and the great nutria and carpincho (capybara) on the river banks."[1]

Mr. Clark further remarks on the desperate struggle for existence which characterises the bordering fertile zones, where rivers and marshy plains permit a more luxuriant and varied vegetable and animal life. After describing how the

[1] *A Visit to South America*, 1878; also *Nature*, vol. xxxi. pp. 263-339.

river sometimes rose 30 feet in eight hours, doing immense destruction, and the abundance of the larger carnivora and large reptiles on its banks, he goes on: "But it was among the flora that the principle of natural selection was most prominently displayed. In such a district—overrun with rodents and escaped cattle, subject to floods that carried away whole islands of botany, and especially to droughts that dried up the lakes and almost the river itself—no ordinary plant could live, even on this rich and watered alluvial debris. The only plants that escaped the cattle were such as were either poisonous, or thorny, or resinous, or indestructibly tough. Hence we had only a great development of solanums, talas, acacias, euphorbias, and laurels. The buttercup is replaced by the little poisonous yellow oxalis with its viviparous buds; the passion-flowers, asclepiads, bignonias, convolvuluses, and climbing leguminous plants escape both floods and cattle by climbing the highest trees and towering overhead in a flood of bloom. The ground plants are the portulacas, turneras, and œnotheras, bitter and ephemeral, on the bare rock, and almost independent of any other moisture than the heavy dews. The pontederias, alismas, and plantago, with grasses and sedges, derive protection from the deep and brilliant pools; and though at first sight the 'monte' doubtless impresses the traveller as a scene of the wildest confusion and ruin, yet, on closer examination, we found it far more remarkable as a manifestation of harmony and law, and a striking example of the marvellous power which plants, like animals, possess, of adapting themselves to the local peculiarities of their habitat, whether in the fertile shades of the luxuriant 'monte' or on the arid, parched-up plains of the treeless pampas."

A curious example of the struggle between plants has been communicated to me by Mr. John Ennis, a resident in New Zealand. The English water-cress grows so luxuriantly in that country as to completely choke up the rivers, sometimes leading to disastrous floods, and necessitating great outlay to keep the stream open. But a natural remedy has now been found in planting willows on the banks. The roots of these trees penetrate the bed of the stream in every direction, and the water-cress, unable to obtain the requisite amount of nourishment, gradually disappears.

Increase of Organisms in a Geometrical Ratio.

The facts which have now been adduced, sufficiently prove that there is a continual competition, and struggle, and war going on in nature, and that each species of animal and plant affects many others in complex and often unexpected ways. We will now proceed to show the fundamental cause of this struggle, and to prove that it is ever acting over the whole field of nature, and that no single species of animal or plant can possibly escape from it. This results from the fact of the rapid increase, in a geometrical ratio, of all the species of animals and plants. In the lower orders this increase is especially rapid, a single flesh-fly (Musca carnaria) producing 20,000 larvæ, and these growing so quickly that they reach their full size in five days; hence the great Swedish naturalist, Linnæus, asserted that a dead horse would be devoured by three of these flies as quickly as by a lion. Each of these larvæ remains in the pupa state about five or six days, so that each parent fly may be increased ten thousand-fold in a fortnight. Supposing they went on increasing at this rate during only three months of summer, there would result one hundred millions of millions of millions for each fly at the commencement of summer,—a number greater probably than exists at any one time in the whole world. And this is only one species, while there are thousands of other species increasing also at an enormous rate; so that, if they were unchecked, the whole atmosphere would be dense with flies, and all animal food and much of animal life would be destroyed by them. To prevent this tremendous increase there must be incessant war against these insects, by insectivorous birds and reptiles as well as by other insects, in the larva as well as in the perfect state, by the action of the elements in the form of rain, hail, or drought, and by other unknown causes; yet we see nothing of this ever-present war, though by its means alone, perhaps, we are saved from famine and pestilence.

Let us now consider a less extreme and more familiar case. We possess a considerable number of birds which, like the redbreast, sparrow, the four common titmice, the thrush, and the blackbird, stay with us all the year round. These lay on an average six eggs, but, as several of them have two or more broods a year, ten will be below the average of the year's increase. Such birds as these often live from fifteen to twenty years in confinement, and we cannot suppose them to live shorter lives in a state of nature, if unmolested; but to avoid possible exaggeration we will take only ten years as the average duration of their lives. Now, if we start with a single pair, and these are allowed to live and breed, unmolested, till they die at the end of ten years,—as they might do if turned loose into a good-sized island with ample vegetable and insect food, but no other competing or destructive birds or quadrupeds—their numbers would amount to more than twenty millions. But we know very well that our bird population is no greater, on the average, now than it was ten years ago. Year by year it may fluctuate a little according as the winters are more or less severe, or from other causes, but on the whole there is no increase. What, then, becomes of the enormous surplus population annually produced? It is evident they must all die or be killed, somehow; and as the increase is, on the average, about five to one, it follows that, if the average number of birds of all kinds in our islands is taken at ten millions—and this is probably far under the mark—then about fifty millions of birds, including eggs as possible birds, must annually die or be destroyed. Yet we see nothing, or almost nothing, of this tremendous slaughter of the innocents going on all around us. In severe winters a few birds are found dead, and a few feathers or mangled remains show us where a wood-pigeon or some other bird has been destroyed by a hawk, but no one would imagine that five times as many birds as the total number in the country in early spring die every year. No doubt a considerable proportion of these do not die here but during or after migration to other countries, but others which are bred in distant countries come here, and thus balance the account. Again, as the average number of young produced is four or five times that of the parents, we ought to have at least five times as many birds in the country at the end of summer as at the beginning, and there is certainly no such enormous disproportion as this. The fact is, that the destruction commences, and is probably most severe, with nestling birds, which are often killed by heavy rains or blown away by severe storms, or left to die of hunger if either of the parents is killed; while they offer a defenceless prey to jackdaws, jays, and magpies, and not a few are ejected from their nests by their foster-brothers the cuckoos. As soon as they are fledged and begin to leave the nest great numbers are destroyed by buzzards, sparrow-hawks, and shrikes. Of those which migrate in autumn a considerable proportion are probably lost at sea or otherwise destroyed before they reach a place of safety; while those which remain with us are greatly thinned by cold and starvation during severe winters. Exactly the same thing goes on with every species of wild animal and plant from the lowest to the highest. All breed at such a rate, that in a few years the progeny of any one species would, if allowed to increase unchecked, alone monopolise the land; but all alike are kept within bounds by various destructive agencies, so that, though the numbers of each may fluctuate, they can never permanently increase except at the expense of some others, which must proportionately decrease.

Cases showing the Great Powers of Increase of Animals.

As the facts now stated are the very foundation of the theory we are considering, and the enormous increase and perpetual destruction continually going on require to be kept ever present in the mind, some direct evidence of actual cases of increase must be adduced. That even the larger animals, which breed comparatively slowly, increase enormously when placed under favourable conditions in new countries, is shown by the rapid spread of cattle and horses in America. Columbus, in his second voyage, left a few black cattle at St. Domingo, and these ran wild and increased so much that, twenty-seven years afterwards, herds of from 4000 to 8000 head were not uncommon. Cattle were afterwards taken from this island to Mexico and to other parts of America, and in 1587, sixty-five years after the conquest of Mexico, the Spaniards exported 64,350 hides from that country and 35,444 from St. Domingo, an indication of the vast numbers of these animals which must then have existed there, since those captured and killed could have been only a small portion of the whole. In the pampas of Buenos Ayres there were, at the end of the last century, about twelve million cows and three million horses, besides great numbers in all other parts of America where open pastures offered suitable conditions. Asses, about fifty years after their introduction, ran wild and multiplied so amazingly in Quito, that the Spanish traveller Ulloa describes them as being a nuisance. They grazed together in great herds, defending themselves with their mouths, and if a horse strayed among them they all fell upon him and did not cease biting and kicking till they left him dead. Hogs were turned out in St. Domingo by Columbus in 1493, and the Spaniards took them to other places where they settled, the result being, that in about half a century these animals were found in great numbers over a large part of America, from 25° north to 40° south latitude. More recently, in New Zealand, pigs have multiplied so greatly in a wild state as to be a serious nuisance and injury to agriculture. To give some idea of their numbers, it is stated that in the province of Nelson there were killed in twenty months 25,000 wild pigs.[1] Now, in the case of all these animals, we know that in their native countries, and even in America at the present time, they do not increase at all in numbers; therefore the whole normal increase must be kept down, year by year, by natural or artificial means of destruction.

Rapid Increase and Wide Spread of Plants.

In the case of plants, the power of increase is even greater and its effects more distinctly visible. Hundreds of square miles of the plains of La Plata are now covered with two or three species of European thistle, often to the exclusion of almost every other plant; but in the native countries of these thistles they occupy, except in cultivated or waste ground, a very subordinate part in the vegetation. Some American plants, like the cotton-weed (Asclepias curassavica), have now become common weeds over a large portion of the tropics. White clover (Trifolium repens) spreads over all the temperate regions of the world, and in New Zealand is exterminating many native species, including even the native flax (Phormium

[1] Still more remarkable is the increase of rabbits both in New Zealand and Australia. No less than seven millions of rabbit-skins have been exported from the former country in a single year, their value being £67,000. In both countries, sheep-runs have been greatly deteriorated in value by the abundance of rabbits, which destroy the herbage; and in some cases they have had to be abandoned altogether.

tenax), a large plant with iris-like leaves 5 or 6 feet high. Mr. W. L. Travers has paid much attention to the effects of introduced plants in New Zealand, and notes the following species as being especially remarkable. The common knot-grass (Polygonum aviculare) grows most luxuriantly, single plants covering a space 4 or 5 feet in diameter, and sending their roots 3 or 4 feet deep. A large sub-aquatic dock (Rumex obtusifolius) abounds in every river-bed, even far up among the mountains. The common sow-thistle (Sonchus oleraceus) grows all over the country up to an elevation of 6000 feet. The water-cress (Nasturtium officinale) grows with amazing vigour in many of the rivers, forming stems 12 feet long and $\frac{3}{4}$ inch in diameter, and completely choking them up. It cost £300 a year to keep the Avon at Christchurch free from it. The sorrel (Rumex acetosella) covers hundreds of acres with a sheet of red. It forms a dense mat, exterminating other plants, and preventing cultivation. It can, however, be itself exterminated by sowing the ground with red clover, which will also vanquish the Polygonum aviculare. The most noxious weed in New Zealand appears, however, to be the Hypochæris radicata, a coarse yellow-flowered composite not uncommon in our meadows and waste places. This has been introduced with grass seeds from England, and is very destructive. It is stated that excellent pasture was in three years destroyed by this weed, which absolutely displaced every other plant on the ground. It grows in every kind of soil, and is said even to drive out the white clover, which is usually so powerful in taking possession of the soil.

In Australia another composite plant, called there the Cape-weed (Cryptostemma calendulaceum), did much damage, and was noticed by Baron Von Hugel in 1833 as "an unexterminable weed"; but, after forty years' occupation, it was found to give way to the dense herbage formed by lucerne and choice grasses.

In Ceylon we are told by Mr. Thwaites, in his *Enumeration of Ceylon Plants*, that a plant introduced into the island less than fifty years ago is helping to alter the character of the vegetation up to an elevation of 3000 feet. This is the Lantana mixta, a verbenaceous plant introduced from the West Indies, which appears to have found in Ceylon a soil and climate exactly suited to it. It now covers thousands of acres with its dense masses of foliage, taking complete possession of land where cultivation has been neglected or abandoned, preventing the growth of any other plants, and even destroying small trees, the tops of which its subscandent stems are able to reach. The fruit of this plant is so acceptable to frugivorous birds of all kinds that, through their instrumentality, it is spreading rapidly, to the complete exclusion of the indigenous vegetation where it becomes established.

Great Fertility not essential to Rapid Increase.

The not uncommon circumstance of slow-breeding animals being very numerous, shows that it is usually the amount of destruction which an animal or plant is exposed to, not its rapid multiplication, that determines its numbers in any country. The passenger-pigeon (Ectopistes migratorius) is, or rather was, excessively abundant in a certain area in North America, and its enormous migrating flocks darkening the sky for hours have often been described; yet this bird lays only two eggs. The fulmar petrel is supposed to be one of the most numerous birds in the world, yet it lays only one egg. On the other hand the great shrike, the tree-creeper, the nut-hatch, the nut-cracker, the hoopoe, and many other birds, lay from four to six or seven eggs, and yet are never abundant. So in plants, the abundance of a species bears little or no relation to its seed-producing power. Some of the grasses and sedges, the wild hyacinth, and many buttercups occur in immense profusion over extensive areas, although each plant produces comparatively few seeds; while several species of bell-flowers, gentians, pinks, and mulleins, and even some of the compositæ, which produce an abundance of minute seeds, many of which are easily scattered by the wind, are yet rare species that never spread beyond a very limited area.

The above-mentioned passenger-pigeon affords such an excellent example of an enormous bird-population kept up by a comparatively slow rate of increase, and in spite of its complete helplessness and the great destruction which it suffers from its numerous enemies, that the following account of one of its breeding-places and migrations by the celebrated

American naturalist, Alexander Wilson, will be read with interest:—

"Not far from Shelbyville, in the State of Kentucky, about five years ago, there was one of these breeding-places, which stretched through the woods in nearly a north and south direction, was several miles in breadth, and was said to be upwards of 40 miles in extent. In this tract almost every tree was furnished with nests wherever the branches could accommodate them. The pigeons made their first appearance there about the 10th of April, and left it altogether with their young before the 25th of May. As soon as the young were fully grown and before they left the nests, numerous parties of the inhabitants from all parts of the adjacent country came with waggons, axes, beds, cooking utensils, many of them accompanied by the greater part of their families, and encamped for several days at this immense nursery. Several of them informed me that the noise was so great as to terrify their horses, and that it was difficult for one person to hear another without bawling in his ear. The ground was strewed with broken limbs of trees, eggs, and young squab pigeons, which had been precipitated from above, and on which herds of hogs were fattening. Hawks, buzzards, and eagles were sailing about in great numbers, and seizing the squabs from the nests at pleasure; while, from 20 feet upwards to the top of the trees, the view through the woods presented a perpetual tumult of crowding and fluttering multitudes of pigeons, their wings roaring like thunder, mingled with the frequent crash of falling timber; for now the axemen were at work cutting down those trees that seemed most crowded with nests, and contrived to fell them in such a manner, that in their descent they might bring down several others; by which means the falling of one large tree sometimes produced 200 squabs little inferior in size to the old birds, and almost one heap of fat. On some single trees upwards of a hundred nests were found, each containing one squab only; a circumstance in the history of the bird not generally known to naturalists.[1] It was dangerous to walk

[1] Later observers have proved that two eggs are laid and usually two young produced, but it may be that in most cases only one of these comes to maturity.

under these flying and fluttering millions, from the frequent fall of large branches, broken down by the weight of the multitudes above, and which in their descent often destroyed numbers of the birds themselves; while the clothes of those engaged in traversing the woods were completely covered with the excrements of the pigeons.

"These circumstances were related to me by many of the most respectable part of the community in that quarter, and were confirmed in part by what I myself witnessed. I passed for several miles through this same breeding-place, where every tree was spotted with nests, the remains of those above described. In many instances I counted upwards of ninety nests on a single tree; but the pigeons had abandoned this place for another, 60 or 80 miles off, towards Green River, where they were said at that time to be equally numerous. From the great numbers that were constantly passing over our heads to or from that quarter, I had no doubt of the truth of this statement. The mast had been chiefly consumed in Kentucky; and the pigeons, every morning a little before sunrise, set out for the Indiana territory, the nearest part of which was about sixty miles distant. Many of these returned before ten o'clock, and the great body generally appeared on their return a little after noon. I had left the public road to visit the remains of the breeding-place near Shelbyville, and was traversing the woods with my gun, on my way to Frankfort, when about ten o'clock the pigeons which I had observed flying the greater part of the morning northerly, began to return in such immense numbers as I never before had witnessed. Coming to an opening by the side of a creek, where I had a more uninterrupted view, I was astonished at their appearance: they were flying with great steadiness and rapidity, at a height beyond gunshot, in several strata deep, and so close together that, could shot have reached them, one discharge could not have failed to bring down several individuals. From right to left, as far as the eye could reach, the breadth of this vast procession extended, seeming everywhere equally crowded. Curious to determine how long this appearance would continue, I took out my watch to note the time, and sat down to observe them. It was then half-past one; I sat for more than an hour, but instead of a diminution of this prodigious procession, it seemed rather to increase, both in numbers and rapidity; and anxious to reach Frankfort before night, I rose and went on. About four o'clock in the afternoon I crossed Kentucky River, at the town of Frankfort, at which time the living torrent above my head seemed as numerous and as extensive as ever. Long after this I observed them in large bodies that continued to pass for six or eight minutes, and these again were followed by other detached bodies, all moving in the same south-east direction, till after six o'clock in the evening. The great breadth of front which this mighty multitude preserved would seem to intimate a corresponding breadth of their breeding-place, which, by several gentlemen who had lately passed through part of it, was stated to me at several miles."

From these various observations, Wilson calculated that the number of birds contained in the mass of pigeons which he saw on this occasion was at least two thousand millions, while this was only one of many similar aggregations known to exist in various parts of the United States. The picture here given of these defenceless birds, and their still more defenceless young, exposed to the attacks of numerous rapacious enemies, brings vividly before us one of the phases of the unceasing struggle for existence ever going on; but when we consider the slow rate of increase of these birds, and the enormous population they are nevertheless able to maintain, we must be convinced that in the case of the majority of birds which multiply far more rapidly, and yet are never able to attain such numbers, the struggle against their numerous enemies and against the adverse forces of nature must be even more severe or more continuous.

Struggle for Life between closely allied Animals and Plants often the most severe.

The struggle we have hitherto been considering has been mainly that between an animal or plant and its direct enemies, whether these enemies are other animals which devour it, or the forces of nature which destroy it. But there is another kind of struggle often going on at the same time between closely related species, which almost always terminates in the destruction of one of them. As an example of what is meant, the missel-thrush has increased in numbers in Scotland during the last thirty years, and this has caused a decrease in the numbers of the closely allied song-thrush in the same country. The black rat (Mus rattus) was the common rat of Europe till, in the beginning of the eighteenth century, the large brown rat (Mus decumanus) appeared on the Lower Volga, and thence spread more or less rapidly till it overran all Europe, and generally drove out the black rat, which in most parts is now comparatively rare or quite extinct. This invading rat has now been carried by commerce all over the world, and in New Zealand has completely extirpated a native rat, which the Maoris allege they brought with them from their home in the Pacific; and in the same country a native fly is being supplanted by the European house-fly. In Russia the small Asiatic cockroach has driven away a larger native species; and in Australia the imported hive-bee is exterminating the small stingless native bee.

The reason why this kind of struggle goes on is apparent if we consider that the allied species fill nearly the same place in the economy of nature. They require nearly the same kind of food, are exposed to the same enemies and the same dangers. Hence, if one has ever so slight an advantage over the other in procuring food or in avoiding danger, in its rapidity of multiplication or its tenacity of life, it will increase more rapidly, and by that very fact will cause the other to decrease and often become altogether extinct. In some cases, no doubt, there is actual war between the two, the stronger killing the weaker; but this is by no means necessary, and there may be cases in which the weaker species, physically, may prevail, by its power of more rapid multiplication, its better withstanding vicissitudes of climates, or its greater cunning in escaping the attacks of the common enemies. The same principle is seen at work in the fact that certain mountain varieties of sheep will starve out other mountain varieties, so that the two cannot be kept together. In plants the same thing occurs. If several distinct varieties of wheat are sown together, and the mixed seed resown, some of the varieties which best suit the soil and climate, or are naturally the most fertile, will beat the others and so yield more seed, and will consequently in a few years supplant the other varieties.

As an effect of this principle, we seldom find closely allied species of animals or plants living together, but often in distinct though adjacent districts where the conditions of life are somewhat different. Thus we may find cowslips (Primula veris) growing in a meadow, and primroses (P. vulgaris) in an adjoining wood, each in abundance, but not often intermingled. And for the same reason the old turf of a pasture or heath consists of a great variety of plants matted together, so much so that in a patch little more than a yard square Mr. Darwin found twenty distinct species, belonging to eighteen distinct genera and to eight natural orders, thus showing their extreme diversity of organisation. For the same reason a number of distinct grasses and clovers are sown in order to make a good lawn instead of any one species; and the quantity of hay produced has been found to be greater from a variety of very distinct grasses than from any one species of grass.

It may be thought that forests are an exception to this rule, since in the north-temperate and arctic regions we find extensive forests of pines or of oaks. But these are, after all, exceptional, and characterise those regions only where the climate is little favourable to forest vegetation. In the tropical and all the warm temperate parts of the earth, where there is a sufficient supply of moisture, the forests present the same variety of species as does the turf of our old pastures; and in the equatorial virgin forests there is so great a variety of forms, and they are so thoroughly intermingled, that the traveller often finds it difficult to discover a second specimen of any particular species which he has noticed. Even the forests of the temperate zones, in all favourable situations, exhibit a considerable variety of trees of distinct genera and families, and it is only when we approach the outskirts of forest vegetation, where either drought or winds or the severity of the winter is adverse to the existence of most trees, that we find extensive tracts monopolised by one or two species. Even Canada has more than sixty different forest trees, and the Eastern United States a hundred and fifty; Europe is rather poor, containing about eighty trees only; while the forests of Eastern Asia, Japan, and Manchuria are exceedingly rich, about a hundred and seventy species being already known. And in all these countries the trees grow intermingled, so that in every extensive forest we have a considerable variety, as may be seen in the few remnants of our primitive woods in some parts of Epping Forest and the New Forest.

Among animals the same law prevails, though, owing to their constant movements and power of concealment, it is not so readily observed. As illustrations we may refer to the wolf, ranging over Europe and Northern Asia, while the jackal inhabits Southern Asia and Northern Africa; the tree-porcupines, of which there are two closely allied species, one inhabiting the eastern, the other the western half of North America; the common hare (Lepus timidus) in Central and Southern Europe, while all Northern Europe is inhabited by the variable hare (Lepus variabilis); the common jay (Garrulus glandarius) inhabiting all Europe, while another species (Garrulus Brandti) is found all across Asia from the Urals to Japan; and many species of birds in the Eastern United States are replaced by closely allied species in the west. Of course there are also numbers of closely related species in the same country, but it will almost always be found that they frequent different stations and have somewhat different habits, and so do not come into direct competition with each other; just as closely allied plants may inhabit the same districts, when one prefers meadows the other woods, one a chalky soil the other sand, one a damp situation the other a dry one. With plants, fixed as they are to the earth, we easily note these peculiarities of station; but with wild animals, which we see only on rare occasions, it requires close and long-continued observation to detect the peculiarities in their mode of life which may prevent all direct competition between closely allied species dwelling in the same area.

The Ethical Aspect of the Struggle for Existence.

Our exposition of the phenomena presented by the struggle for existence may be fitly concluded by a few remarks on its ethical aspect. Now that the war of nature is better known, it has been dwelt upon by many writers as presenting so vast an amount of cruelty and pain as to be revolting to our instincts of humanity, while it has proved a stumbling-block in the way of those who would fain believe in an all-wise and benevolent ruler of the universe. Thus, a brilliant writer says: "Pain, grief, disease, and death, are these the inventions of a loving God? That no animal shall rise to excellence except by being fatal to the life of others, is this the law of a kind Creator? It is useless to say that pain has its benevolence, that massacre has its mercy. Why is it so ordained that bad should be the raw material of good? Pain is not the less pain because it is useful; murder is not less murder because it is conducive to development. Here is blood upon the hand still, and all the perfumes of Arabia will not sweeten it."[1]

Even so thoughtful a writer as Professor Huxley adopts similar views. In a recent article on "The Struggle for Existence" he speaks of the myriads of generations of herbivorous animals which "have been tormented and devoured by carnivores"; of the carnivores and herbivores alike "subject to all the miseries incidental to old age, disease, and over-multiplication"; and of the "more or less enduring suffering," which is the meed of both vanquished and victor. And he concludes that, since thousands of times a minute, were our ears sharp enough, we should hear sighs and groans of pain like those heard by Dante at the gate of hell, the world cannot be governed by what we call benevolence.[2]

Now there is, I think, good reason to believe that all this is greatly exaggerated; that the supposed "torments" and "miseries" of animals have little real existence, but are the reflection of the imagined sensations of cultivated men and women in similar circumstances; and that the amount of actual suffering caused by the struggle for existence among animals is altogether insignificant. Let us, therefore, endeavour to ascertain what are the real facts on which these tremendous accusations are founded.

In the first place, we must remember that animals are entirely spared the pain we suffer in the anticipation of death—a pain far greater, in most cases, than the reality. This leads, probably, to an almost perpetual enjoyment of their lives; since their constant watchfulness against danger, and even their actual flight from an enemy, will be the enjoyable

[1] Winwood Reade's *Martyrdom of Man*, p. 520.
[2] *Nineteenth Century*, February 1888, pp. 162, 163.

exercise of the powers and faculties they possess, unmixed with any serious dread. There is, in the next place, much evidence to show that violent deaths, if not too prolonged, are painless and easy; even in the case of man, whose nervous system is in all probability much more susceptible to pain than that of most animals. In all cases in which persons have escaped after being seized by a lion or tiger, they declare that they suffered little or no pain, physical or mental. A well-known instance is that of Livingstone, who thus describes his sensations when seized by a lion: "Starting and looking half round, I saw the lion just in the act of springing on me. I was upon a little height; he caught my shoulder as he sprang, and we both came to the ground below together. Growling horribly close to my ear, he shook me as a terrier-dog does a rat. The shock produced a stupor similar to that which seems to be felt by a mouse after the first shake of the cat. It causes a sort of dreaminess, *in which there was no sense of pain or feeling of terror*, though I was quite conscious of all that was happening. It was like what patients partially under the influence of chloroform describe, who see all the operation, but feel not the knife. This singular condition was not the result of any mental process. The shake annihilated fear, and allowed no sense of horror in looking round at the beast."

This absence of pain is not peculiar to those seized by wild beasts, but is equally produced by any accident which causes a general shock to the system. Mr. Whymper describes an accident to himself during one of his preliminary explorations of the Matterhorn, when he fell several hundred feet, bounding from rock to rock, till fortunately embedded in a snow-drift near the edge of a tremendous precipice. He declares that while falling and feeling blow after blow, he neither lost consciousness nor suffered pain, merely thinking, calmly, that a few more blows would finish him. We have therefore a right to conclude, that when death follows soon after any great shock it is as easy and painless a death as possible; and this is certainly what happens when an animal is seized by a beast of prey. For the enemy is one which hunts for food, not for pleasure or excitement; and it is doubtful whether any carnivorous animal in a state of nature begins to seek after prey till driven to do so by hunger. When an animal is caught, therefore, it is very soon devoured, and thus the first shock is followed by an almost painless death. Neither do those which die of cold or hunger suffer much. Cold is generally severest at night and has a tendency to produce sleep and painless extinction. Hunger, on the other hand, is hardly felt during periods of excitement, and when food is scarce the excitement of seeking for it is at its greatest. It is probable, also, that when hunger presses, most animals will devour anything to stay their hunger, and will die of gradual exhaustion and weakness not necessarily painful, if they do not fall an earlier prey to some enemy or to cold.[1]

Now let us consider what are the enjoyments of the lives of most animals. As a rule they come into existence at a time of year when food is most plentiful and the climate most suitable, that is in the spring of the temperate zone and at the commencement of the dry season in the tropics. They grow vigorously, being supplied with abundance of food; and when they reach maturity their lives are a continual round of healthy excitement and exercise, alternating with complete repose. The daily search for the daily food employs all their faculties and exercises every organ of their bodies, while this exercise leads to the satisfaction of all their physical needs. In our own case, we can give no more perfect definition of happiness, than this exercise and this satisfaction; and we must therefore conclude that animals, as a rule, enjoy all the happiness of which they are capable. And this normal state of happiness is not alloyed, as with us, by long periods—whole lives often—of poverty or ill-health, and of the unsatisfied longing for pleasures which others enjoy but to which we cannot attain. Illness, and what answers to poverty in animals—continued hunger—are quickly followed by unanticipated and almost painless extinction. Where we err is, in giving to animals feelings and emotions which they do not possess. To us the very sight of blood and of torn or mangled limbs is painful, while the idea of the suffering implied by it

[1] The Kestrel, which usually feeds on mice, birds, and frogs, sometimes stays its hunger with earthworms, as do some of the American buzzards. The Honey-buzzard sometimes eats not only earthworms and slugs, but even corn; and the Buteo borealis of North America, whose usual food is small mammals and birds, sometimes eats crayfish.

is heartrending. We have a horror of all violent and sudden death, because we think of the life full of promise cut short, of hopes and expectations unfulfilled, and of the grief of mourning relatives. But all this is quite out of place in the case of animals, for whom a violent and a sudden death is in every way the best. Thus the poet's picture of

> "Nature red in tooth and claw
> With ravine"

is a picture the evil of which is read into it by our imaginations, the reality being made up of full and happy lives, usually terminated by the quickest and least painful of deaths.

On the whole, then, we conclude that the popular idea of the struggle for existence entailing misery and pain on the animal world is the very reverse of the truth. What it really brings about, is, the maximum of life and of the enjoyment of life with the minimum of suffering and pain. Given the necessity of death and reproduction—and without these there could have been no progressive development of the organic world,—and it is difficult even to imagine a system by which a greater balance of happiness could have been secured. And this view was evidently that of Darwin himself, who thus concludes his chapter on the struggle for existence: "When we reflect on this struggle, we may console ourselves with the full belief that the war of nature is not incessant, that no fear is felt, that death is generally prompt, and that the vigorous, the healthy, and the happy survive and multiply."

CHAPTER III

THE VARIABILITY OF SPECIES IN A STATE OF NATURE

Importance of variability—Popular ideas regarding it—Variability of the lower animals—The variability of insects—Variation among lizards—Variation among birds—Diagrams of bird-variation—Number of varying individuals—Variation in the mammalia—Variation in internal organs—Variations in the skull—Variations in the habits of Animals—The Variability of plants—Species which vary little—Concluding remarks.

THE foundation of the Darwinian theory is the variability of species, and it is quite useless to attempt even to understand that theory, much less to appreciate the completeness of the proof of it, unless we first obtain a clear conception of the nature and extent of this variability. The most frequent and the most misleading of the objections to the efficacy of natural selection arise from ignorance of this subject, an ignorance shared by many naturalists, for it is only since Mr. Darwin has taught us their importance that varieties have been systematically collected and recorded; and even now very few collectors or students bestow upon them the attention they deserve. By the older naturalists, indeed, varieties—especially if numerous, small, and of frequent occurrence—were looked upon as an unmitigated nuisance, because they rendered it almost impossible to give precise definitions of species, then considered the chief end of systematic natural history. Hence it was the custom to describe what was supposed to be the "typical form" of species, and most collectors were satisfied if they possessed this typical form in their cabinets. Now, however, a collection is valued in proportion as it contains illustrative specimens of all the varieties that occur in each species, and in some cases these have been carefully described, so that we possess a considerable mass of information on the subject. Utilising this information we will now endeavour to give some idea of the nature and extent of variation in the species of animals and plants.

It is very commonly objected that the widespread and constant variability which is admitted to be a characteristic of domesticated animals and cultivated plants is largely due to the unnatural conditions of their existence, and that we have no proof of any corresponding amount of variation occurring in a state of nature. Wild animals and plants, it is said, are usually stable, and when variations occur these are alleged to be small in amount and to affect superficial characters only; or if larger and more important, to occur so rarely as not to afford any aid in the supposed formation of new species.

This objection, as will be shown, is utterly unfounded; but as it is one which goes to the very root of the problem, it is necessary to enter at some length into the various proofs of variation in a state of nature. This is the more necessary because the materials collected by Mr. Darwin bearing on this question have never been published, and comparatively few of them have been cited in *The Origin of Species;* while a considerable body of facts has been made known since the publication of the last edition of that work.

Variability of the Lower Animals.

Among the lowest and most ancient marine organisms are the Foraminifera, little masses of living jelly, apparently structureless, but which secrete beautiful shelly coverings, often perfectly symmetrical, as varied in form as those of the mollusca and far more complicated. These have been studied with great care by many eminent naturalists, and the late Dr. W. B. Carpenter in his great work—the *Introduction to the Study of the Foraminifera*—thus refers to their variability: "There is not a single species of plant or animal of which the range of variation has been studied by the collocation and comparison of so large a number of specimens as have passed under the review of Messrs. Williamson, Parker, Rupert Jones, and myself in our studies of the types of this group;" and he states as the result of this extensive comparison of specimens: "The range of variation is so great among the Foraminifera as to include not merely those differential characters which have been usually accounted *specific*, but also those upon which the greater part of the *genera* of this group have been founded, and even in some instances those of its *orders*."[1]

Coming now to a higher group—the Sea-Anemones—Mr. P. H. Gosse and other writers on these creatures often refer to variations in size, in the thickness and length of the tentacles, the form of the disc and of the mouth, and the character of surface of the column, while the colour varies enormously in a great number of the species. Similar variations occur in all the various groups of marine invertebrata, and in the great sub-kingdom of the mollusca they are especially numerous. Thus, Dr. S. P. Woodward states that many present a most perplexing amount of variation, resulting (as he supposes) from supply of food, variety of depth and of saltness of the water; but we know that many variations are quite independent of such causes, and we will now consider a few cases among the land-mollusca in which they have been more carefully studied.

In the small forest region of Oahu, one of the Sandwich Islands; there have been found about 175 species of land-shells represented by 700 or 800 varieties; and we are told by the Rev. J. T. Gulick, who studied them carefully, that "we frequently find a genus represented in several successive valleys by allied species, sometimes feeding on the same, sometimes on different plants. In every such case the valleys that are nearest to each other furnish the most nearly allied forms; *and a full set of the varieties of each species presents a minute gradation of forms between the more divergent types found in the more widely separated localities.*"

In most land-shells there is a considerable amount of variation in colour, markings, size, form, and texture or striation of the surface, even in specimens collected in the same locality. Thus, a French author has enumerated no less than 198 varieties of the common wood-snail (Helix nemoralis), while of the equally common garden-snail (Helix hortensis) ninety varieties have been described. Fresh-water shells are also

[1] *Foraminifera*, preface, p. x.

subject to great variation, so that there is much uncertainty as to the number of species; and variations are especially frequent in the Planorbidæ, which exhibit many eccentric deviations from the usual form of the species—deviations which must often affect the form of the living animal. In Mr. Ingersoll's Report on the Recent Mollusca of Colorado many of these extraordinary variations are referred to, and it is stated that a shell (Helisonia trivolvis) abundant in some small ponds and lakes, had scarcely two specimens alike, and many of them closely resembled other and altogether distinct species.[1]

The Variability of Insects.

Among Insects there is a large amount of variation, though very few entomologists devote themselves to its investigation. Our first examples will be taken from the late Mr. T. Vernon Wollaston's book, *On the Variation of Species*, and they must be considered as indications of very widespread though little noticed phenomena. He speaks of the curious little carabideous beetles of the genus Notiophilus as being "extremely unstable both in their sculpture and hue;" of the common Calathus mollis as having "the hind wings at one time ample, at another rudimentary, and at a third nearly obsolete;" and of the same irregularity as to the wings being characteristic of many Orthoptera and of the Homopterous Fulgoridæ. Mr. Westwood in his *Modern Classification of Insects* states that "the species of Gerris, Hydrometra, and Velia are mostly found perfectly apterous, though occasionally with full-sized wings."

It is, however, among the Lepidoptera (butterflies and moths) that the most numerous cases of variation have been observed, and every good collection of these insects affords striking examples. I will first adduce the testimony of Mr. Bates, who speaks of the butterflies of the Amazon valley exhibiting innumerable local varieties or races, while some species showed great individual variability. Of the beautiful Mechanitis Polymnia he says, that at Ega on the Upper Amazons, "it varies not only in general colour and pattern, but also very considerably in the shape of the wings, especially in the male sex." Again, at St. Paulo, Ithomia

[1] *United States Geological Survey of the Territories*, 1874.

Orolina exhibits four distinct varieties, all occurring together, and these differ not only in colour but in form, one variety being described as having the fore wings much elongated in the male, while another is much larger and has "the hind wings in the male different in shape." Of Heliconius Numata Mr. Bates says: "This species is so variable that it is difficult to find two examples exactly alike," while "it varies in structure as well as in colours. The wings are sometimes broader, sometimes narrower; and their edges are simple in some examples and festooned in others." Of another species of the same genus, H. melpomene, ten distinct varieties are described all more or less connected by intermediate forms, and four of these varieties were obtained at one locality, Serpa on the north bank of the Amazon. Ceratina Ninonia is another of these very unstable species exhibiting many local varieties which are, however, incomplete and connected by intermediate forms; while the several species of the genus Lycorea all vary to such an extent as almost to link them together, so that Mr. Bates thinks they might all fairly be considered as varieties of one species only.

Turning to the Eastern Hemisphere we have in Papilo Severus a species which exhibits a large amount of simple variation, in the presence or absence of a pale patch on the upper wings, in the brown submarginal marks on the lower wings, in the form and extent of the yellow band, and in the size of the specimens. The most extreme forms, as well as the intermediate ones, are often found in one locality and in company with each other. A small butterfly (Terias hecabe) ranges over the whole of the Indian and Malayan regions to Australia, and everywhere exhibits great variations, many of which have been described as distinct species; but a gentleman in Australia bred two of these distinct forms (T. hecabe and T. Æsiope), with several intermediates, from one batch of caterpillars found feeding together on the same plant.[1] It is therefore very probable that a considerable number of supposed distinct species are only individual varieties.

Cases of variation similar to those now adduced among butterflies might be increased indefinitely, but it is as well to note that such important characters as the neuration of the

[1] *Proceedings of the Entomological Society of London*, 1875, p. vii.

wings, on which generic and family distinctions are often established, are also subject to variation. The Rev. R. P. Murray, in 1872, laid before the Entomological Society examples of such variation in six species of butterflies, and other cases have been since described. The larvæ of butterflies and moths are also very variable, and one observer recorded in the *Proceedings of the Entomological Society for* 1870 no less than sixteen varieties of the caterpillar of the bedstraw hawk-moth (Deilephela galii).

Variation among Lizards.

Passing on from the lower animals to the vertebrata, we find more abundant and more definite evidence as to the extent and amount of individual variation. I will first give a case among the Reptilia from some of Mr. Darwin's unpublished MSS., which have been kindly lent me by Mr. Francis Darwin.

"M. Milne Edwards (*Annales des Sci. Nat.*, 1 ser., tom. xvi. p. 50) has given a curious table of measurements of fourteen specimens of Lacerta muralis; and, taking the length of the head as a standard, he finds the neck, trunk, tail, front and hind legs, colour, and femoral pores, all varying wonderfully; and so it is more or less with other species. So apparently trifling a character as the scales on the head affording almost the only constant characters."

As the table of measurements above referred to would give no clear conception of the nature and amount of the variation without a laborious study and comparison of the figures, I have endeavoured to find a method of presenting the facts to the eye, so that they may be easily grasped and appreciated. In the diagram opposite, the comparative variations of the different organs of this species are given by means of variously bent lines. The head is represented by a straight line because it presented (apparently) no variation. The body is next given, the specimens being arranged in the order of their size from No. 1, the smallest, to No. 14, the largest, the actual lengths being laid down from a base line at a suitable distance below, in this case two inches below the centre, the mean length of the body of the fourteen specimens being two inches. The respective lengths of the neck, legs, and toe of

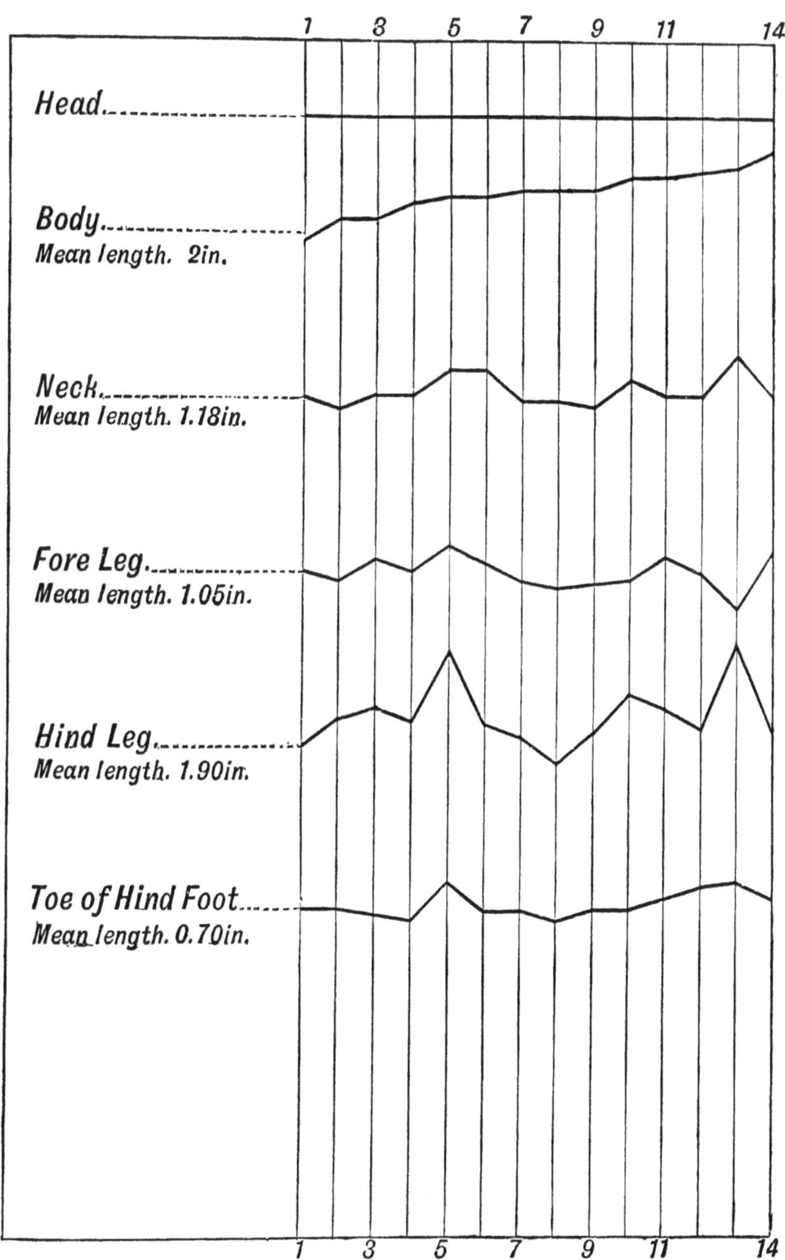

The lengths in the table are given in millimetres, which are here reduced to inches for the means.

FIG. 1.—Variations of Lacerta muralis.

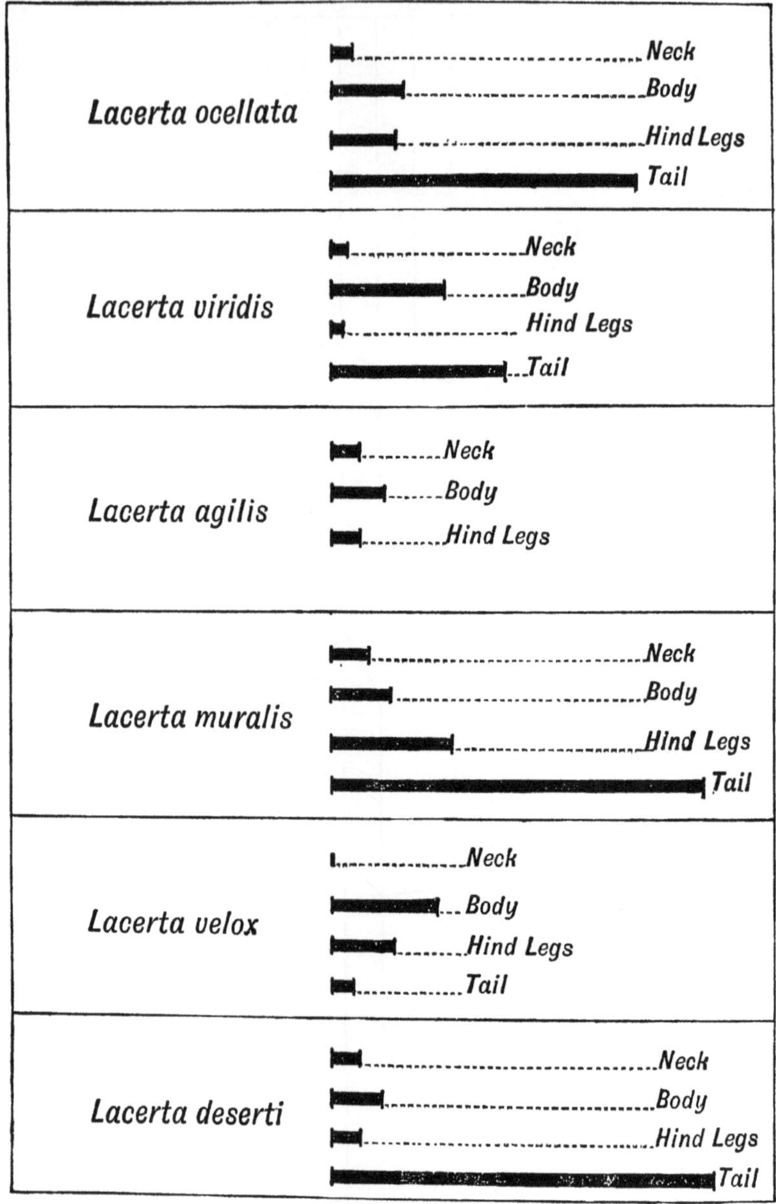

FIG. 2.—Variation of Lizards.

each specimen are then laid down in the same manner at convenient distances apart for comparison; and we see that their variations bear no definite relation to those of the body, and not much to those of each other. With the exception of No. 5, in which all the parts agree in being large, there is a marked independence of each part, shown by the lines often curving in opposite directions; which proves that in those specimens one part is large while the other is small. The actual amount of the variation is very great, ranging from one-sixth of the mean length in the neck to considerably more than a fourth in the hind leg, and this among only fourteen examples which happen to be in a particular museum.

To prove that this is not an isolated case, Professor Milne Edwards also gives a table showing the amount of variation in the museum specimens of six common species of lizards, also taking the head as the standard, so that the comparative variation of each part to the head is given. In the accompanying diagram (Fig. 2) the variations are exhibited by means of lines of varying length. It will be understood that, however much the specimens varied in *size*, if they had kept the same *proportions*, the variation line would have been in every case reduced to a point, as in the neck of L. velox which exhibits no variation. The different proportions of the variation lines for each species may show a distinct mode of variation, or may be merely due to the small and differing number of specimens; for it is certain that whatever amount of variation occurs among a few specimens will be greatly increased when a much larger number of specimens are examined. That the amount of variation is large, may be seen by comparing it with the actual length of the head (given below the diagram) which was used as a standard in determining the variation, but which itself seems not to have varied.[1]

Variation among Birds.

Coming now to the class of Birds, we find much more copious evidence of variation. This is due partly to the fact that Ornithology has perhaps a larger body of devotees than any other branch of natural history (except entomology); to the moderate size of the majority of birds; and to the circum-

[1] *Ann. des Sci. Nat.*, tom. xvi. p. 50.

stance that the form and dimensions of the wings, tail, beak, and feet offer the best generic and specific characters and can all be easily measured and compared. The most systematic observations on the individual variation of birds have been made by Mr. J. A. Allen, in his remarkable memoir: "On the Mammals and Winter Birds of East Florida, with an examination of certain assumed specific characters in Birds, and a sketch of the Bird Faunæ of Eastern North America," published in the *Bulletin of the Museum of Comparative Zoology* at Harvard College, Cambridge, Massachusetts, in 1871. In this work exact measurements are given of all the chief external parts of a large number of species of common American birds, from twenty to sixty or more specimens of each species being measured, so that we are able to determine with some precision the nature and extent of the variation that usually occurs. Mr. Allen says: "The facts of the case show that a variation of from 15 to 20 per cent in general size, and an equal degree of variation in the relative size of different parts, may be ordinarily expected among specimens of the same species and sex, taken at the same locality, while in some cases the variation is even greater than this." He then goes on to show that each part varies to a considerable extent independently of the other parts; so that when the size varies, the proportions of all the parts vary, often to a much greater amount. The wing and tail, for example, besides varying in length, vary in the proportionate length of each feather, and this causes their outline to vary considerably in shape. The bill also varies in length, width, depth, and curvature. The tarsus varies in length, as does each toe separately and independently; and all this not to a minute degree requiring very careful measurement to detect it at all, but to an amount easily seen without any measurement, as it averages one-sixth of the whole length and often reaches one-fourth. In twelve species of common perching birds the wing varied (in from twenty-five to thirty specimens) from 14 to 21 per cent of the mean length, and the tail from 13·8 to 23·4 per cent. The variation of the form of the wing can be very easily tested by noting which feather is longest, which next in length, and so on, the respective feathers being indicated by the numbers 1, 2, 3, etc., commencing with the outer one. As an example of the irregular variation constantly met with, the following occurred among twenty-five specimens of Dendræca coronata. Numbers bracketed imply that the corresponding feathers were of equal length.[1]

Relative Lengths of Primary Wing Feathers of Dendræca coronata.

Longest.	Second in Length.	Third in Length.	Fourth in Length.	Fifth in Length.	Sixth in Length.
2	3	1	4	5	6
3	2	4	1	5	6
3	{2, 4}	1	5	6	7
{2, 3}	4	1	5	6	7
{2, 1, 3, 4}	5	6	7	8	9

Here we have five very distinct proportionate lengths of the wing feathers, any one of which is often thought sufficient to characterise a distinct species of bird; and though this is rather an extreme case, Mr. Allen assures us that "the comparison, extended in the table to only a few species, has been carried to scores of others with similar results."

Along with this variation in size and proportions there occurs a large amount of variation in colour and markings. "The difference in intensity of colour between the extremes of a series of fifty or one hundred specimens of any species, collected at a single locality, and nearly at the same season of the year, is often as great as occurs between truly distinct species." But there is also a great amount of individual variability in the markings of the same species. Birds having the plumage varied with streaks and spots differ exceedingly in different individuals of the same species in respect to the size, shape, and number of these marks, and in the general aspect of the plumage resulting from such variations. "In the common

[1] See *Winter Birds of Florida*, p. 206, Table F.

song sparrow (Melospiza melodia), the fox-coloured sparrow (Passerella iliaca), the swamp sparrow (Melospiza palustris), the black and white creeper (Mniotilta varia), the water-wagtail (Seiurus novæboracencis), in Turdus fuscescens and its allies, the difference in the size of the streaks is often very considerable. In the song sparrow they vary to such an extent that in some cases they are reduced to narrow lines; in others so enlarged as to cover the greater part of the breast and sides of the body, sometimes uniting on the middle of the breast into a nearly continuous patch."

Mr. Allen then goes on to particularise several species in which such variations occur, giving cases in which two specimens taken at the same place on the same day exhibited the two extremes of coloration. Another set of variations is thus described: "The white markings so common on the wings and tails of birds, as the bars formed by the white tips of the greater wing-coverts, the white patch occasionally present at the base of the primary quills, or the white band crossing them, and the white patch near the end of the outer tail-feathers are also extremely liable to variation in respect to their extent and the number of feathers to which, in the same species, these markings extend." It is to be especially noted that all these varieties are distinct from those which depend on season, on age, or on sex, and that they are such as have in many other species been considered to be of specific value.

These variations of colour could not be presented to the eye without a series of carefully engraved plates, but in order to bring Mr. Allen's *measurements*, illustrating variations of size and proportion, more clearly before the reader, I have prepared a series of diagrams illustrating the more important facts and their bearings on the Darwinian theory.

The first of these is intended, mainly, to show the actual amount of the variation, as it gives the true length of the wing and tail in the extreme cases among thirty specimens of each of three species. The shaded portion shows the minimum length, the unshaded portion the additional length in the maximum. The point to be specially noted here is, that in each of these common species there is about the same amount of variation, and that it is so great as to be obvious at a glance.

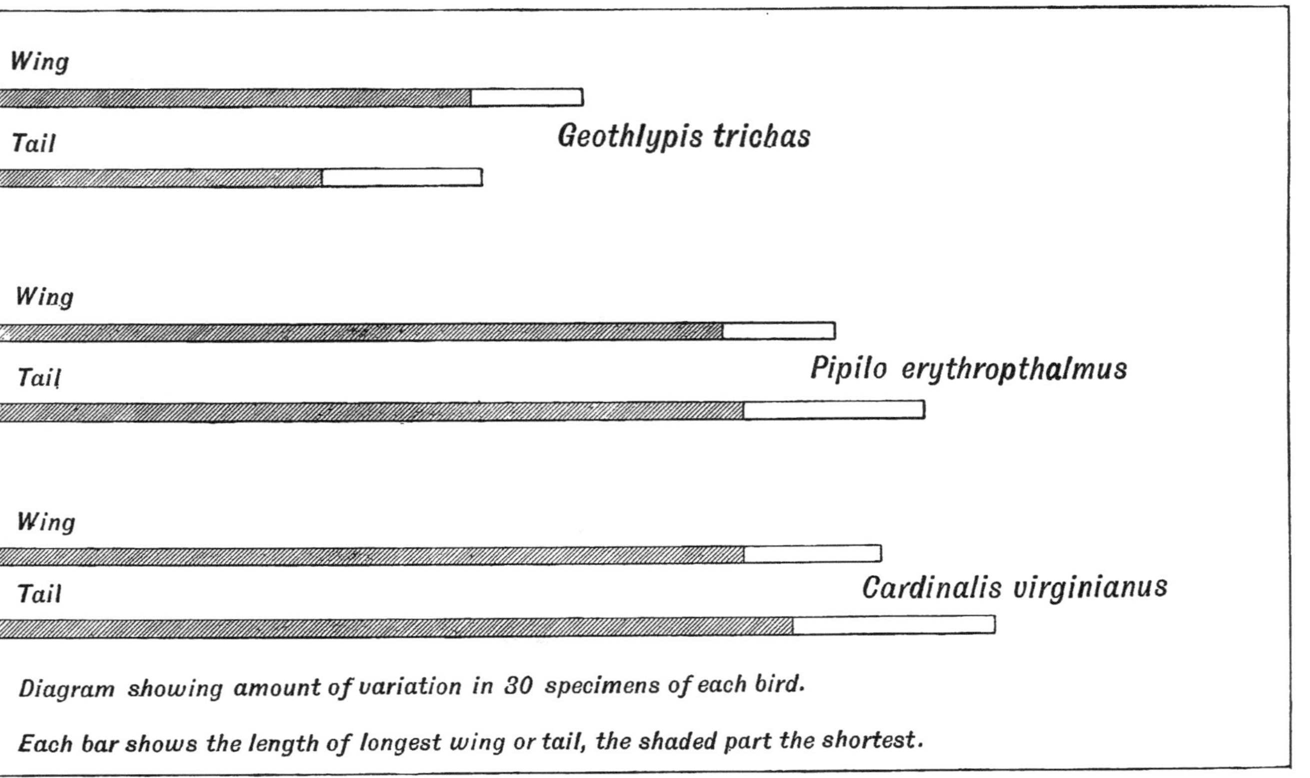

Fig. 3.—Variation of Wings and Tail.

There is here no question of "minute" or "infinitesimal" variation, which many people suppose to be the only kind of variation that exists. It cannot even be called small; yet from all the evidence we now possess it seems to be the amount which characterises most of the common species of birds.

It may be said, however, that these are the extreme variations, and only occur in one or two individuals, while the great majority exhibit little or no difference. Other diagrams will show that this is not the case; but even if it were so, it would be no objection at all, because these are the extremes among thirty specimens only. We may safely assume that these thirty specimens, taken by chance, are not, in the case of all these species, exceptional lots, and therefore we might expect at least two similarly varying specimens in each additional thirty. But the number of individuals, even in a very rare species, is probably thirty thousand or more, and in a common species thirty, or even three hundred, millions. Even one individual in each thirty, varying to the amount shown in the diagram, would give at least a million in the total population of any common bird, and among this million many would vary much more than the extreme among thirty only. We should thus have a vast body of individuals varying to a large extent in the length of the wings and tail, and offering ample material for the modification of these organs by natural selection. We will now proceed to show that other parts of the body vary, simultaneously, but independently, to an equal amount.

The first bird taken is the common Bob-o-link or Rice-bird (Dolichonyx oryzivorus), and the Diagram, Fig. 4, exhibits the variations of seven important characters in twenty male adult specimens.[1] These characters are—the lengths of the body, wing, tail, tarsus, middle toe, outer toe, and hind toe, being as many as can be conveniently exhibited in one diagram. The length of the body is not given by Mr. Allen, but as it forms a convenient standard of comparison, it has been obtained by deducting the length of the tail from the total length of the birds as given by him. The diagram has been constructed as follows:—The twenty specimens are first arranged in a series according to the body-lengths (which may be con-

[1] See Table I, p. 211, of Allen's *Winter Birds of Florida.*

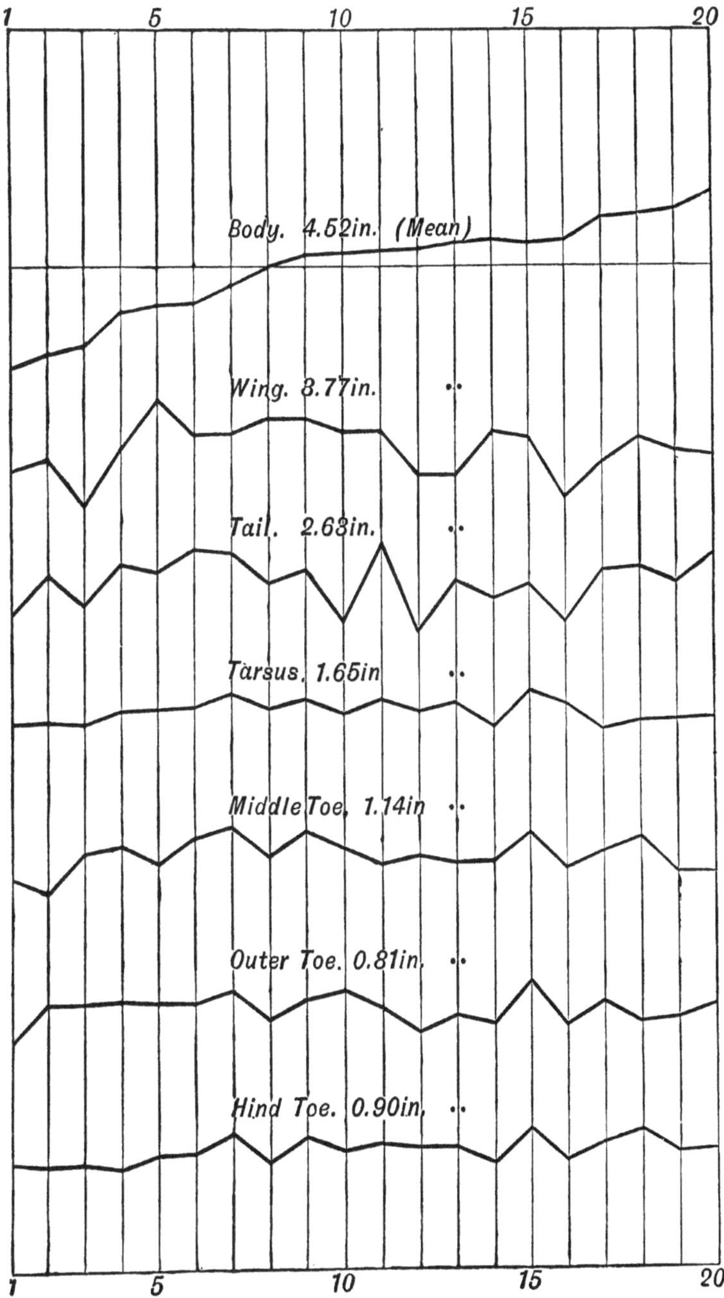

Fig. 4.—Dolichonyx oryzivorus. 20 Males.

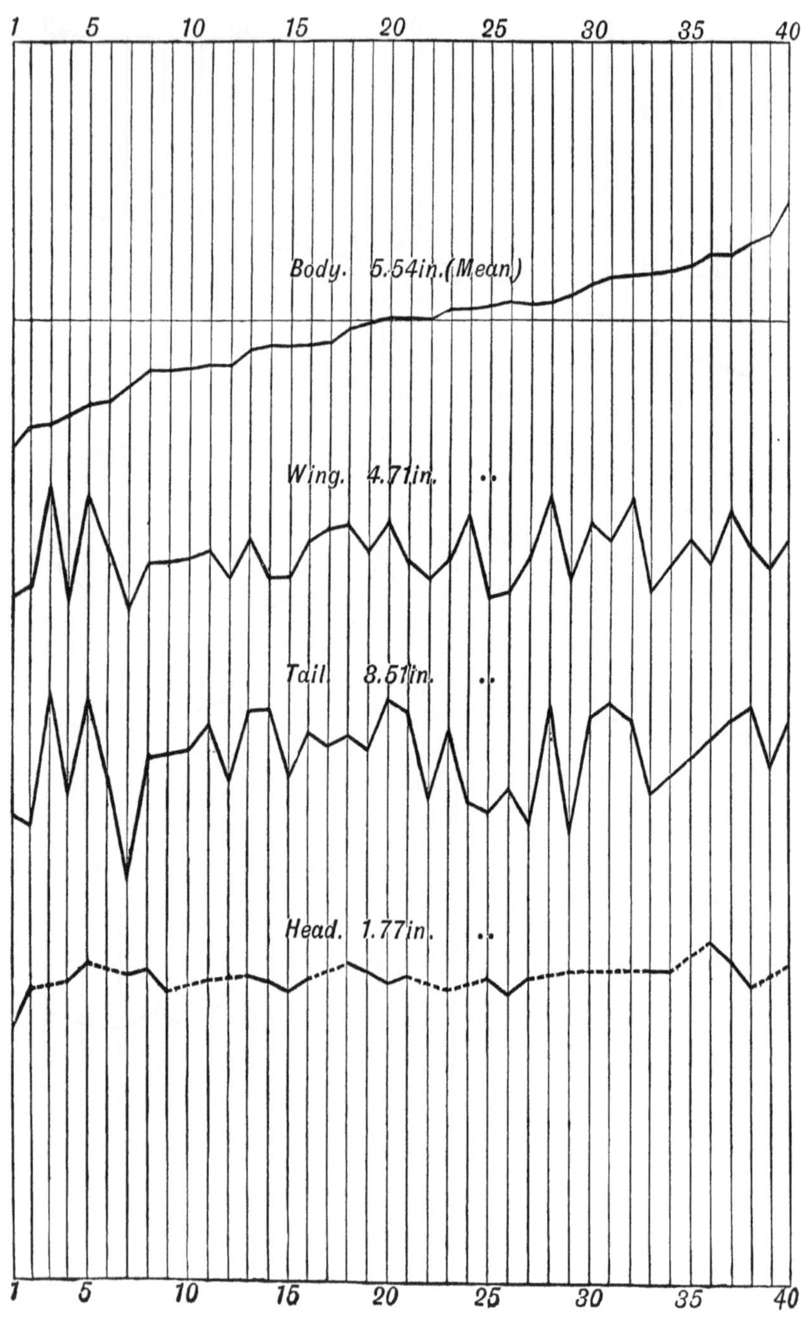

Fig. 5.—Agelæus phœniceus. 40 Males.

sidered to give the size of the bird), from the shortest to the longest, and the same number of vertical lines are drawn, numbered from one to twenty. In this case (and wherever practicable) the body-length is measured from the lower line of the diagram, so that the actual length of the bird is exhibited as well as the actual variations of length. These can be well estimated by means of the horizontal line drawn at the mean between the two extremes, and it will be seen that one-fifth of the total number of specimens taken on either side exhibits a very large amount of variation, which would of course be very much greater if a hundred or more specimens were compared. The lengths of the wing, tail, and other parts are then laid down, and the diagram thus exhibits at a glance the comparative variation of these parts in every specimen as well as the actual amount of variation in the twenty specimens; and we are thus enabled to arrive at some important conclusions.

We note, first, that the variations of none of the parts follow the variations of the body, but are sometimes almost in an opposite direction. Thus the longest wing corresponds to a rather small body, the longest tail to a medium body, while the longest leg and toes belong to only a moderately large body. Again, even related parts do not constantly vary together but present many instances of independent variation, as shown by the want of parallelism in their respective variation-lines. In No. 5 (see Fig. 4) the wing is very long, the tail moderately so; while in No. 6 the wing is much shorter while the tail is considerably longer. The tarsus presents comparatively little variation; and although the three toes may be said to vary in general together, there are many divergencies; thus, in passing from No. 9 to No. 10, the outer toe becomes longer, while the hind toe becomes considerably shorter; while in Nos. 3 and 4 the middle toe varies in an opposite way to the outer and the hind toes.

In the next diagram (Fig. 5) we have the variations in forty males of the Red-winged Blackbird (Agelæus phœniceus), and here we see the same general features. One-fifth of the whole number of specimens offer a large amount of variation either below or above the mean; while the wings, tail, and head vary quite independently of the body. The wing and tail too,

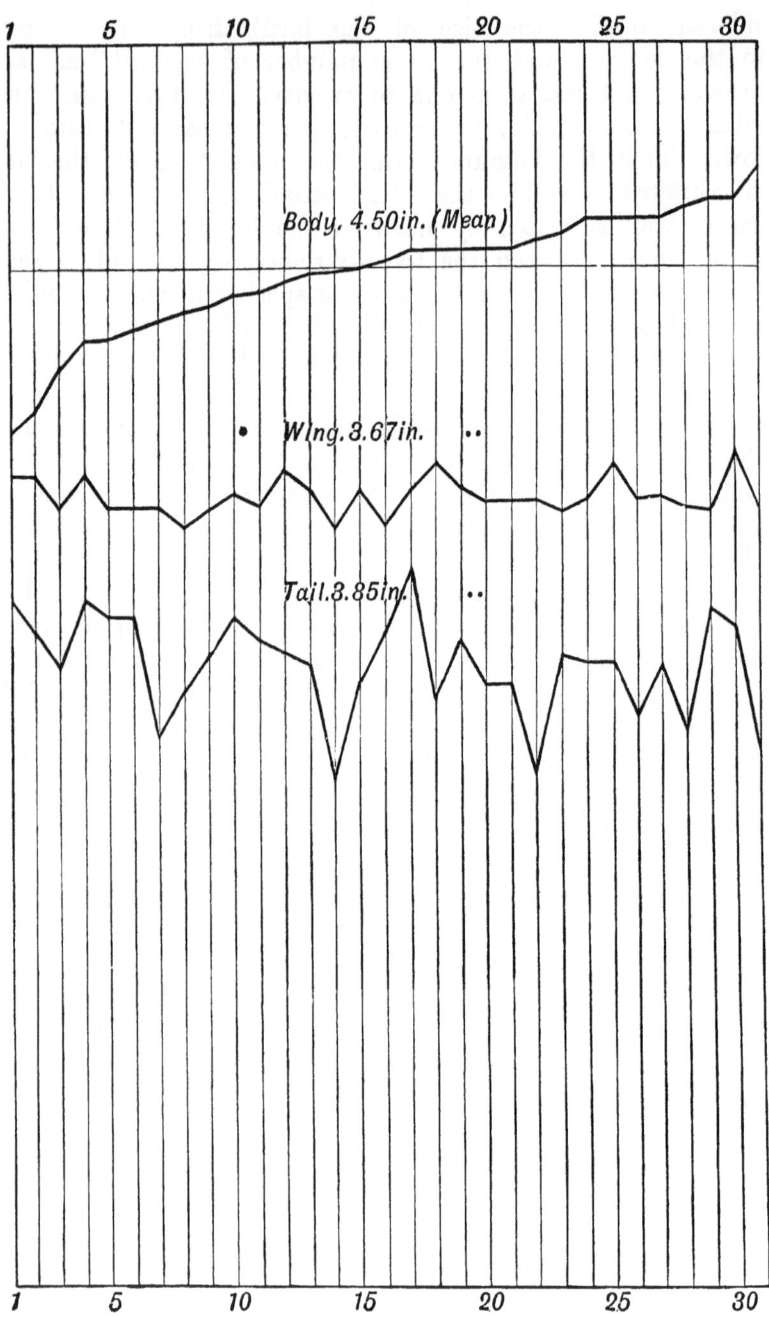

FIG. 6.—Cardinalis virginianus. 31 Males.

though showing some amount of correlated variation, yet in no less than nine cases vary in opposite directions as compared with the preceding species.

The next diagram (Fig. 6), showing the variations of thirty-one males of the Cardinal bird (Cardinalis virginianus), exhibits these features much more strongly. The amount of variation in proportion to the size of the bird is very much greater; while the variations of the wing and tail not only have no correspondence with that of the body but very little with each other. In no less than twelve or thirteen instances they vary in opposite directions, while even where they correspond in direction the amount of the variation is often very disproportionate.

As the proportions of the tarsi and toes of birds have great influence on their mode of life and habits and are often used as specific or even generic characters, I have prepared a diagram (Fig. 7) to show the variation in these parts only, among twenty specimens of each of four species of birds, four or five of the most variable alone being given. The extreme divergence of each of the lines in a vertical direction shows the actual amount of variation; and if we consider the small length of the toes of these small birds, averaging about three-quarters of an inch, we shall see that the variation is really very large; while the diverging curves and angles show that each part varies, to a great extent, independently. It is evident that if we compared some thousands of individuals instead of only twenty, we should have an amount of independent variation occurring each year which would enable almost any modification of these important organs to be rapidly effected.

In order to meet the objection that the large amount of variability here shown depends chiefly on the observations of one person and on the birds of a single country, I have examined Professor Schlegel's Catalogue of the Birds in the Leyden Museum, in which he usually gives the range of variation of the specimens in the museum (which are commonly less than a dozen and rarely over twenty) as regards some of their more important dimensions. These fully support the statement of Mr. Allen, since they show an equal amount of variability when the numbers compared are

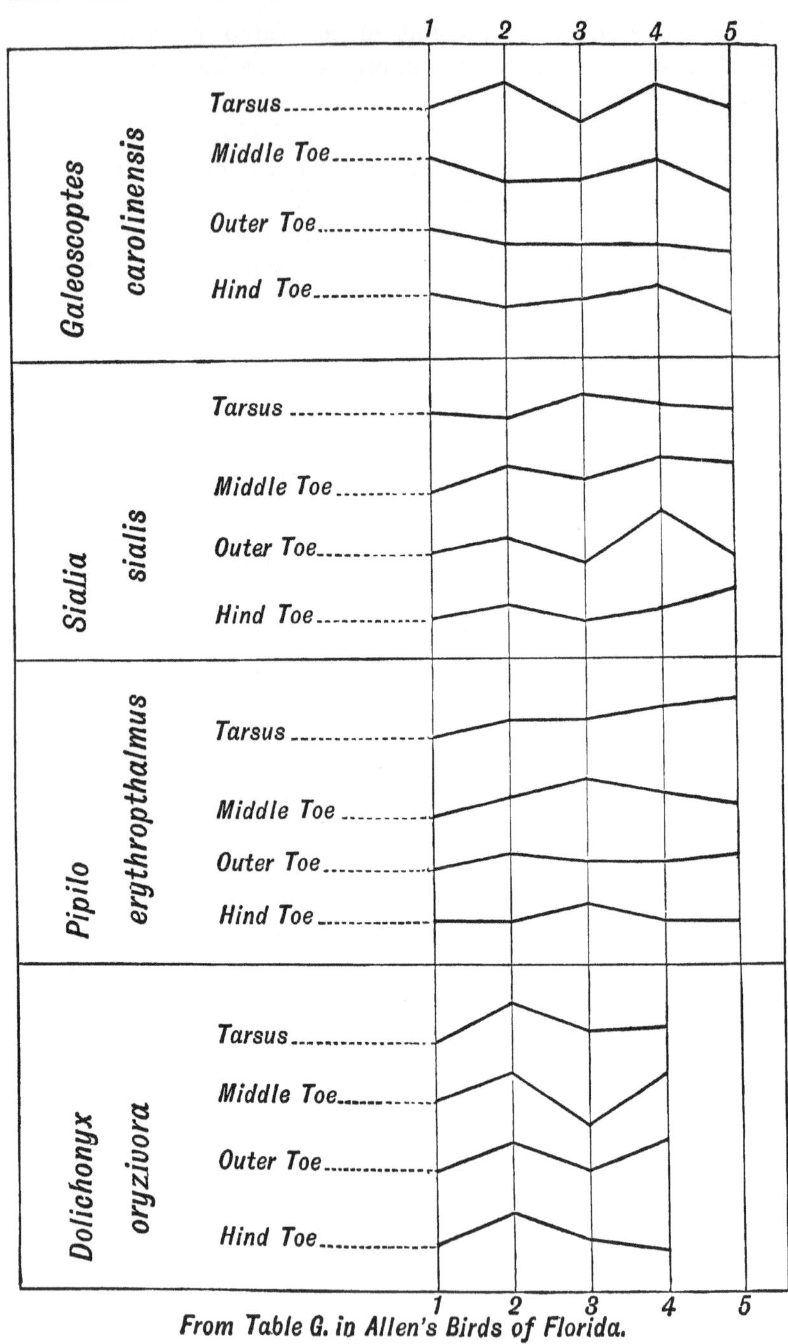

FIG. 7.—Variation of Tarsus and Toes.

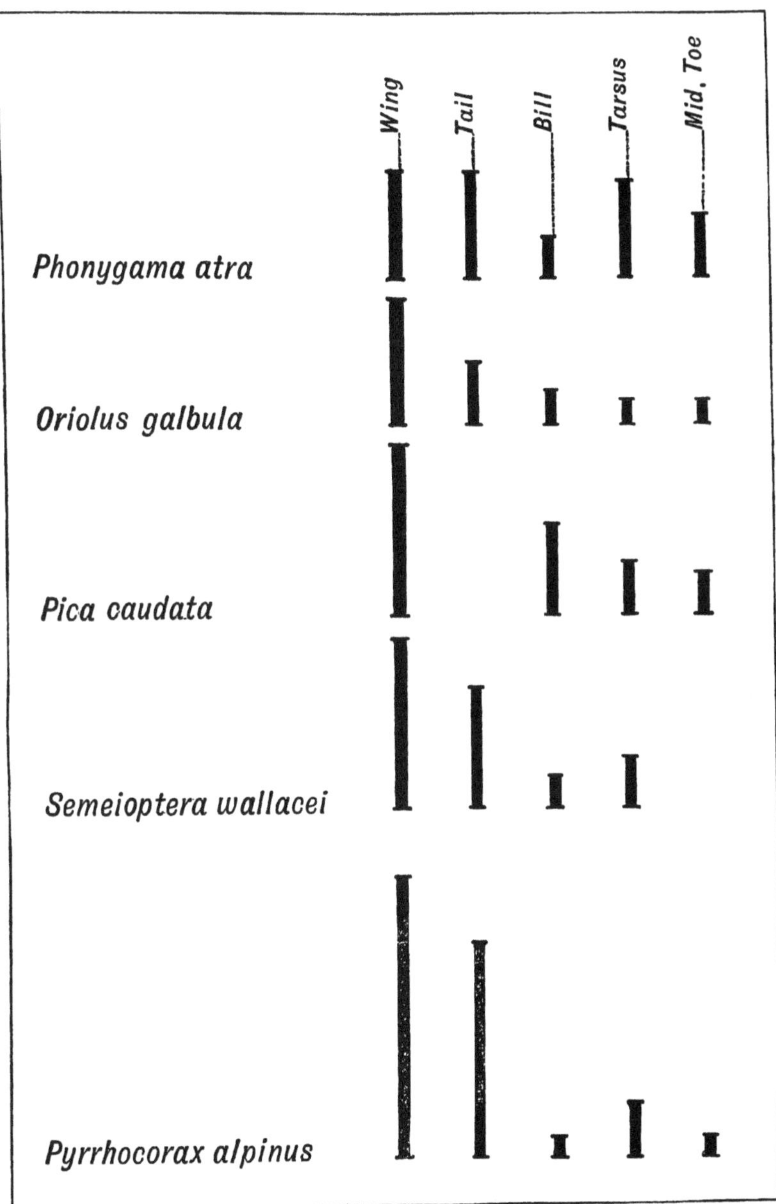

FIG. 8.—Variation of Birds in Leyden Museum.

sufficient, which, however, is not often the case. The accompanying diagram exhibits the actual differences of size in five organs which occur in five species taken almost at random from this catalogue. Here, again, we perceive that the variation is decidedly large, even among a very small number of specimens; while the facts all show that there is no ground whatever for the common assumption that natural species consist of individuals which are nearly all alike, or that the variations which occur are "infinitesimal" or even "small."

The proportionate Number of Individuals which present a considerable amount of Variation.

The notion that variation is a comparatively exceptional phenomenon, and that in any case considerable variations occur very rarely in proportion to the number of individuals which do not vary, is so deeply rooted that it is necessary to show by every possible method of illustration how completely opposed it is to the facts of nature. I have therefore prepared some diagrams in which each of the individual birds measured is represented by a spot, placed at a proportionate distance, right and left, from the median line accordingly as it varies in excess or defect of the mean length as regards the particular part compared. As the object in this set of diagrams is to show the number of individuals which vary considerably in proportion to those which vary little or not at all, the scale has been enlarged in order to allow room for placing the spots without overlapping each other.

In the diagram opposite twenty males of Icterus Baltimore are registered, so as to exhibit to the eye the proportionate number of specimens which vary, to a greater or less amount, in the length of the tail, wing, tarsus, middle toe, hind toe, and bill. It will be noticed that there is usually no very great accumulation of dots about the median line which shows the average dimensions, but that a considerable number are spread at varying distances on each side of it.

In the next diagram (Fig. 10), showing the variation among forty males of Agelæus phœniceus, this approach to an equable spreading of the variations is still more apparent; while in Fig. 12, where fifty-eight specimens of Cardinalis virginianus are registered, we see a remarkable spreading out of the spots, showing in some of the characters a tendency to segregation into two or more groups of individuals, each varying considerably from the mean.

In order fully to appreciate the teaching of these diagrams,

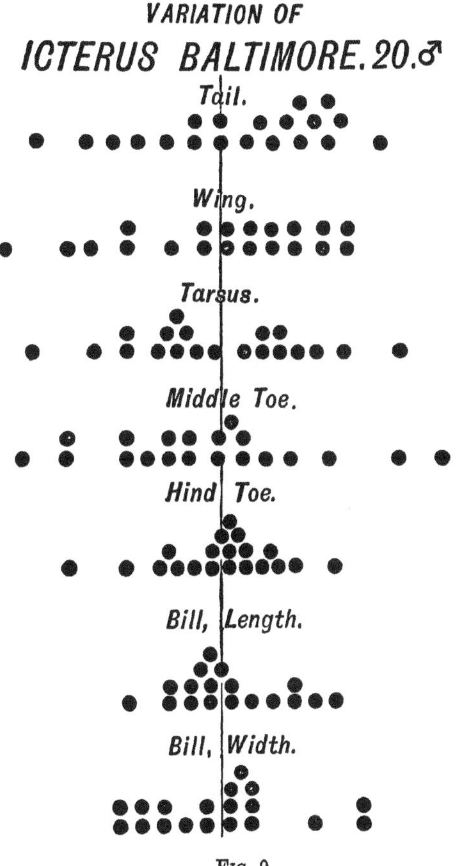

Fig. 9.

we must remember, that, whatever kind and amount of variations are exhibited by the few specimens here compared, would be greatly extended and brought into symmetrical form if large numbers—thousands or millions—were subjected to the same process of measurement and registration. We know, from the general law which governs variations from a mean value, that with increasing numbers the range

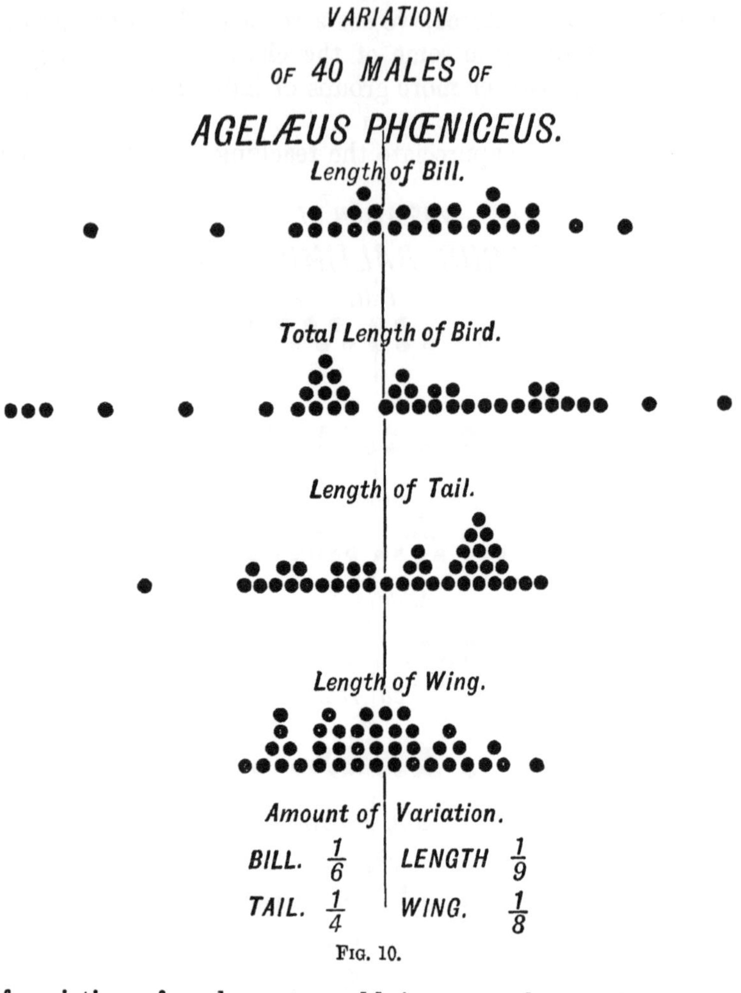

VARIATION
OF 40 MALES OF
AGELÆUS PHŒNICEUS.

Length of Bill.

Total Length of Bird.

Length of Tail.

Length of Wing.

Amount of Variation.

BILL.	$\frac{1}{6}$	LENGTH	$\frac{1}{9}$
TAIL.	$\frac{1}{4}$	WING.	$\frac{1}{8}$

Fig. 10.

of variation of each part would increase also, at first rather rapidly and then more slowly; while gaps and irregularities

Curves of Variation

Fig. 11.

would be gradually filled up, and at length the distribution of the dots would indicate a tolerably regular curve of double curvature like those shown in Fig. 11. The great divergence of the dots, when even a few specimens are compared, shows that the curve, with high numbers, would be a flat one like the lower curve in the illustration here given. This being the case it would follow that a very large proportion of the total number of individuals constituting a species would diverge considerably from its average condition as regards each part or organ; and as we know from the previous diagrams of variation (Figs. 1 to 7) that each part varies to a considerable extent, *independently*, the materials constantly ready for natural selection

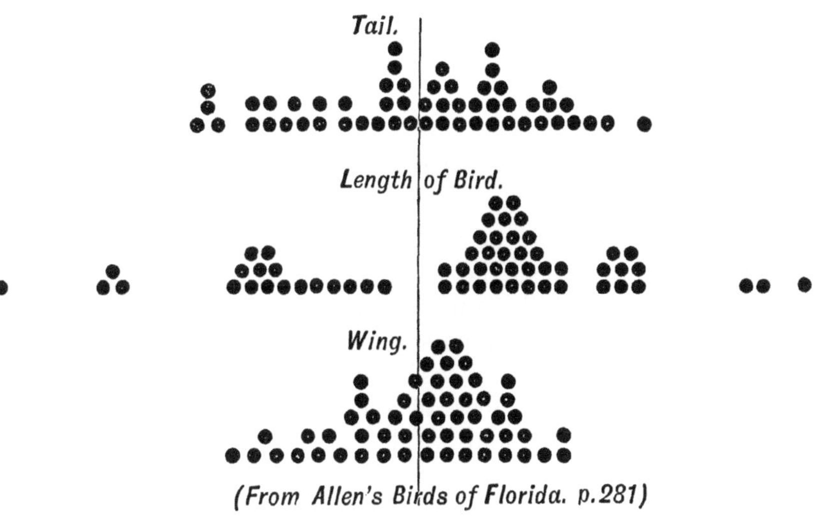

Fig. 12

to act upon are abundant in quantity and very varied in kind. Almost any combination of variations of distinct parts will be available, where required; and this, as we shall see further on, obviates one of the most weighty objections which have been urged against the efficiency of natural selection in producing new species, genera, and higher groups.

Variation in the Mammalia.

Owing to the generally large size of this class of animals, and the comparatively small number of naturalists who study them, large series of specimens are only occasionally examined and compared, and thus the materials for determining the question of their variability in a state of nature are comparatively scanty. The fact that our domestic animals belonging to this group, especially dogs, present extreme varieties not surpassed even by pigeons and poultry among birds, renders it almost certain that an equal amount of variability exists in the wild state; and this is confirmed by the example of a species of squirrel (Sciurus carolinensis), of which sixteen specimens, all males and all taken in Florida, were measured and tabulated by Mr. Allen. The diagram here given shows, that, both the general amount of the variation and the independent variability of the several members of the body, accord completely with the variations so common in the class of birds; while their amount and their independence of each other are even greater than usual.

Variation in the Internal Organs of Animals.

In case it should be objected that the cases of variation hitherto adduced are in the external parts only, and that there is no proof that the internal organs vary in the same manner, it will be advisable to show that such varieties also occur. It is, however, impossible to adduce the same amount of evidence in this class of variation, because the great labour of dissecting large numbers of specimens of the same species is rarely undertaken, and we have to trust to the chance observations of anatomists recorded in their regular course of study.

It must, however, be noted that a very large proportion of the variations already recorded in the external parts of animals necessarily imply corresponding internal variations. When feet and legs vary in size, it is because the bones vary; when the head, body, limbs, and tail change their proportions, the bony skeleton must also change; and even when the wing or tail feathers of birds become longer or more numerous, there is sure to be a corresponding change in the bones which support and the muscles which move them. I will, however, give a few cases of variations which have been directly observed.

Mr. Frank E. Beddard has kindly communicated to me some remarkable variations he has observed in the internal

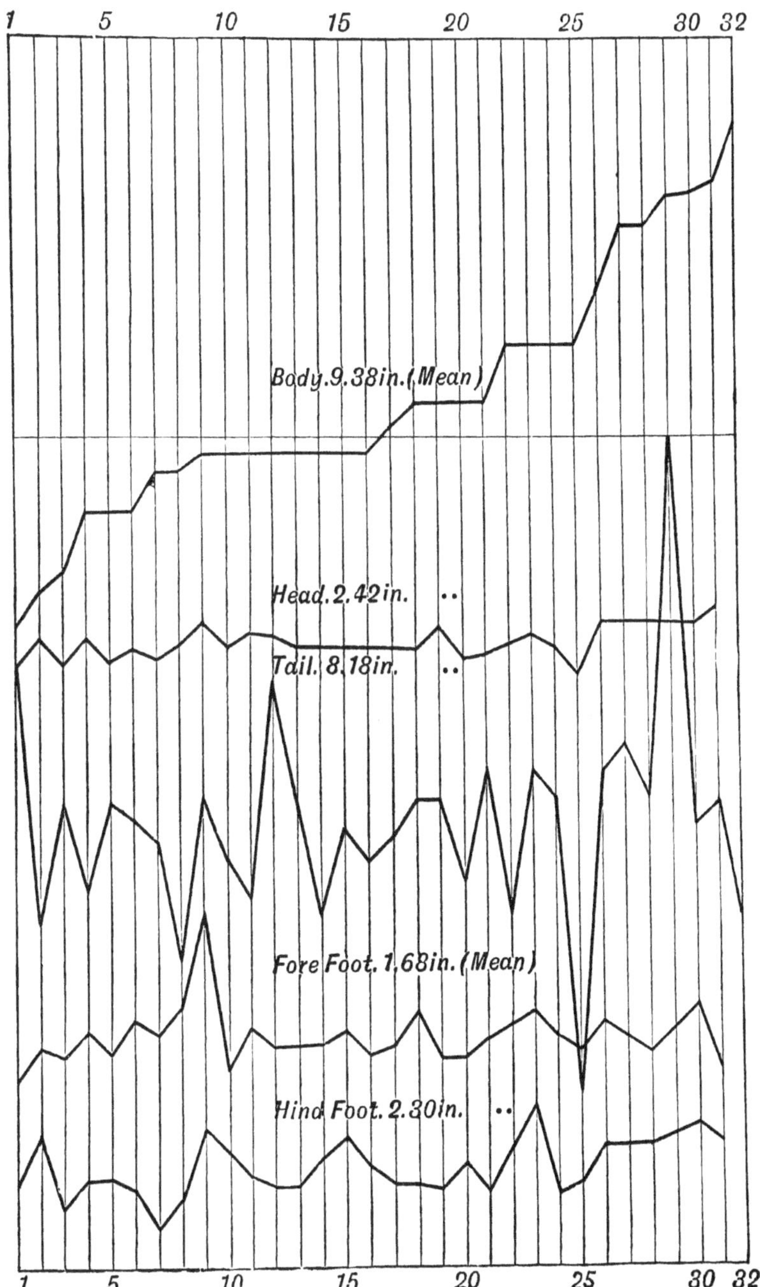

FIG. 13.—Sciurus carolinensis. 32 specimens. Florida.

organs of a species of earthworm (Perionyx excavatus). The normal characters of this species are—

Setæ forming a complete row round each segment.

Two pairs of spermathecæ—spherical pouches without diverticulæ—in segments 8 and 9.

Two pairs of testes in segments 11 and 12.

Ovaries, a single pair in segment 13.

Oviducts open by a common pore in the middle of segment 14.

Vasa deferentia open separately in segment 18, each furnished at its termination with a large prostate gland.

Between two and three hundred specimens were examined, and among them thirteen specimens exhibited the following marked variations :—

(1) The number of the spermathecæ varied from two to three or four pairs, their position also varying.

(2) There were occasionally two pairs of ovaries, each with its own oviduct; the external apertures of these varied in position, being upon segments 13 and 14, 14 and 15, or 15 and 16. Occasionally when there was only the normal single oviduct pore present it varied in position, once occurring on the 10th, and once on the 11th segment.

(3) The male generative pores varied in position from segments 14 to 20. In one instance there were two pairs instead of the normal single pair, and in this case each of the four apertures had its own prostate gland.

Mr. Beddard remarks that all, or nearly all, the above variations are found *normally* in other genera and species.

When we consider the enormous number of earthworms and the comparatively very small number of individuals examined, we may be sure, not only that such variations as these occur with considerable frequency, but also that still more extraordinary deviations from the normal structure may often exist.

The next example is taken from Mr. Darwin's unpublished MSS.

"In some species of Shrews (Sorex) and in some field-mice (Arvicola), the Rev. L. Jenyns (*Ann. Nat. Hist.*, vol. vii. pp. 267, 272) found the proportional length of the intestinal canal to vary considerably. He found the same variability in the number of the caudal vertebræ. In three specimens of an Arvicola he found the gall-bladder having a very different degree of development, and there is reason to believe it is sometimes absent. Professor Owen has shown that this is the case with the gall-bladder of the giraffe."

Dr. Crisp (*Proc. Zool. Soc.*, 1862, p. 137) found the gall-bladder present in some specimens of Cervus superciliaris while absent in others; and he found it to be absent in three giraffes which he dissected. A double gall-bladder was found in a sheep, and in a small mammal preserved in the Hunterian Museum there are three distinct gall-bladders.

The length of the alimentary canal varies greatly. In three adult giraffes described by Professor Owen it was from 124 to 136 feet long; one dissected in France had this canal 211 feet long; while Dr. Crisp measured one of the extraordinary length of 254 feet, and similar variations are recorded in other animals.[1]

The number of ribs varies in many animals. Mr. St. George Mivart says: "In the highest forms of the Primates, the number of true ribs is seven, but in Hylobates there are sometimes eight pairs. In Semnopithecus and Colobus there are generally seven, but sometimes eight pairs of true ribs. In the Cebidæ there are generally seven or eight pairs, but in Ateles sometimes nine" (*Proc. Zool. Soc.*, 1865, p. 568). In the same paper it is stated that the number of dorsal vertebræ in man is normally twelve, very rarely thirteen. In the Chimpanzee there are normally thirteen dorsal vertebræ, but occasionally there are fourteen or only twelve.

Variations in the Skull.

Among the nine adult male Orang-utans, collected by myself in Borneo, the skulls differed remarkably in size and proportions. The orbits varied in width and height, the cranial ridge was either single or double, either much or little developed, and the zygomatic aperture varied considerably in

[1] *Proc. Zool. Soc.*, 1864, p. 64.

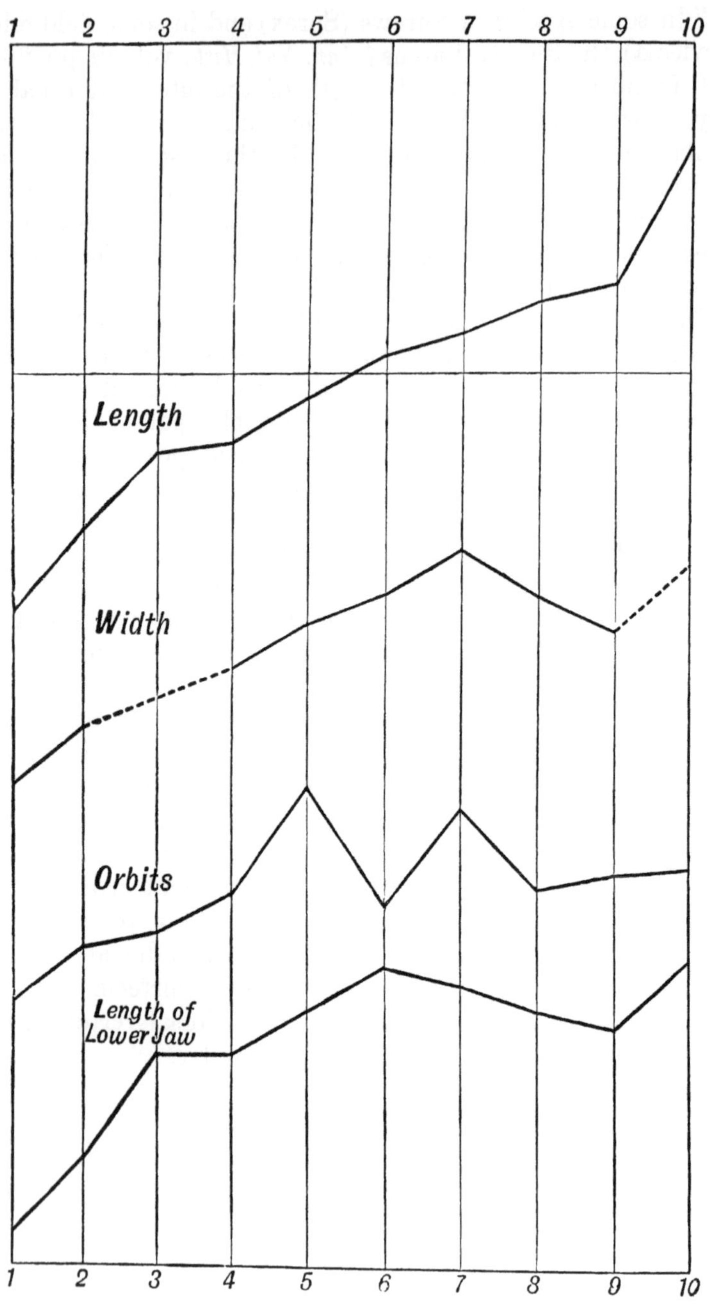

FIG. 14.—Variation of Skull of Wolf. 10 specimens.

size. I noted particularly that these variations bore no necessary relation to each other, so that a large temporal muscle and zygomatic aperture might exist either with a large or a small cranium; and thus was explained the curious difference between the single-crested and the double-crested skulls, which had been supposed to characterise distinct species. As an instance of the amount of variation in the skulls of fully adult male orangs, I found the width between the orbits externally to be only 4 inches in one specimen and fully 5 inches in another.

Exact measurements of large series of comparable skulls of the mammalia are not easily found, but from those available I have prepared three diagrams (Figs. 14, 15, and 16), in order to exhibit the facts of variation in this very important organ. The first shows the variation in ten specimens of the common wolf (Canis lupus) from one district in North America, and we see that it is not only large in amount, but that each part exhibits a considerable independent variability.[1]

In Diagram 15 we have the variations of eight skulls of the Indian Honey-bear (Ursus labiatus), as tabulated by the late Dr. J. E. Gray of the British Museum. For such a small number of specimens the amount of variation is very large—from one-eighth to one-fifth of the mean size,—while there are an extraordinary number of instances of independent variability. In Diagram 16 we have the length and width of twelve skulls of adult males of the Indian wild boar (Sus cristatus), also given by Dr. Gray, exhibiting in both sets of measurements a variation of more than one-sixth, combined with a very considerable amount of independent variability.[2]

The few facts now given, as to variations of the internal parts of animals, might be multiplied indefinitely by a search through the voluminous writings of comparative anatomists. But the evidence already adduced, taken in conjunction with the much fuller evidence of variation in all external organs, leads us to the conclusion that wherever variations are looked for among a considerable number of individuals of the more

[1] J. A. Allen, on Geographical Variation among North American Mammals, *Bull. U. S. Geol. and Geog. Survey*, vol. ii. p. 314 (1876).

[2] *Proc. Zool. Soc. Lond.*, 1864, p. 700, and 1868, p. 28.

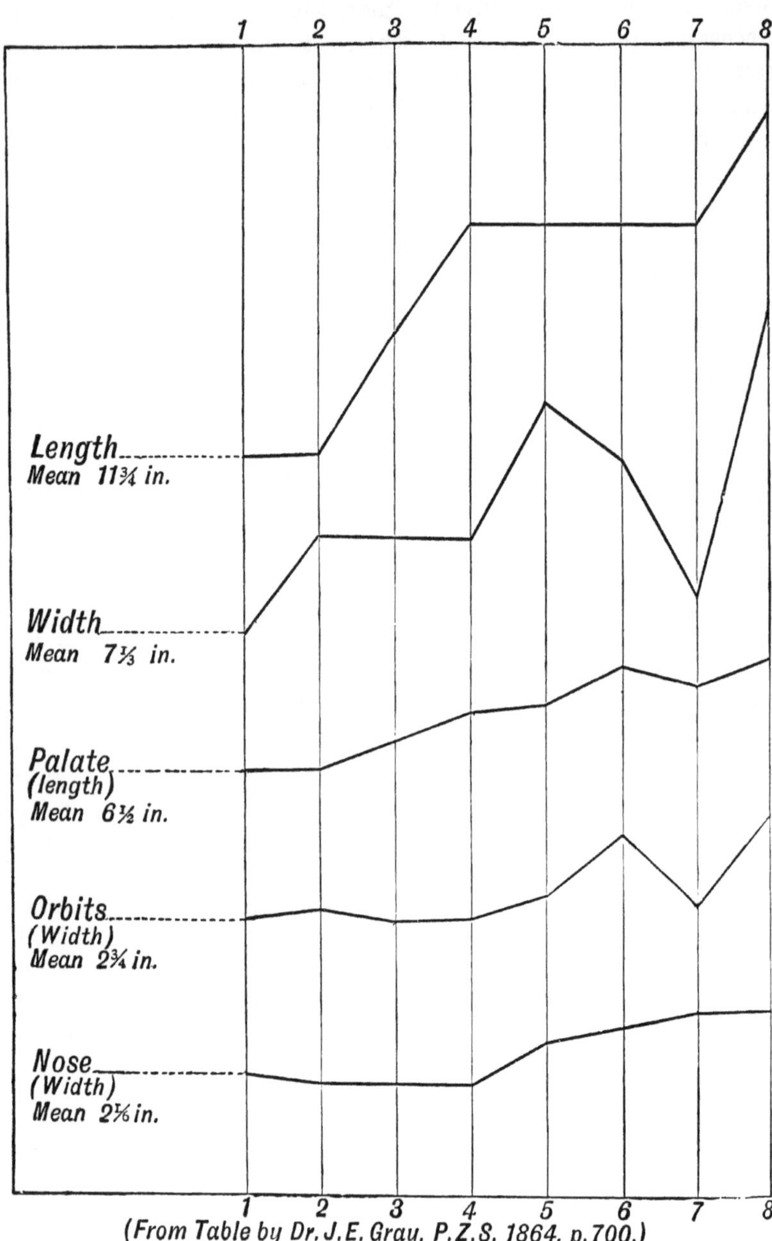

Fig. 15.—Variation of 8 skulls (Ursus labiatus).

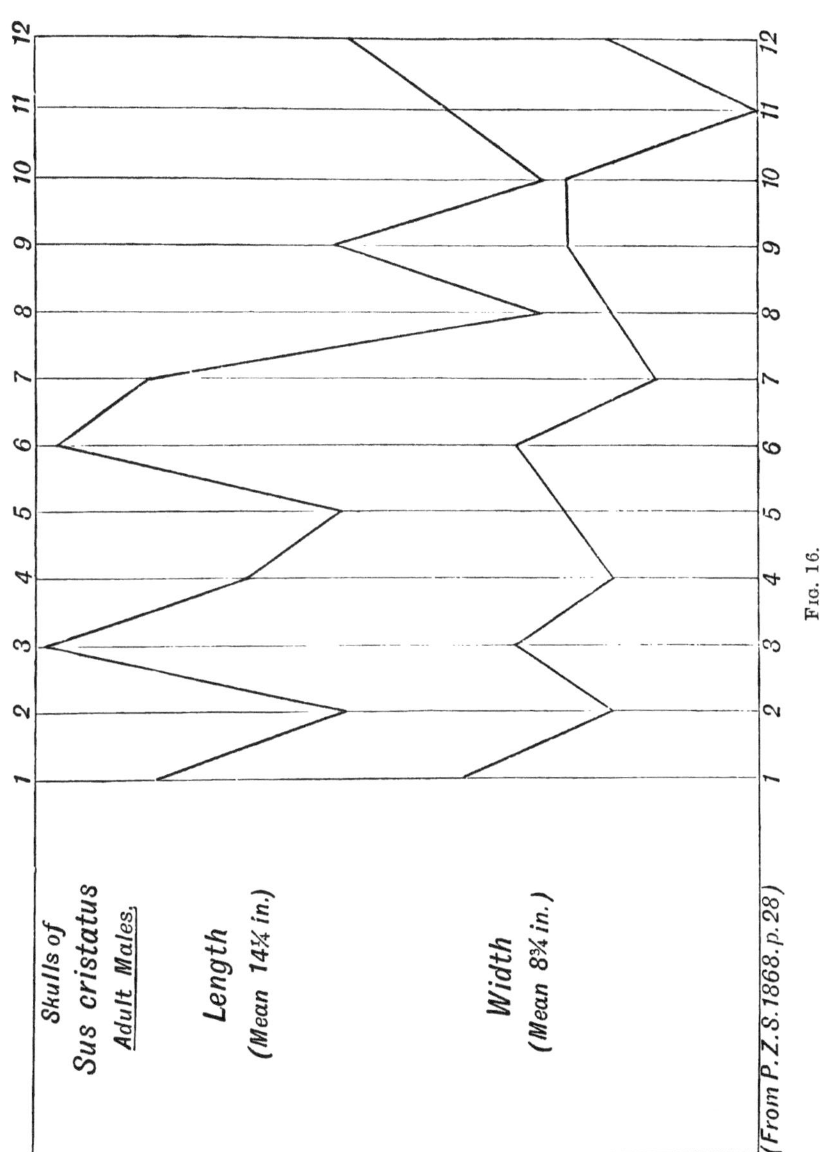
Skulls of
Sus cristatus
Adult Males.
Length
(Mean 14¼ in.)
Width
(Mean 8¾ in.)
(From P.Z.S. 1868. p. 28)
1 2 3 4 5 6 7 8 9 10 11 12

Fig. 16.

common species they are sure to be found; that they are everywhere of considerable amount, often reaching 20 per cent of the size of the part implicated; and that they are to a great extent independent of each other, and thus afford almost any combination of variations that may be needed.

It must be particularly noticed that the whole series of variation-diagrams here given (except the three which illustrate the number of varying individuals) in every case represent the actual amount of the variation, not on any reduced or enlarged scale, but as it were life-size. Whatever number of inches or decimals of an inch the species varies in any of its parts is marked on the diagrams, so that with the help of an ordinary divided rule or a pair of compasses the variation of the different parts can be ascertained and compared just as if the specimens themselves were before the reader, but with much greater ease.

In my lectures on the Darwinian theory in America and in this country I used diagrams constructed on a different plan, equally illustrating the large amount of independent variability, but less simple and less intelligible. The present method is a modification of that used by Mr. Francis Galton in his researches on the theory of variability, the upper line (showing the variability of the body) in Diagrams 4, 5, 6, and 13, being laid down on the method he has used in his experiments with sweet-peas and in pedigree moth-breeding.[1] I believe, after much consideration, and many tedious experiments in diagram-making, that no better method can be adopted for bringing before the eye, both the amount and the peculiar features of individual variability.

Variations of the Habits of Animals.

Closely connected with those variations of internal and external structure which have been already described, are the changes of habits which often occur in certain individuals or in whole species, since these must necessarily depend upon some corresponding change in the brain or in other parts of the organism; and as these changes are of great importance in relation to the theory of instinct, a few examples of them will be now adduced.

[1] See *Trans. Entomological Society of London*, 1887, p. 24.

The Kea (Nestor notabilis) is a curious parrot inhabiting the mountain ranges of the Middle Island of New Zealand. It belongs to the family of Brush-tongued parrots, and naturally feeds on the honey of flowers and the insects which frequent them, together with such fruits or berries as are found in the region. Till quite recently this comprised its whole diet, but since the country it inhabits has become occupied by Europeans it has developed a taste for a carnivorous diet, with alarming results. It began by picking the sheepskins hung out to dry or the meat in process of being cured. About 1868 it was first observed to attack living sheep, which had frequently been found with raw and bleeding wounds on their backs. Since then it is stated that the bird actually burrows into the living sheep, eating its way down to the kidneys, which form its special delicacy. As a natural consequence, the bird is being destroyed as rapidly as possible, and one of the rare and curious members of the New Zealand fauna will no doubt shortly cease to exist. The case affords a remarkable instance of how the climbing feet and powerful hooked beak developed for one set of purposes can be applied to another altogether different purpose, and it also shows how little real stability there may be in what appear to us the most fixed habits of life. A somewhat similar change of diet has been recorded by the Duke of Argyll, in which a goose, reared by a golden eagle, was taught by its foster-parent to eat flesh, which it continued to do regularly and apparently with great relish.[1]

Change of habits appears to be often a result of imitation, of which Mr. Tegetmeier gives some good examples. He states that if pigeons are reared exclusively with small grain, as wheat or barley, they will starve before eating beans. But when they are thus starving, if a bean-eating pigeon is put among them, they follow its example, and thereafter adopt the habit. So fowls sometimes refuse to eat maize, but on seeing others eat it, they do the same and become excessively fond of it. Many persons have found that their yellow crocuses were eaten by sparrows, while the blue, purple, and white coloured varieties were left untouched; but Mr. Tegetmeier, who grows only these latter colours, found that after

[1] *Nature*, vol. xix. p. 554.

two years the sparrows began to attack them, and thereafter destroyed them quite as readily as the yellow ones; and he believes it was merely because some bolder sparrow than the rest set the example. On this subject Mr. Charles C. Abbott well remarks: "In studying the habits of our American birds—and I suppose it is true of birds everywhere—it must at all times be remembered that there is less stability in the habits of birds than is usually supposed; and no account of the habits of any one species will exactly detail the various features of its habits as they really are, in every portion of the territory it inhabits." [1]

Mr. Charles Dixon has recorded a remarkable change in the mode of nest-building of some common chaffinches which were taken to New Zealand and turned out there. He says: "The cup of the nest is small, loosely put together, apparently lined with feathers, and the walls of the structure are prolonged for about 18 inches, and hang loosely down the side of the supporting branch. The whole structure bears some resemblance to the nests of the hangnests (Icteridæ), with the exception that the cavity is at the top. Clearly these New Zealand chaffinches were at a loss for a design when fabricating their nest. They had no standard to work by, no nests of their own kind to copy, no older birds to give them any instruction, and the result is the abnormal structure I have just described." [2]

These few examples are sufficient to show that both the habits and instincts of animals are subject to variation; and had we a sufficient number of detailed observations we should probably find that these variations were as numerous, as diverse in character, as large in amount, and as independent of each other as those which we have seen to characterise their bodily structure.

The Variability of Plants.

The variability of plants is notorious, being proved not only by the endless variations which occur whenever a species is largely grown by horticulturists, but also by the great difficulty that is felt by botanists in determining the limits of species in

[1] *Nature*, vol. xvi. p. 163; and vol. xi. p. 227.
[2] *Ibid.*, vol. xxxi. (1885), p. 533.

many large genera. As examples we may take the roses, the brambles, and the willows as well illustrating this fact. In Mr. Baker's *Revision of the British Roses* (published by the Linnean Society in 1863), he includes under the single species, Rosa canina—the common dog-rose—no less than twenty-eight named *varieties* distinguished by more or less constant characters and often confined to special localities, and to these are referred about seventy of the *species* of British and continental botanists. Of the genus Rubus or bramble, *five* British species are given in Bentham's *Handbook of the British Flora*, while in the fifth edition of Babington's *Manual of British Botany*, published about the same time, no less than *forty-five* species are described. Of willows (Salix) the same two works enumerate *fifteen* and *thirty-one* species respectively. The hawkweeds (Hieracium) are equally puzzling, for while Mr. Bentham admits only seven British species, Professor Babington describes no less than thirty-two, besides several named varieties.

A French botanist, Mons. A. Jordan, has collected numerous forms of a common little plant, the spring whitlow-grass (Draba verna); he has cultivated these for several successive years, and declares that they preserve their peculiarities unchanged; he also says that they each come true from seed, and thus possess all the characteristics of true species. He has described no less than fifty-two such species or permanent varieties, all found in the south of France; and he urges botanists to follow his example in collecting, describing, and cultivating all such varieties as may occur in their respective districts. Now, as the plant is very common almost all over Europe and ranges from North America to the Himalayas, the number of similar forms over this wide area would probably have to be reckoned by hundreds if not by thousands.

The class of facts now adduced must certainly be held to prove that in many large genera and in some single species there is a very large amount of variation, which renders it quite impossible for experts to agree upon the limits of species. We will now adduce a few striking cases of individual variation.

The distinguished botanist, Alp. de Candolle, made a special study of the oaks of the whole world, and has stated some remarkable facts as to their variability. He declares that on the same branch of oak he has noted the following variations: (1) In the length of the petiole, as one to three; (2) in the form of the leaf, being either elliptical or obovoid; (3) in the margin being entire, or notched, or even pinnatifid; (4) in the extremity being acute or blunt; (5) in the base being sharp, blunt, or cordate; (6) in the surface being pubescent or smooth; (7) the perianth varies in depth and lobing; (8) the stamens vary in number, independently; (9) the anthers are mucronate or blunt; (10) the fruit stalks vary greatly in length, often as one to three; (11) the number of fruits varies; (12) the form of the base of the cup varies; (13) the scales of the cup vary in form; (14) the proportions of the acorns vary; (15) the times of the acorns ripening and falling vary.

Besides this, many species exhibit well-marked varieties which have been described and named, and these are most numerous in the best-known species. Our British oak (Quercus robur) has twenty-eight varieties; Quercus Lusitanica has eleven; Quercus calliprinos has ten; and Quercus coccifera eight.

A most remarkable case of variation in the parts of a common flower has been given by Dr. Hermann Müller. He examined two hundred flowers of Myosurus minimus, among which he found *thirty-five* different proportions of the sepals, petals, and anthers, the first varying from four to seven, the second from two to five, and the third from two to ten. Five sepals occurred in one hundred and eighty-nine out of the two hundred, but of these one hundred and five had three petals, forty-six had four petals, and twenty-six had five petals; but in each of these sets the anthers varied in number from three to eight, or from two to nine. We have here an example of the same amount of "independent variability" that, as we have seen, occurs in the various dimensions of birds and mammals; and it may be taken as an illustration of the kind and degree of variability that may be expected to occur among small and little specialised flowers.[1]

In the common wind-flower (Anemone nemorosa) an almost equal amount of variation occurs; and I have myself gathered

[1] *Nature*, vol. xxvi. p. 81.

in one locality flowers varying from $\frac{7}{8}$ inch to $1\frac{3}{4}$ inch in diameter; the bracts varying from $1\frac{1}{2}$ inch to 4 inches across; and the petaloid sepals either broad or narrow, and varying in number from five to ten. Though generally pure white on their upper surface, some specimens are a full pink, while others have a decided bluish tinge.

Mr. Darwin states that he carefully examined a large number of plants of Geranium phæum and G. pyrenaicum (not perhaps truly British but frequently found wild), which had escaped from cultivation, and had spread by seed in an open plantation; and he declares that "the seedlings varied in almost every single character, both in their flowers and foliage, to a degree which I have never seen exceeded; yet they could not have been exposed to any great change of their conditions."[1]

The following examples of variation in important parts of plants were collected by Mr. Darwin and have been copied from his unpublished MSS.:—

"De Candolle (*Mem. Soc. Phys. de Genève*, tom. ii. part ii. p. 217) states that Papaver bracteatum and P. orientale present indifferently two sepals and four petals, or three sepals and six petals, which is sufficiently rare with other species of the genus."

"In the Primulaceæ and in the great class to which this family belongs the unilocular ovarium is free, but M. Dubury (*Mem. Soc. Phys. de Genève*, tom. ii. p. 406) has often found individuals in Cyclamen hederæfolium, in which the base of the ovary was connected for a third part of its length with the inferior part of the calyx."

"M. Aug. St. Hilaire (Sur la Gynobase, *Mem. des Mus. d'Hist. Nat.*, tom. x. p. 134), speaking of some bushes of the Gomphia oleæfolia, which he at first thought formed a quite distinct species, says: 'Voilà donc dans un même individu des loges et un style qui se rattachent tantôt a un axe vertical, et tantôt a un gynobase; donc celui-ci n'est qu'un axe veritable; mais cet axe est deprimé au lieu d'être vertical." He adds (p. 151), 'Does not all this indicate that nature has tried, in a manner, in the family of Rutaceæ to produce from a single multilocular ovary, one-styled and symmetrical, several unilocular ovaries, each with its own style.' And he

[1] *Animals and Plants under Domestication*, vol. ii. p. 258.

subsequently shows that, in Xanthoxylum monogynum, 'it often happens that on the same plant, on the same panicle, we find flowers with one or with two ovaries;' and that this is an important character is shown by the Rutaceæ (to which Xanthoxylum belongs), being placed in a group of natural orders characterised by having a solitary ovary."

"De Candolle has divided the Cruciferæ into five sub-orders in accordance with the position of the radicle and cotyledons, yet Mons. T. Gay (*Ann. des Scien. Nat.*, ser. i. tom. vii. p. 389) found in sixteen seeds of Petrocallis Pyrenaica the form of the embryo so uncertain that he could not tell whether it ought to be placed in the sub-orders 'Pleurorhizée' or 'Notorhizée'; so again (p. 400) in Cochlearia saxatilis M. Gay examined twenty-nine embryos, and of these sixteen were vigorously 'pleurorhizées,' nine had characters intermediate between pleuro- and notor- hizées, and four were pure notorhizées."

"M. Raspail asserts (*Ann. des Scien. Nat.*, ser. i. tom. v. p. 440) that a grass (Nostus Borbonicus) is so eminently variable in its floral organisation, that the varieties might serve to make a family with sufficiently numerous genera and tribes—a remark which shows that important organs must be here variable."

Species which vary little.

The preceding statements, as to the great amount of variation occurring in animals and plants, do not prove that all species vary to the same extent, or even vary at all, but, merely, that a considerable number of species in every class, order, and family do so vary. It will have been observed that the examples of great variability have all been taken from common species, or species which have a wide range and are abundant in individuals. Now Mr. Darwin concludes, from an elaborate examination of the floras and faunas of several distinct regions, that common, wide ranging species, as a rule, vary most, while those that are confined to special districts and are therefore comparatively limited in number of individuals vary least. By a similar comparison it is shown that species of large genera vary more than species of small genera. These facts explain, to some extent, why the opinion has been so prevalent that variation is very limited in amount and exceptional in character. For naturalists of the old school, and all mere collectors, were interested in species in proportion to their rarity, and would often have in their collections a larger number of specimens of a rare species than of a species that was very common. Now as these rare species do really vary much less than the common species, and in many cases hardly vary at all, it was very natural that a belief in the fixity of species should prevail. It is not, however, as we shall see presently, the rare, but the common and widespread species which become the parents of new forms, and thus the non-variability of any number of rare or local species offers no difficulty whatever in the way of the theory of evolution.

Concluding Remarks.

We have now shown in some detail, at the risk of being tedious, that individual variability is a general character of all common and widespread species of animals or plants; and, further, that this variability extends, so far as we know, to every part and organ, whether external or internal, as well as to every mental faculty. Yet more important is the fact that each part or organ varies to a considerable extent independently of other parts. Again, we have shown, by abundant evidence, that the variation that occurs is very large in amount—usually reaching 10 or 20, and sometimes even 25 per cent of the average size of the varying part; while not one or two only, but from 5 to 10 per cent of the specimens examined exhibit nearly as large an amount of variation. These facts have been brought clearly before the reader by means of numerous diagrams, drawn to scale and exhibiting the actual variations in inches, so that there can be no possibility of denying either their generality or their amount. The importance of this full exposition of the subject will be seen in future chapters, when we shall frequently have to refer to the facts here set forth, especially when we deal with the various theories of recent writers and the criticisms that have been made of the Darwinian theory.

A full exposition of the facts of variation among wild animals and plants is the more necessary, because comparatively few of them were published in Mr. Darwin's works, while the more important have only been made known since the last edition of *The Origin of Species* was prepared; and it is clear that Mr. Darwin himself did not fully recognise the enormous amount of variability that actually exists. This is indicated by his frequent reference to the extreme slowness of the changes for which variation furnishes the materials, and also by his use of such expressions as the following: "A variety when once formed must again, *perhaps after a long interval of time*, vary or present individual differences of the same favourable nature as before" (*Origin*, p. 66). And again, after speaking of changed conditions "affording a better chance of the occurrence of favourable variations," he adds: "*Unless such occur natural selection can do nothing*" (*Origin*, p. 64). These expressions are hardly consistent with the fact of the constant and large amount of variation, of every part, in all directions, which evidently occurs in each generation of all the more abundant species, and which must afford an ample supply of favourable variations whenever required; and they have been seized upon and exaggerated by some writers as proofs of the extreme difficulties in the way of the theory. It is to show that such difficulties do not exist, and in the full conviction that an adequate knowledge of the facts of variation affords the only sure foundation for the Darwinian theory of the origin of species, that this chapter has been written.

CHAPTER IV

VARIATION OF DOMESTICATED ANIMALS AND CULTIVATED PLANTS

The facts of variation and artificial selection—Proofs of the generality of variation—Variations of apples and melons—Variations of flowers—Variations of domestic animals—Domestic pigeons—Acclimatisation—Circumstances favourable to selection by man—Conditions favourable to variation—Concluding remarks.

HAVING so fully discussed variation under nature it will be unnecessary to devote so much space to domesticated animals and cultivated plants, especially as Mr. Darwin has published two remarkable volumes on the subject where those who desire it may obtain ample information. A general sketch of the more important facts will, however, be given, for the purpose of showing how closely they correspond with those described in the preceding chapter, and also to point out the general principles which they illustrate. It will also be necessary to explain how these variations have been increased and accumulated by artificial selection, since we are thereby better enabled to understand the action of natural selection, to be discussed in the succeeding chapter.

The facts of Variation and Artificial Selection.

Every one knows that in each litter of kittens or of puppies no two are alike. Even in the case in which several are exactly alike in colours, other differences are always perceptible to those who observe them closely. They will differ in size, in the proportions of their bodies and limbs, in the length or texture of their hairy covering, and notably in their disposition. They each possess, too, an individual countenance, almost as varied when closely studied as that of a human being; not only can a shepherd distinguish every sheep in his flock, but we all know that each kitten in the successive families of our old favourite cat has a face of its own, with an expression and individuality distinct from all its brothers and sisters. Now this individual variability exists among all creatures whatever, which we can closely observe, even when the two parents are very much alike and have been matched in order to preserve some special breed. The same thing occurs in the vegetable kingdom. All plants raised from seed differ more or less from each other. In every bed of flowers or of vegetables we shall find, if we look closely, that there are countless small differences, in the size, in the mode of growth, in the shape or colour of the leaves, in the form, colour, or markings of the flowers, or in the size, form, colour, or flavour of the fruit. These differences are usually small, but are yet easily seen, and in their extremes are very considerable; and they have this important quality, that they have a tendency to be reproduced, and thus by careful breeding any particular variation or group of variations can be increased to an enormous extent—apparently to any extent not incompatible with the life, growth, and reproduction of the plant or animal.

The way this is done is by artificial selection, and it is very important to understand this process and its results. Suppose we have a plant with a small edible seed, and we want to increase the size of that seed. We grow as large a quantity of it as possible, and when the crop is ripe we carefully choose a few of the very largest seeds, or we may by means of a sieve sort out a quantity of the largest seeds. Next year we sow only these large seeds, taking care to give them suitable soil and manure, and the result is found to be that the *average* size of the seeds is larger than in the first crop, and that the largest seeds are now somewhat larger and more numerous. Again sowing these, we obtain a further slight increase of size, and in a very few years we obtain a greatly improved race, which will always produce larger seeds than the unimproved race, even if cultivated without any special care. In this way all our fine sorts of vegetables, fruits, and flowers have been obtained, all our choice breeds of cattle or of poultry, our wonderful race-horses, and our endless varieties of dogs. It is a very common but mistaken idea that this improvement is due to crossing and feeding in the case of animals, and to improved cultivation in the case of plants. Crossing is occasionally used in order to obtain a combination of qualities found in two distinct breeds, and also because it is found to increase the constitutional vigour; but every breed possessing any exceptional quality is the result of the selection of variations occurring year after year and accumulated in the manner just described. Purity of breed, with repeated selection of the best varieties of that breed, is the foundation of all improvement in our domestic animals and cultivated plants.

Proofs of the Generality of Variation.

Another very common error is, that variation is the exception, and rather a rare exception, and that it occurs only in one direction at a time—that is, that only one or two of the numerous possible modes of variation occur at the same time. The experience of breeders and cultivators, however, proves that variation is the rule instead of the exception, and that it occurs, more or less, in almost every direction. This is shown by the fact that different species of plants and animals have required different *kinds* of modification to adapt them to our use, and we have never failed to meet with variation *in that particular direction*, so as to enable us to accumulate it and so to produce ultimately a large amount of change in the required direction. Our gardens furnish us with numberless examples of this property of plants. In the cabbage and lettuce we have found variation in the size and mode of growth of the leaf, enabling us to produce by selection the almost innumerable varieties, some with solid heads of foliage quite unlike any plant in a state of nature, others with curiously wrinkled leaves like the savoy, others of a deep purple colour used for pickling. From the very same species as the cabbage (Brassica oleracea) have arisen the broccoli and cauliflower, in which the leaves have undergone little alteration, while the branching heads of flowers grow into a compact mass forming one of our most delicate vegetables. The brussels sprouts are another form of the same plant, in which the whole mode of growth has been altered, numerous little heads of leaves being produced on the stem. In other varieties the ribs of the leaves are thickened so as to become themselves a culinary vegetable; while, in the Kohlrabi, the stem grows into a turnip-like mass just above ground. Now all these extraordinarily distinct plants come from one original species which still grows wild on our coasts; and it must have varied in all these directions, otherwise variations could not have been accumulated to the extent we now see them. The flowers and seeds of all these plants have remained nearly stationary, because no attempt has been made to accumulate the slight variations that no doubt occur in them.

If now we turn to another set of plants, the turnips, radishes, carrots, and potatoes, we find that the roots or underground tubers have been wonderfully enlarged and improved, and also altered in shape and colour, while the stems, leaves, flowers, and fruits have remained almost unchanged. In the various kinds of peas and beans it is the pod or fruit and the seed that has been subjected to selection, and therefore greatly modified; and it is here very important to notice that while all these plants have undergone cultivation in a great variety of soils and climates, with different manures and under different systems, yet the flowers have remained but little altered, those of the broad bean, the scarlet-runner, and the garden-pea, being nearly the same in all the varieties. This shows us how little change is produced by mere cultivation, or even by variety of soil and climate, if there is no *selection* to preserve and accumulate the small variations that are continually occurring. When, however, a great amount of modification has been effected in one country, change to another country produces a decided effect. Thus it has been found that some of the numerous varieties of maize produced and cultivated in the United States change considerably, not only in their size and colour, but even in the shape of the seed when grown for a few successive years in Germany.[1] In all our cultivated fruit trees the fruits vary immensely in shape, size, colour, flavour, time of ripening, and other qualities, while the leaves and flowers usually differ so little that they are hardly distinguishable except to a very close observer.

[1] Darwin, *Animals and Plants under Domestication*, vol. i. p. 322.

Variations of Apples and of Melons.

The most remarkable varieties are afforded by the apple and the melon, and some account of these will be given as illustrating the effects of slight variations accumulated by selection. All our apples are known to have descended from the common crab of our hedges (Pyrus malus), and from this at least a thousand distinct varieties have been produced. These differ greatly in the size and form of the fruit, in its colour, and in the texture of the skin. They further differ in the time of ripening, in their flavour, and in their keeping properties; but apple trees also differ in many other ways. The foliage of the different varieties can often be distinguished by peculiarities of form and colour, and it varies considerably in the time of its appearance; in some hardly a leaf appears till the tree is in full bloom, while others produce their leaves so early as almost to hide the flowers. The flowers differ in size and colour, and in one case in structure also, that of the St. Valery apple having a double calyx with ten divisions, and fourteen styles with oblique stigmas, but without stamens or corolla. The flowers, therefore, have to be fertilised with the pollen from other varieties in order to produce fruit. The pips or seeds differ also in shape, size, and colour; some varieties are liable to canker more than others, while the Winter Majetin and one or two others have the strange constitutional peculiarity of never being attacked by the mealy bug even when all the other trees in the same orchard are infested with it.

All the cucumbers and gourds vary immensely, but the melon (Cucumis melo) exceeds them all. A French botanist, M. Naudin, devoted six years to their study. He found that previous botanists had described thirty distinct species, as they thought, which were really only varieties of melons. They differ chiefly in their fruits, but also very much in foliage and mode of growth. Some melons are only as large as small plums, others weigh as much as sixty-six pounds. One variety has a scarlet fruit. Another is not more than an inch in diameter, but sometimes more than a yard in length, twisting about in all directions like a serpent. Some melons are exactly like cucumbers; and an Algerian variety, when ripe, cracks and falls to pieces, just as occurs in a wild gourd (C. momordica).[1]

Variations of Flowers.

Turning to flowers, we find that in the same genus as our currant and gooseberry, which we have cultivated for their fruits, there are some ornamental species, as the Ribes sanguinea, and in these the flowers have been selected so as to produce deep red, pink, or white varieties. When any particular flower becomes fashionable and is grown in large quantities, variations are always met with sufficient to produce great varieties of tint or marking, as shown by our roses, auriculas, and geraniums. When varied leaves are required, it is found that a number of plants vary sufficiently in this direction also, and we have zonal geraniums, variegated ivies, gold and silver marked hollies, and many others.

Variations of Domestic Animals.

Coming now to our domesticated animals, we find still more extraordinary cases ; and it appears as if any special quality or modification in an animal can be obtained if we only breed it in sufficient quantity, watch carefully for the required variations, and carry on selection with patience and skill for a sufficiently long period. Thus, in sheep we have enormously increased the wool, and have obtained the power of rapidly forming flesh and fat ; in cows we have increased the production of milk ; in horses we have obtained strength, endurance, or speed, and have greatly modified size, form, and colour ; in poultry we have secured various colours of plumage, increase of size, and almost perpetual egg-laying. But it is in dogs and pigeons that the most marvellous changes have been effected, and these require our special attention.

Our various domestic dogs are believed to have originated from several distinct wild species, because in every part of the world the native dogs resemble some wild dogs or wolves of the same country. Thus perhaps several species of wolves and jackals were domesticated in very early times, and from breeds derived from these, crossed and improved by selection,

[1] These facts are taken from Darwin's *Domesticated Animals and Cultivated Plants*, vol. i. pp. 359, 360, 392-401 ; vol. ii. pp. 231, 275, 330.

our existing dogs have descended. But this intermixture of distinct species will go a very little way in accounting for the peculiarities of the different breeds of dogs, many of which are totally unlike any wild animal. Such is the case with greyhounds, bloodhounds, bulldogs, Blenheim spaniels, terriers, pugs, turnspits, pointers, and many others; and these differ so greatly in size, shape, colour, and habits, as well as in the form and proportions of all the different parts of the body, that it seems impossible that they could have descended from any of the known wild dogs, wolves, or allied animals, none of which differ nearly so much in size, form, and proportions. We have here a remarkable proof that variation is not confined to superficial characters—to the colour, hair, or external appendages, when we see how the entire skeletons of such forms as the greyhound and the bulldog have been gradually changed in opposite directions till they are both completely unlike that of any known wild animal, recent or extinct. These changes have been the result of some thousands of years of domestication and selection, different breeds being used and preserved for different purposes; but some of the best breeds are known to have been improved and perfected in modern times. About the middle of the last century a new and improved kind of foxhound was produced; the greyhound was also greatly improved at the end of the last century, while the true bulldog was brought to perfection about the same period. The Newfoundland dog has been so much changed since it was first imported that it is now quite unlike any existing native dog in that island.[1]

Domestic Pigeons.

The most remarkable and instructive example of variation produced by human selection is afforded by the various races and breeds of domestic pigeons, not only because the variations produced are often most extraordinary in amount and diverse in character, but because in this case there is no doubt whatever that all have been derived from one wild species, the common rock-pigeon (Columba livia). As this is a very important point it is well to state the evidence on which the belief is founded. The wild rock-pigeon is of a slaty-blue

[1] See Darwin's *Animals and Plants under Domestication*, vol. i. pp. 40-42.

colour, the tail has a dark band across the end, the wings have two black bands, and the outer tail-feathers are edged with white at the base. No other wild pigeon in the world has this combination of characters. Now in every one of the domestic varieties, even the most extreme, all the above marks, even to the white edging of the outer tail-feathers, are sometimes found perfectly developed. When birds belonging to two distinct breeds are crossed one or more times, neither of the parents being blue, or having any of the above-named marks, the mongrel offspring are very apt to acquire some of these characters. Mr. Darwin gives instances which he observed himself. He crossed some white fantails with some black barbs, and the mongrels were black, brown, or mottled. He also crossed a barb with a spot, which is a white bird with a red tail and red spot on the forehead, and the mongrel offspring were dusky and mottled. On now crossing these two sets of mongrels with each other, he obtained a bird of a beautiful blue colour, with the barred and white edged tail, and double-banded wings, so as almost exactly to resemble a wild rock-pigeon. This bird was descended in the second generation from a pure white and pure black bird, both of which when unmixed breed their kind remarkably true. These facts, well known to experienced pigeon-fanciers, together with the habits of the birds, which all like to nest in holes, or dovecots, not in trees like the great majority of wild pigeons, have led to the general belief in the single origin of all the different kinds.

In order to afford some idea of the great differences which exist among domesticated pigeons, it will be well to give a brief abstract of Mr. Darwin's account of them. He divides them into eleven distinct races, most of which have several sub-races.

RACE I. *Pouters.*—These are especially distinguished by the enormously enlarged crop, which can be so inflated in some birds as almost to conceal the beak. They are very long in the body and legs and stand almost upright, so as to present a very distinct appearance. Their skeleton has become modified, the ribs being broader and the vertebræ more numerous than in other pigeons.

RACE II. *Carriers.*—These are large, long-necked birds, with a long pointed beak, and the eyes surrounded with a naked carunculated skin or wattle, which is also largely developed at the base of the beak. The opening of the mouth is unusually wide. There are several sub-races, one being called Dragons.

RACE III. *Runts.*—These are very large-bodied, long-beaked pigeons, with naked skin round the eyes. The wings are usually very long, the legs long, and the feet large, and the skin of the neck is often red. There are several sub-races, and these differ very much, forming a series of links between the wild rock-pigeon and the carrier.

RACE IV. *Barbs.*—These are remarkable for their very short and thick beak, so unlike that of most pigeons that fanciers compare it with that of a bullfinch. They have also a naked carunculated skin round the eyes, and the skin over the nostrils swollen.

RACE V. *Fantails.*—Short-bodied and rather small-beaked pigeons, with an enormously developed tail, consisting usually of from fourteen to forty feathers instead of twelve, the regular number in all other pigeons, wild and tame. The tail spreads out like a fan and is usually carried erect, and the bird bends back its slender neck, so that in highly-bred varieties the head touches the tail. The feet are small, and they walk stiffly.

RACE VI. *Turbits and Owls.*—These are characterised by the feathers of the middle of neck and breast in front spreading out irregularly so as to form a frill. The Turbits also have a crest on the head, and both have the beak exceedingly short.

RACE VII. *Tumblers.*—These have a small body and short beak, but they are specially distinguished by the singular habit of tumbling over backwards during flight. One of the sub-races, the Indian Lotan or Ground tumbler, if slightly shaken and placed on the ground, will immediately begin tumbling head over heels until taken up and soothed. If not taken up, some of them will go on tumbling till they die.

Some English tumblers are almost equally persistent. A writer, quoted by Mr. Darwin, says that these birds generally begin to tumble almost as soon as they can fly; "at three months old they tumble well, but still fly strong; at five or six months they tumble excessively; and in the second year they mostly give up flying, on account of their tumbling so much and so close to the ground. Some fly round with the flock, throwing a clean summersault every few yards till they are obliged to settle from giddiness and exhaustion. These are called Air-tumblers, and they commonly throw from twenty to thirty summersaults in a minute, each clear and clean. I have one red cock that I have on two or three occasions timed by my watch, and counted forty summersaults in the minute. At first they throw a single summersault, then it is double, till it becomes a continuous roll, which puts an end to flying, for if they fly a few yards over they go, and roll till they reach the ground. Thus I had one kill herself, and another broke his leg. Many of them turn over only a few inches from the ground, and will tumble two or three times in flying across their loft. These are called House-tumblers from tumbling in the house. The act of tumbling seems to be one over which they have no control, an involuntary movement which they seem to try to prevent. I have seen a bird sometimes in his struggles fly a yard or two straight upwards, the impulse forcing him backwards while he struggles to go forwards." [1]

The Short-faced tumblers are an improved sub-race which have almost lost the power of tumbling, but are valued for possessing some other characteristics in an extreme degree. They are very small, have almost globular heads, and a very minute beak, so that fanciers say the head of a perfect bird should resemble a cherry with a barleycorn stuck in it. Some of these weigh less than seven ounces, whereas the wild rock-pigeon weighs about fourteen ounces. The feet, too, are very short and small, and the middle toe has twelve or thirteen instead of fourteen or fifteen scutellæ. They have often only nine primary wing-feathers instead of ten as in all other pigeons.

[1] Mr. Brent in *Journal of Horticulture*, 1861, p. 76; quoted by Darwin, *Animals and Plants under Domestication*, vol. i. p. 151.

RACE VIII. *Indian Frill-back.*—In these birds the beak is very short, and the feathers of the whole body are reversed or turn backwards.

RACE IX. *Jacobin.*—These curious birds have a hood of feathers almost enclosing the head and meeting in front of the neck. The wings and tail are unusually long.

RACE X. *Trumpeter.*—Distinguished by a tuft of feathers curling forwards over the beak, and the feet very much feathered. They obtain their name from the peculiar voice unlike that of any other pigeon. The coo is rapidly repeated, and is continued for several minutes. The feet are covered with feathers so large as often to appear like little wings.

RACE XI. comprises *Laughers, Frill-backs, Nuns, Spots, and Swallows.*—They are all very like the common rock-pigeon, but have each some slight peculiarity. The Laughers have a peculiar voice, supposed to resemble a laugh. The Nuns are white, with the head, tail, and primary wing-feathers black or red. The Spots are white, with the tail and a spot on the forehead red. The Swallows are slender, white in colour, with the head and wings of some darker colour.

Besides these races and sub-races a number of other kinds have been described, and about one hundred and fifty varieties can be distinguished. It is interesting to note that almost every part of the bird, whose variations can be noted and selected, has led to variations of a considerable extent, and many of these have necessitated changes in the plumage and in the skeleton quite as great as any that occur in the numerous distinct species of large genera. The form of the skull and beak varies enormously, so that the skulls of the Short-faced tumbler and some of the Carriers differ more than any wild pigeons, even those classed in distinct genera. The breadth and number of the ribs vary, as well as the processes on them; the number of the vertebræ and the length of the sternum also vary; and the perforations in the sternum vary in size and shape. The oil gland varies in development, and is sometimes absent. The number of the wing-feathers varies, and those of the tail to an enormous extent. The proportions of the leg and feet and the number of the scutellæ also vary. The eggs also vary somewhat in size and shape; and the amount of downy clothing on the young bird, when first hatched, differs very considerably. Finally, the attitude of the body, the manner of walking, the mode of flight, and the voice, all exhibit modifications of the most remarkable kind.[1]

Acclimatisation.

A very important kind of variation is that constitutional change termed acclimatisation, which enables any organism to become gradually adapted to a different climate from the parent stock. As closely allied species often inhabit different countries possessing very different climates, we should expect to find cases illustrating this change among our domesticated animals and cultivated plants. A few examples will therefore be adduced showing that such constitutional variation does occur.

Among animals the cases are not numerous, because no systematic attempt has been made to select varieties for this special quality. It has, however, been observed that, though no European dogs thrive well in India, the Newfoundland dog, originating from a severe climate, can hardly be kept alive. A better case, perhaps, is furnished by merino sheep, which, when imported directly from England, do not thrive, while those which have been bred in the intermediate climate of the Cape of Good Hope do much better. When geese were first introduced into Bogota, they laid few eggs at long intervals, and few of the young survived. By degrees, however, the fecundity improved, and in about twenty years became equal to what it is in Europe. According to Garcilaso, when fowls were first introduced into Peru they were not fertile, whereas now they are as much so as in Europe.

Plants furnish much more important evidence. Our nurserymen distinguish in their catalogues varieties of fruit-trees which are more or less hardy, and this is especially the case in America, where certain varieties only will stand the severe climate of Canada. There is one variety of pear, the Forelle, which both in England and France withstood frosts

[1] This account of domestic pigeons is greatly condensed from Mr. Darwin's work already referred to.

that killed the flowers and buds of all other kinds of pears. Wheat, which is grown over so large a portion of the world, has become adapted to special climates. Wheat imported from India and sown in good wheat soil in England produced the most meagre ears; while wheat taken from France to the West Indian Islands produced either wholly barren spikes or spikes furnished with two or three miserable seeds, while West Indian seed by its side yielded an enormous harvest. The orange was very tender when first introduced into Italy, and continued so as long as it was propagated by grafts, but when trees were raised from seed many of these were found to be hardier, and the orange is now perfectly acclimatised in Italy. Sweet-peas (Lathyrus odoratus) imported from England to the Calcutta Botanic Gardens produced few blossoms and no seed; those from France flowered a little better, but still produced no seed, but plants raised from seed brought from Darjeeling in the Himalayas, but originally derived from England, flower and seed profusely in Calcutta.[1]

An observation by Mr. Darwin himself is perhaps even more instructive. He says: "On 24th May 1864 there was a severe frost in Kent, and two rows of scarlet runners (Phaseolus multiflorus) in my garden, containing 390 plants of the same age and equally exposed, were all blackened and killed except about a dozen plants. In an adjoining row of Fulmer's dwarf bean (Phaseolus vulgaris) one single plant escaped. A still more severe frost occurred four days afterwards, and of the dozen plants which had previously escaped only three survived; these were not taller or more vigorous than the other young plants, but they escaped completely, with not even the tips of their leaves browned. It was impossible to behold these three plants, with their blackened, withered, and dead brethren all around them, and not see at a glance that they differed widely in their constitutional power of resisting frost."

The preceding sketch of the variation that occurs among domestic animals and cultivated plants shows how wide it is in range and how great in amount; and we have good reason to believe that similar variation extends to all organised beings. In the class of fishes, for example, we have one kind which has

[1] *Animals and Plants under Domestication*, vol. ii. pp. 307-311.

been long domesticated in the East, the gold and silver carps; and these present great variation, not only of colour but in the form and structure of the fins and other external organs. In like manner, the only domesticated insects, hive bees and silk-worm moths, present numbers of remarkable varieties which have been produced by the selection of chance variations just as in the case of plants and the higher animals.

Circumstances favourable to Selection by Man.

It may be supposed, that the systematic selection which has been employed for the purpose of improving the races of animals or plants useful to man is of comparatively recent origin, though some of the different races are known to have been in existence in very early times. But Mr. Darwin has pointed out, that unconscious selection must have begun to produce an effect as soon as plants were cultivated or animals domesticated by man. It would have been very soon observed that animals and plants produced their like, that seed of early wheat produced early wheat, that the offspring of very swift dogs were also swift, and as every one would try to have a good rather than a bad sort this would necessarily lead to the slow but steady improvement of all useful plants and animals subject to man's care. Soon there would arise distinct breeds, owing to the varying uses to which the animals and plants were put. Dogs would be wanted chiefly to hunt one kind of game in one part of the country and another kind elsewhere; for one purpose scent would be more important, for another swiftness, for another strength and courage, for yet another watchfulness and intelligence, and this would soon lead to the formation of very distinct races. In the case of vegetables and fruits, different varieties would be found to succeed best in certain soils and climates; some might be preferred on account of the quantity of food they produced, others for their sweetness and tenderness, while others might be more useful on account of their ripening at a particular season, and thus again distinct varieties would be established. An instance of unconscious selection leading to distinct results in modern times is afforded by two flocks of Leicester sheep which both originated from the same stock, and were then bred pure for upwards of fifty years by two gentlemen, Mr. Buckley and Mr. Burgess. Mr. Youatt, one of the greatest authorities on breeding domestic animals, says: "There is not a suspicion existing in the mind of any one at all acquainted with the subject that the owner of either of them has deviated in any one instance from the pure blood of Mr. Bakewell's original flock, and yet the difference between the sheep possessed by these two gentlemen is so great that they have the appearance of being quite different varieties." In this case there was no desire to deviate from the original breed, and the difference must have arisen from some slight difference of taste or judgment in selecting, each year, the parents for the next year's stock, combined perhaps with some direct effect of the slight differences of climate and soil on the two farms.

Most of our domesticated animals and cultivated plants have come to us from the earliest seats of civilisation in Western Asia or Egypt, and have therefore been the subjects of human care and selection for some thousands of years, the result being that, in many cases, we do not know the wild stock from which they originally sprang. The horse, the camel, and the common bull and cow are nowhere found in a wild state, and they have all been domesticated from remote antiquity. The original of the domestic fowl is still wild in India and the Malay Islands, and it was domesticated in India and China before 1400 B.C. It was introduced into Europe about 600 B.C. Several distinct breeds were known to the Romans about the commencement of the Christian era, and they have since spread all over the civilised world and been subjected to a vast amount of conscious and unconscious selection, to many varieties of climate and to differences of food; the result being seen in the wonderful diversity of breeds which differ quite as remarkably as do the different races of pigeons already described.

In the vegetable kingdom, most of the cereals—wheat, barley, etc.—are unknown as truly wild plants; and the same is the case with many vegetables, for De Candolle states that out of 157 useful cultivated plants thirty-two are quite unknown in a wild state, and that forty more are of doubtful origin. It is not improbable that most of these do exist wild, but they have been so profoundly changed by thousands of years of cultivation as to be quite unrecognisable. The peach is unknown in a wild state, unless it is derived from the common almond, on which point there is much difference of opinion among botanists and horticulturists.

The immense antiquity of most of our cultivated plants sufficiently explains the apparent absence of such useful productions in Australia and the Cape of Good Hope, notwithstanding that they both possess an exceedingly rich and varied flora. These countries having been, until a comparatively recent period, inhabited only by uncivilised men, neither cultivation nor selection has been carried on for a sufficiently long time. In North America, however, where there was evidently a very ancient if low form of civilisation, as indicated by the remarkable mounds, earthworks, and other prehistoric remains, maize was cultivated, though it was probably derived from Peru; and the ancient civilisation of that country and of Mexico has given rise to no fewer than thirty-three useful cultivated plants.

Conditions favourable to the production of Variations.

In order that plants and animals may be improved and modified to any considerable extent, it is of course essential that suitable variations should occur with tolerable frequency. There seem to be three conditions which are especially favourable to the production of variations: (1) That the particular species or variety should be kept in very large numbers; (2) that it should be spread over a wide area and thus subjected to a considerable diversity of physical conditions; and (3) that it should be occasionally crossed with some distinct but closely allied race. The first of these conditions is perhaps the most important, the chance of variations of any particular kind being increased in proportion to the quantity of the original stock and of its annual offspring. It has been remarked that only those breeders who keep large flocks can effect much improvement; and it is for the same reason that pigeons and fowls, which can be so easily and rapidly increased, and which have been kept in such large numbers by so great a number of persons, have produced such strange and numerous varieties. In like manner, nurserymen who grow fruit and flowers in large quantities have a great advantage over private amateurs in the production of new varieties.

Although I believe, for reasons which will be given further on, that some amount of variability is a constant and necessary property of all organisms, yet there appears to be good evidence to show that changed conditions of life tend to increase it, both by a direct action on the organisation and by indirectly affecting the reproductive system. Hence the extension of civilisation, by favouring domestication under altered conditions, facilitates the process of modification. Yet this change does not seem to be an essential condition, for nowhere has the production of extreme varieties of plants and flowers been carried farther than in Japan, where careful selection continued for many generations must have been the chief factor. The effect of occasional crosses often results in a great amount of variation, but it also leads to instability of character, and is therefore very little employed in the production of fixed and well-marked races. For this purpose, in fact, it has to be carefully avoided, as it is only by isolation and pure breeding that any specially desired qualities can be increased by selection. It is for this reason that among savage peoples, whose animals run half wild, little improvement takes place; and the difficulty of isolation also explains why distinct and pure breeds of cats are so rarely met with. The wide distribution of useful animals and plants from a very remote epoch has, no doubt, been a powerful cause of modification, because the particular breed first introduced into each country has often been kept pure for many years, and has also been subjected to slight differences of conditions. It will also usually have been selected for a somewhat different purpose in each locality, and thus very distinct races would soon originate.

The important physiological effects of crossing breeds or strains, and the part this plays in the economy of nature, will be explained in a future chapter.

Concluding Remarks.

The examples of variation now adduced—and these might have been almost indefinitely increased—will suffice to show that there is hardly an organ or a quality in plants or animals which has not been observed to vary; and further, that whenever any of these variations have been useful to man he has been able to increase them to a marvellous extent by the simple process of always preserving the best varieties to breed from. Along with these larger variations others of smaller amount occasionally appear, sometimes in external, sometimes in internal characters, the very bones of the skeleton often changing slightly in form, size, or number; but as these secondary characters have been of no use to man, and have not been specially selected by him, they have, usually, not been developed to any great amount except when they have been closely dependent on those external characters which he has largely modified.

As man has considered only utility to himself, or the satisfaction of his love of beauty, of novelty, or merely of something strange or amusing, the variations he has thus produced have something of the character of monstrosities. Not only are they often of no use to the animals or plants themselves, but they are not unfrequently injurious to them. In the Tumbler pigeons, for instance, the habit of tumbling is sometimes so excessive as to injure or kill the bird; and many of our highly-bred animals have such delicate constitutions that they are very liable to disease, while their extreme peculiarities of form or structure would often render them quite unfit to live in a wild state. In plants, many of our double flowers, and some fruits, have lost the power of producing seed, and the race can thus be continued only by means of cuttings or grafts. This peculiar character of domestic productions distinguishes them broadly from wild species and varieties, which, as will be seen by and by, are necessarily adapted in every part of their organisation to the conditions under which they have to live. Their importance for our present inquiry depends on their demonstrating the occurrence of incessant slight variations in all parts of an organism, with the transmission to the offspring of the special characteristics of the parents; and also, that all such slight variations are capable of being accumulated by selection till they present very large and important divergencies from the ancestral stock.

We thus see, that the evidence as to variation afforded by animals and plants under domestication strikingly accords with that which we have proved to exist in a state of nature.

And it is not at all surprising that it should be so, since all the species were in a state of nature when first domesticated or cultivated by man, and whatever variations occur must be due to purely natural causes. Moreover, on comparing the variations which occur in any one generation of domesticated animals with those which we know to occur in wild animals, we find no evidence of greater individual variation in the former than in the latter. The results of man's selection are more striking to us because we have always considered the varieties of each domestic animal to be essentially identical, while those which we observe in a wild state are held to be essentially diverse. The greyhound and the spaniel seem wonderful, as varieties of one animal produced by man's selection; while we think little of the diversities of the fox and the wolf, or the horse and the zebra, because we have been accustomed to look upon them as radically distinct animals, not as the results of nature's selection of the varieties of a common ancestor.

CHAPTER V

NATURAL SELECTION BY VARIATION AND SURVIVAL OF THE FITTEST

Effect of struggle for existence under unchanged conditions—The effect under change of conditions—Divergence of character—In insects—In birds—In mammalia—Divergence leads to a maximum of life in each area—Closely allied species inhabit distinct areas—Adaptation to conditions at various periods of life—The continued existence of low forms of life—Extinction of low types among the higher animals—Circumstances favourable to the origin of new species—Probable origin of the dippers—The importance of isolation—On the advance of organisation by natural selection—Summary of the first five chapters.

In the preceding chapters we have accumulated a body of facts and arguments which will enable us now to deal with the very core of our subject—the formation of species by means of natural selection. We have seen how tremendous is the struggle for existence always going on in nature owing to the great powers of increase of all organisms; we have ascertained the fact of variability extending to every part and organ, each of which varies simultaneously and for the most part independently; and we have seen that this variability is both large in its amount in proportion to the size of each part, and usually affects a considerable proportion of the individuals in the large and dominant species. And, lastly, we have seen how similar variations, occurring in cultivated plants and domestic animals, are capable of being perpetuated and accumulated by artificial selection, till they have resulted in all the wonderful varieties of our fruits, flowers, and vegetables, our domestic animals and household pets, many of which differ from each other far more in external characters, habits, and instincts than do species in a state of nature. We have now to inquire whether there is any analogous process in nature, by which wild animals and plants can be permanently modified and new races or new species produced.

Effect of Struggle for Existence under Unchanged Conditions.

Let us first consider what will be the effect of the struggle for existence upon the animals and plants which we see around us, under conditions which do not perceptibly vary from year to year or from century to century. We have seen that every species is exposed to numerous and varied dangers throughout its entire existence, and that it is only by means of the exact adaptation of its organisation—including its instincts and habits—to its surroundings that it is enabled to live till it produces offspring which may take its place when it ceases to exist. We have seen also that, of the whole annual increase only a very small fraction survives; and though the survival in individual cases may sometimes be due rather to accident than to any real superiority, yet we cannot doubt that, in the long run, those survive which are best fitted by their perfect organisation to escape the dangers that surround them. This "survival of the fittest" is what Darwin termed "natural selection," because it leads to the same results in nature as are produced by man's selection among domestic animals and cultivated plants. Its primary effect will, clearly, be to keep each species in the most perfect health and vigour, with every part of its organisation in full harmony with the conditions of its existence. It prevents any possible deterioration in the organic world, and produces that appearance of exuberant life and enjoyment, of health and beauty, that affords us so much pleasure, and which might lead a superficial observer to suppose that peace and quietude reigned throughout nature.

The Effect under changed Conditions.

But the very same process which, so long as conditions remain substantially the same, secures the continuance of each species of animal or plant in its full perfection, will usually, under changed conditions, bring about whatever change of structure or habits may be necessitated by them. The changed conditions to which we refer are such as we know have occurred throughout all geological time and in every part of the world. Land and water have been continually shifting their positions; some regions are undergoing subsidence with diminution of area, others elevation with extension of area; dry land has been converted into marshes, while marshes have been drained or have even been elevated into plateaux. Climate too has changed again and again, either through the elevation of mountains in high latitudes leading to the accumulation of snow and ice, or by a change in the direction of winds and ocean currents produced by the subsidence or elevation of lands which connected continents and divided oceans. Again, along with all these changes have come not less important changes in the distribution of species. Vegetation has been greatly modified by changes of climate and of altitude; while every union of lands before separated has led to extensive migrations of animals into new countries, disturbing the balance that before existed among its forms of life, leading to the extermination of some species and the increase of others.

When such physical changes as these have taken place, it is evident that many species must either become modified or cease to exist. When the vegetation has changed in character the herbivorous animals must become able to live on new and perhaps less nutritious food; while the change from a damp to a dry climate may necessitate migration at certain periods to escape destruction by drought. This will expose the species to new dangers, and require special modifications of structure to meet them. Greater swiftness, increased cunning, nocturnal habits, change of colour, or the power of climbing trees and living for a time on their foliage or fruit, may be the means adopted by different species to bring themselves into harmony with the new conditions; and by the continued survival of those individuals, only, which varied sufficiently in the right direction, the necessary modifications of structure or of function would be brought about, just as surely as man has been able to breed the greyhound to hunt by sight and the foxhound by scent, or has produced from the same wild plant such distinct forms as the cauliflower and the brussels sprouts.

We will now consider the special characteristics of the changes in species that are likely to be effected, and how far they agree with what we observe in nature.

Divergence of Character.

In species which have a wide range the struggle for existence will often cause some individuals or groups of individuals to adopt new habits in order to seize upon vacant places in nature where the struggle is less severe. Some, living among extensive marshes, may adopt a more aquatic mode of life; others, living where forests abound, may become more arboreal. In either case we cannot doubt that the changes of structure needed to adapt them to their new habits would soon be brought about, because we know that variations in all the external organs and all their separate parts are very abundant and are also considerable in amount. That such divergence of character has actually occurred we have some direct evidence. Mr. Darwin informs us that in the Catskill Mountains in the United States there are two varieties of wolves, one with a light greyhound-like form which pursues deer, the other more bulky with shorter legs, which more frequently attacks sheep.[1] Another good example is that of the insects in the island of Madeira, many of which have either lost their wings or have had them so much reduced as to be useless for flight, while the very same species on the continent of Europe possess fully developed wings. In other cases the wingless Madeira species are distinct from, but closely allied to, winged species of Europe. The explanation of this change is, that Madeira, like many oceanic islands in the temperate zone, is much exposed to sudden gales of wind, and as most of the fertile land is on the coast, insects which flew much would be very liable to be blown out to sea and lost. Year after year, therefore, those individuals which had shorter wings, or which used them least, were preserved; and thus, in time, terrestrial, wingless, or imperfectly winged races or species have been produced. That this is the true explanation of this singular fact is proved by much corroborative evidence. There are some few flower-frequenting insects in Madeira to whom wings are essential, and in these the wings are somewhat larger than in the same species on the mainland. We thus see that there is no general tendency to the abortion of wings in Madeira, but that it is simply a case of adaptation to new conditions. Those insects

[1] *Origin of Species*, p. 71.

to whom wings were not absolutely essential escaped a serious danger by not using them, and the wings therefore became reduced or were completely lost. But when they were essential they were enlarged and strengthened, so that the insect could battle against the winds and save itself from destruction at sea. Many flying insects, not varying fast enough, would be destroyed before they could establish themselves, and thus we may explain the total absence from Madeira of several whole families of winged insects which must have had many opportunities of reaching the islands. Such are the large groups of the tiger-beetles (Cicindelidæ), the chafers (Melolonthidæ), the click-beetles (Elateridæ), and many others.

But the most curious and striking confirmation of this portion of Mr. Darwin's theory is afforded by the case of Kerguelen Island. This island was visited by the *Transit of Venus* expedition. It is one of the stormiest places on the globe, being subject to almost perpetual gales, while, there being no wood, it is almost entirely without shelter. The Rev. A. E. Eaton, an experienced entomologist, was naturalist to the expedition, and he assiduously collected the few insects that were to be found. All were incapable of flight, and most of them entirely without wings. They included a moth, several flies, and numerous beetles. As these insects could hardly have reached the islands in a wingless state, even if there were any other known land inhabited by them—which there is not—we must assume that, like the Madeiran insects, they were originally winged, and lost their power of flight because its possession was injurious to them.

It is no doubt due to the same cause that some butterflies on small and exposed islands have their wings reduced in size, as is strikingly the case with the small tortoise-shell butterfly (Vanessa urticæ) inhabiting the Isle of Man, which is only about half the size of the same species in England or Ireland; and Mr. Wollaston notes that Vanessa callirhoe—a closely allied South European form of our red-admiral butterfly—is permanently smaller in the small and bare island of Porto Santo than in the larger and more wooded adjacent island of Madeira.

A very good example of comparatively recent divergence of character, in accordance with new conditions of life, is afforded by our red grouse. This bird, the Lagopus scoticus of naturalists, is entirely confined to the British Isles. It is, however, very closely allied to the willow grouse (Lagopus albus), a bird which ranges all over Europe, Northern Asia, and North America, but which, unlike our species, changes to white in winter. No difference in form or structure can be detected between the two birds, but as they differ so decidedly in colour—our species being usually rather darker in winter than in summer, while there are also slight differences in the call-note and in habits,—the two species are generally considered to be distinct. The differences, however, are so clearly adaptations to changed conditions that we can hardly doubt that, during the early part of the glacial period, when our islands were united to the continent, our grouse was identical with that of the rest of Europe. But when the cold passed away and our islands became permanently separated from the mainland, with a mild and equable climate and very little snow in winter, the change to white at that season became hurtful, rendering the birds more conspicuous instead of serving as a means of concealment. The colour was, therefore, gradually changed by the process of variation and natural selection; and as the birds obtained ample shelter among the heather which clothes so many of our moorlands, it became useful for them to assimilate with its brown and dusky stems and withered flowers rather than with the snow of the higher mountains. An interesting confirmation of this change having really occurred is afforded by the occasional occurrence in Scotland of birds with a considerable amount of white in the winter plumage. This is considered to be a case of reversion to the ancestral type, just as the slaty colours and banded wings of the wild rock-pigeon sometimes reappear in our fancy breeds of domestic pigeons.[1]

The principle of "divergence of character" pervades all nature from the lowest groups to the highest, as may be well seen in the class of birds. Among our native species we see it well marked in the different species of titmice, pipits, and chats. The great titmouse (Parus major) by its larger size and stronger bill is adapted to feed on larger insects, and is even said sometimes to kill small and weak birds. The smaller and weaker coal titmouse (Parus ater) has adopted a

[1] Yarrell's *British Birds*, fourth edition, vol. iii. p. 77.

more vegetarian diet, eating seeds as well as insects, and feeding on the ground as well as among trees. The delicate little blue titmouse (Parus cœruleus), with its very small bill, feeds on the minutest insects and grubs which it extracts from crevices of bark and from the buds of fruit-trees. The marsh titmouse, again (Parus palustris), has received its name from the low and marshy localities it frequents; while the crested titmouse (Parus cristatus) is a northern bird frequenting especially pine forests, on the seeds of which trees it partially feeds. Then, again, our three common pipits—the tree-pipit (Anthus arboreus), the meadow-pipit (Anthus pratensis), and the rock-pipit or sea-lark (Anthus obscurus) have each occupied a distinct place in nature to which they have become specially adapted, as indicated by the different form and size of the hind toe and claw in each species. So, the stone-chat (Saxicola rubicola), the whin-chat (S. rubetra), and the wheat-ear (S. œnanthe) are all slightly divergent forms of one type, with modifications in the shape of the wing, feet, and bill adapting them to slightly different modes of life. The whin-chat is the smallest, and frequents furzy commons, fields, and lowlands, feeding on worms, insects, small molluscs, and berries; the stone-chat is next in size, and is especially active and lively, frequenting heaths and uplands, and is a permanent resident with us, the two other species being migrants; while the larger and more conspicuous wheat-ear, besides feeding on grubs, beetles, etc., is able to capture flying insects on the wing, something after the manner of true flycatchers.

These examples sufficiently indicate how divergence of character has acted, and has led to the adaptation of numerous allied species, each to a more or less special mode of life, with the variety of food, of habits, and of enemies which must necessarily accompany such diversity. And when we extend our inquiries to higher groups we find the same indications of divergence and special adaptation, often to a still more marked extent. Thus we have the larger falcons, which prey upon birds, while some of the smaller species, like the hobby (Falco subbuteo), live largely on insects. The true falcons capture their prey in the air, while the hawks usually seize it on or near the ground, feeding on hares, rabbits, squirrels, grouse, pigeons, and poultry. Kites and buzzards, on the other hand, seize their prey upon the ground, and the former feed largely on reptiles and offal as well as on birds and quadrupeds. Others have adopted fish as their chief food, and the osprey snatches its prey from the water with as much facility as a gull or a petrel; while the South American caracaras (Polyborus) have adopted the habits of vultures and live altogether on carrion. In every great group there is the same divergence of habits. There are ground-pigeons, rock-pigeons, and wood-pigeons,—seed-eating pigeons and fruit-eating pigeons; there are carrion-eating, insect-eating, and fruit-eating crows. Even kingfishers are, some aquatic, some terrestrial in their habits; some live on fish, some on insects, some on reptiles. Lastly, among the primary divisions of birds we find a purely terrestrial group—the Ratitæ, including the ostriches, cassowaries, etc.; other great groups, including the ducks, cormorants, gulls, penguins, etc., are aquatic; while the bulk of the Passerine birds are aerial and arboreal. The same general facts can be detected in all other classes of animals. In the mammalia, for example, we have in the common rat a fish-eater and flesh-eater as well as a grain-eater, which has no doubt helped to give it the power of spreading over the world and driving away the native rats of other countries. Throughout the Rodent tribe we find everywhere aquatic, terrestrial, and arboreal forms. In the weasel and cat tribes some live more in trees, others on the ground; squirrels have diverged into terrestrial, arboreal, and flying species; and finally, in the bats we have a truly aerial, and in the whales a truly aquatic order of mammals. We thus see that, beginning with different varieties of the same species, we have allied species, genera, families, and orders, with similarly divergent habits, and adaptations to different modes of life, indicating some general principle in nature which has been operative in the development of the organic world. But in order to be thus operative it must be a generally useful principle, and Mr. Darwin has very clearly shown us in what this utility consists.

Divergence leads to a Maximum of Organic Forms in each Area.

Divergence of character has a double purpose and use. In the first place it enables a species which is being overcome by rivals, or is in process of extinction by enemies, to save itself by adopting new habits or by occupying vacant places in nature. This is the immediate and obvious effect of all the numerous examples of divergence of character which we have pointed out. But there is another and less obvious result, which is, that the greater the diversity in the organisms inhabiting a country or district the greater will be the total amount of life that can be supported there. Hence the continued action of the struggle for existence will tend to bring about more and more diversity in each area, which may be shown to be the case by several kinds of evidence. As an example, a piece of turf, three feet by four in size, was found by Mr. Darwin to contain twenty species of plants, and these twenty species belonged to eighteen genera and to eight orders, showing how greatly they differed from each other. Farmers find that a greater quantity of hay is obtained from ground sown with a variety of genera of grasses, clover, etc., than from similar land sown with one or two species only; and the same principle applies to rotation of crops, plants differing very widely from each other giving the best results. So, in small and uniform islands, and in small ponds of fresh water, the plants and insects, though few in number, are found to be wonderfully varied in character.

The same principle is seen in the naturalisation of plants and animals by man's agency in distant lands, for the species that thrive best and establish themselves permanently are not only very varied among themselves but differ greatly from the native inhabitants. Thus, in the Northern United States there are, according to Dr. Asa Gray, 260 naturalised flowering plants which belong to no less than 162 genera; and of these, 100 genera are not natives of the United States. So, in Australia, the rabbit, though totally unlike any native animal, has increased so much that it probably outnumbers in individuals all the native mammals of the country; and in New Zealand the rabbit and the pig have equally multiplied. Darwin remarks that this "advantage of diversification of structure in the inhabitants of the same region is, in fact, the same as that of the physiological division of labour in the organs of the same body. No physiologist doubts that a stomach adapted to digest vegetable matter alone, or flesh alone, draws more nutriment from these substances. So, in the general economy of any land, the more widely and perfectly the animals and plants are diversified for different habits of life, so will a greater number of individuals be capable of there supporting themselves."[1]

The most closely allied Species inhabit distinct Areas.

One of the curious results of the general action of this principle in nature is, that the most closely allied species—those whose differences though often real and important are hardly perceptible to any one but a naturalist—are usually not found in the same but in widely separated countries. Thus, the nearest allies to our European golden plover are found in North America and East Asia; the nearest ally of our European jay is found in Japan, although there are several other species of jays in Western Asia and North Africa; and though we have several species of titmice in England they are not very closely allied to each other. The form most akin to our blue tit is the azure tit of Central Asia (Parus azureus); the Parus ledouci of Algeria is very near our coal tit, and the Parus lugubris of South-Eastern Europe and Asia Minor is nearest to our marsh tit. So, our four species of wild pigeons—the ring-dove, stock-dove, rock-pigeon, and turtle-dove—are not closely allied to each other, but each of them belongs, according to some ornithologists, to a separate genus or subgenus, and has its nearest relatives in distant parts of Asia and Africa. In mammalia the same thing occurs. Each mountain region of Europe and Asia has usually its own species of wild sheep and goat, and sometimes of antelope and deer; so that in each region there is found the greatest diversity in this class of animals, while the closest allies inhabit quite distinct and often distant areas. In plants we find the same phenomenon prevalent. Distinct species of columbine are found in Central Europe (Aguilegia vulgaris), in Eastern Europe, and Siberia (A. glandulosa), in the Alps (A. Alpina), in the Pyrenees (A. pyrenaiea), in the Greek mountains (A. ottonis), and in Corsica (A. Bernardi), but rarely are two

[1] *Origin of Species*, p. 89.

species found in the same area. So, each part of the world has its own peculiar forms of pines, firs, and cedars, but the closely allied species or varieties are in almost every case inhabitants of distinct areas. Examples are the deodar of the Himalayas, the cedar of Lebanon, and that of North Africa, all very closely allied but confined to distinct areas; and the numerous closely allied species of true pine (genus Pinus), which almost always inhabit different countries or occupy different stations. We will now consider some other modes in which natural selection will act, to adapt organisms to changed conditions.

Adaptation to Conditions at Various Periods of Life.

It is found, that, in domestic animals and cultivated plants, variations occurring at any one period of life reappear in the offspring at the same period, and can be perpetuated and increased by selection without modifying other parts of the organisation. Thus, variations in the caterpillar or the cocoon of the silkworm, in the eggs of poultry, and in the seeds or young shoots of many culinary vegetables, have been accumulated till those parts have become greatly modified and, for man's purposes, improved. Owing to this fact it is easy for organisms to become so modified as to avoid dangers that occur at any one period of life. Thus it is that so many seeds have become adapted to various modes of dissemination or protection. Some are winged, or have down or hairs attached to them, so as to enable them to be carried long distances in the air; others have curious hooks and prickles, which cause them to be attached firmly to the fur of mammals or the feathers of birds; while others are buried within sweet or juicy and brightly coloured fruits, which are seen and devoured by birds, the hard smooth seeds passing through their bodies in a fit state for germination. In the struggle for existence it must benefit a plant to have increased means of dispersing its seeds, and of thus having young plants produced in a greater variety of soils, aspects, and surroundings, with a greater chance of some of them escaping their numerous enemies and arriving at maturity. The various differences referred to would, therefore, be brought about by variation and survival of the fittest, just as surely as the length and quality of cotton on the seed of the cotton-plant have been increased by man's selection.

The larvæ of insects have thus been wonderfully modified in order to escape the numerous enemies to whose attacks they are exposed at this period of their existence. Their colours and markings have become marvellously adapted to conceal them among the foliage of the plant they live upon, and this colour often changes completely after the last moult, when the creature has to descend to the ground for its change to the pupa state, during which period a brown instead of a green colour is protective. Others have acquired curious attitudes and large ocelli, which cause them to resemble the head of some reptile, or they have curious horns or coloured ejectile processes which frighten away enemies; while a great number have acquired secretions which render them offensive to the taste of their enemies, and these are always adorned with very conspicuous markings or brilliant colours, which serve as a sign of inedibility and prevent their being needlessly attacked. This, however, is a portion of the very large subject of organic colour and marking, which will be fully discussed and illustrated in a separate chapter.

In this way every possible modification of an animal or plant, whether in colour, form, structure, or habits, which would be serviceable to it or to its progeny at any period of its existence, may be readily brought about. There are some curious organs which are used only once in a creature's life, but which are yet essential to its existence, and thus have very much the appearance of design by an intelligent designer. Such are, the great jaws possessed by some insects, used exclusively for opening the cocoon, and the hard tip to the beak of unhatched birds used for breaking the eggshell. The increase in thickness or hardness of the cocoons or the eggs being useful for protection against enemies or to avoid accidents, it is probable that the change has been very gradual, because it would be constantly checked by the necessity for a corresponding change in the young insects or birds enabling them to overcome the additional obstacle of a tougher cocoon or a harder eggshell. As we have seen, however, that every part of the organism appears to be varying independently, at the same time, though to different amounts, there seems no reason to believe that the necessity for two or more coincident variations would prevent the required change from taking place.

The Continued Existence of Low Forms of Life.

Since species are continually undergoing modifications giving them some superiority over other species or enabling them to occupy fresh places in nature, it may be asked—Why do any low forms continue to exist? Why have they not long since been improved and developed into higher forms? The answer, probably, is, that these low forms occupy places in nature which cannot be filled by higher forms, and that they have few or no competitors; they therefore continue to exist. Thus, earthworms are adapted to their mode of life better than they would be if more highly organised. So, in the ocean, the minute foraminifera and infusoria, and the larger sponges and corals, occupy places which more highly developed creatures could not fill. They form, as it were, the base of the great structure of animal life, on which the next higher forms rest; and though in the course of ages they may undergo some changes, and diversification of form and structure, in accordance with changed conditions, their essential nature has probably remained the same from the very dawn of life on the earth. The low aquatic diatomaceæ and confervæ, together with the lowest fungi and lichens, occupy a similar position in the vegetable kingdom, filling places in nature which would be left vacant if only highly organised plants existed. There is, therefore, no motive power to destroy or seriously to modify them; and they have thus probably persisted, under slightly varying forms, through all geological time.

Extinction of Lower Types among the Higher Animals.

So soon, however, as we approach the higher and more fully developed groups, we see indications of the often repeated extinction of lower by higher forms. This is shown by the great gaps that separate the mammalia, birds, reptiles, and fishes from each other; while the lowest forms of each are always few in number and confined to limited areas. Such are the lowest mammals—the echidna and ornithorhynchus of Australia; the lowest birds—the apteryx of New Zealand and the cassowaries of the New Guinea region; while the lowest fish—the amphioxus or lancelet, is completely isolated, and has apparently survived only by its habit of burrowing in the sand. The great distinctness of the carnivora, ruminants, rodents, whales, bats, and other orders of mammalia; of the accipitres, pigeons, and parrots, among birds; and of the beetles, bees, flies, and moths, among insects, all indicate an enormous amount of extinction among the comparatively low forms by which, on any theory of evolution, these higher and more specialised groups must have been preceded.

Circumstances favourable to the Origin of New Species by Natural Selection.

We have already seen that, when there is no change in the physical or organic conditions of a country, the effect of natural selection is to keep all the species inhabiting it in a state of perfect health and full development, and to preserve the balance that already exists between the different groups of organisms. But, whenever the physical or organic conditions change, to however small an extent, some corresponding change will be produced in the flora and fauna, since, considering the severe struggle for existence and the complex relations of the various organisms, it is hardly possible that the change should not be beneficial to some species and hurtful to others. The most common effect, therefore, will be that some species will increase and others will diminish; and in cases where a species was already small in numbers a further diminution might lead to extinction. This would afford room for the increase of other species, and thus a considerable readjustment of the proportions of the several species might take place. When, however, the change was of a more important character, directly affecting the existence of many species so as to render it difficult for them to maintain themselves without some considerable change in structure or habits, that change would, in some cases, be brought about by variation and natural selection, and thus new varieties or new species might be formed. We have to consider, then, which are the species that would be most likely to be so modified, while others, not becoming modified, would succumb to the changed conditions and become extinct.

The most important condition of all is, undoubtedly, that variations should occur of sufficient amount, of a sufficiently diverse character, and in a large number of individuals, so as to afford ample materials for natural selection to act upon; and this, we have seen, does occur in most, if not in all, large, wide-ranging, and dominant species. From some of these, therefore, the new species adapted to the changed conditions would usually be derived; and this would especially be the case when the change of conditions was rather rapid, and when a correspondingly rapid modification could alone save some species from extinction. But when the change was very gradual, then even less abundant and less widely distributed species might become modified into new forms, more especially if the extinction of many of the rarer species left vacant places in the economy of nature.

Probable Origin of the Dippers.

An excellent example of how a limited group of species has been able to maintain itself by adaptation to one of these "vacant places" in nature, is afforded by the curious little birds called dippers or water-ouzels, forming the genus Cinclus and the family Cinclidæ of naturalists. These birds are something like small thrushes, with very short wings and tail, and very dense plumage. They frequent, exclusively, mountain torrents in the northern hemisphere, and obtain their food entirely in the water, consisting, as it does, of water-beetles, caddis-worms and other insect-larvæ, as well as numerous small fresh-water shells. These birds, although not far removed in structure from thrushes and wrens, have the extraordinary power of flying under water; for such, according to the best observers, is their process of diving in search of their prey, their dense and somewhat fibrous plumage retaining so much air that the water is prevented from touching their bodies or even from wetting their feathers to any great extent. Their powerful feet and long curved claws enable them to hold on to stones at the bottom, and thus to retain their position while picking up insects, shells, etc. As they frequent chiefly the most rapid and boisterous torrents, among rocks, waterfalls, and huge boulders, the water is never frozen over, and they are thus able to live during the severest winters. Only a very few species of dipper are known, all those of the old world being so closely allied to our British bird that some ornithologists consider them to be merely local races of one species; while in North America and the northern Andes there are two other species.

Here then we have a bird, which, in its whole structure, shows a close affinity to the smaller typical perching birds, but which has departed from all its allies in its habits and mode of life, and has secured for itself a place in nature where it has few competitors and few enemies. We may well suppose, that, at some remote period, a bird which was perhaps the common and more generalised ancestor of most of our thrushes, warblers, wrens, etc., had spread widely over the great northern continent, and had given rise to numerous varieties adapted to special conditions of life. Among these some took to feeding on the borders of clear streams, picking out such larvæ and molluscs as they could reach in shallow water. When food became scarce they would attempt to pick them out of deeper and deeper water, and while doing this in cold weather many would become frozen and starved. But any which possessed denser and more hairy plumage than usual, which was able to keep out the water, would survive; and thus a race would be formed which would depend more and more on this kind of food. Then, following up the frozen streams into the mountains, they would be able to live there during the winter; and as such places afforded them much protection from enemies and ample shelter for their nests and young, further adaptations would occur, till the wonderful power of diving and flying under water was acquired by a true land-bird.

That such habits might be acquired under stress of need is rendered highly probable by the facts stated by the well-known American naturalist, Dr. Abbott. He says that "the water-thrushes (Seiurus sp.) all wade in water, and often, seeing minute mollusca on the bottom of the stream, plunge both head and neck beneath the surface, so that often, for several seconds, a large part of the body is submerged. Now these birds still have the plumage pervious to water, and so are liable to be drenched and sodden; but they have also the faculty of giving these drenched feathers such a good shaking that flight is practicable a moment after leaving the water. Certainly the water-thrushes (Seiurus ludovicianus, S. auricapillus; and S. noveboracensis) have taken many preliminary steps to becoming as aquatic as the dipper; and the winter-wren, and even the Maryland yellow-throat are not far behind."[1]

Another curious example of the way in which species have been modified to occupy new places in nature, is afforded by the various animals which inhabit the water-vessels formed by the leaves of many epiphytal species of Bromelia. Fritz Müller has described a caddis-fly larva which lives among these leaves, and which has been modified in the pupa state in accordance with its surroundings. The pupæ of caddis-flies inhabiting streams have fringes of hair on the tarsi to enable them to reach the surface on leaving their cases. But in the species inhabiting bromelia leaves there is no need for swimming, and accordingly we find the tarsi entirely bare. In the same plants are found curious little Entomostraca, very abundant there but found nowhere else. These form a new genus, but are most nearly allied to Cythere, a marine type. It is believed that the transmission of this species from one tree to another must be effected by the young crustacea, which are very minute, clinging to beetles, many of which, both terrestrial and aquatic, also inhabit the bromelia leaves; and as some water-beetles are known to frequent the sea, it is perhaps by these means that the first emigrants established themselves in this strange new abode. Bromeliæ are often very abundant on trees growing on the water's edge, and this would facilitate the transition from a marine to an arboreal habitat. Fritz Müller has also found, among the bromelia leaves, a small frog bearing its eggs on its back, and having some other peculiarities of structure. Several beautiful little aquatic plants of the genus Utricularia or bladder-wort also inhabit bromelia leaves; and these send runners out to neighbouring plants and thus spread themselves with great rapidity.

[1] *Nature*, vol. xxx. p. 30.

The Importance of Isolation.

Isolation is no doubt an important aid to natural selection, as shown by the fact that islands so often present a number of peculiar species; and the same thing is seen on the two sides of a great mountain range or on opposite coasts of a continent. The importance of isolation is twofold. In the first place, it leads to a body of individuals of each species being limited in their range and thus subjected to uniform conditions for long spaces of time. Both the direct action of the environment and the natural selection of such varieties only as are suited to the conditions, will, therefore, be able to produce their full effect. In the second place, the process of change will not be interfered with by intercrossing with other individuals which are becoming adapted to somewhat different conditions in an adjacent area. But this question of the swamping effects of intercrossing will be considered in another chapter.

Mr. Darwin was of opinion that, on the whole, the largeness of the area occupied by a species was of more importance than isolation, as a factor in the production of new species, and in this I quite agree with him. It must, too, be remembered, that isolation will often be produced in a continuous area whenever a species becomes modified in accordance with varied conditions or diverging habits. For example, a wide-ranging species may in the northern and colder part of its area become modified in one direction, and in the southern part in another direction; and though for a long time an intermediate form may continue to exist in the intervening area, this will be likely soon to die out, both because its numbers will be small, and it will be more or less pressed upon in varying seasons by the modified varieties, each better able to endure extremes of climate. So, when one portion of a terrestrial species takes to a more arboreal or to a more aquatic mode of life, the change of habit itself leads to the isolation of each portion. Again, as will be more fully explained in a future chapter, any difference of habits or of haunts usually leads to some modification of colour or marking, as a means of concealment from enemies; and there is reason to believe that this difference will be intensified by natural selection as a means of identification and recognition by members of the same variety or incipient species. It has also been observed that each differently coloured variety of wild animals, or of domesticated animals which have run wild, keep together, and refuse to pair with individuals of the other colours; and this must of itself act to keep the races separate as completely as physical isolation.

On the Advance of Organisation by Natural Selection.

As natural selection acts solely by the preservation of useful variations, or those which are beneficial to the organism under the conditions to which it is exposed, the result must necessarily be that each species or group tends to become more and more improved in relation to its conditions. Hence we should expect that the larger groups in each class of animals and plants—those which have persisted and have been abundant throughout geological ages—would, almost necessarily, have arrived at a high degree of organisation, both physical and mental. Illustrations of this are to be seen everywhere. Among mammalia we have the carnivora, which from Eocene times have been becoming more and more specialised, till they have culminated in the cat and dog tribes, which have reached a degree of perfection both in structure and intelligence fully equal to that of any other animals. In another line of development, the herbivora have been specialised for living solely on vegetable food till they have culminated in the sheep, the cattle, the deer, and the antelopes. The horse tribe, commencing with an early four-toed ancestor in the Eocene age, has increased in size and in perfect adaptation of feet and teeth to a life on open plains, and has reached its highest perfection in the horse, the ass, and the zebra. In birds, also, we see an advance from the imperfect tooth-billed and reptile-tailed birds of the secondary epoch, to the wonderfully developed falcons, crows, and swallows of our time. So, the ferns, lycopods, conifers, and monocotyledons of the palæozoic and mesozoic rocks, have developed into the marvellous wealth of forms of the higher dicotyledons that now adorn the earth.

But this remarkable advance in the higher and larger groups does not imply any universal law of progress in organisation, because we have at the same time numerous examples (as has been already pointed out) of the persistence of lowly organised forms, and also of absolute degradation or degeneration. Serpents, for example, have been developed from some lizard-like type which has lost its limbs; and though this loss has enabled them to occupy fresh places in nature and to increase and flourish to a marvellous extent, yet it must be considered to be a retrogression rather than an advance in organisation. The same remark will apply to the whale tribe among mammals; to the blind amphibia and insects of the great caverns; and among plants to the numerous cases in which flowers, once specially adapted to be fertilised by insects, have lost their gay corollas and their special adaptations, and have become degraded into wind-fertilised forms. Such are our plantains, our meadow burnet, and even, as some botanists maintain, our rushes, sedges, and grasses. The causes which have led to this degeneration will be discussed in a future chapter; but the facts are undisputed, and they show us that although variation and the struggle for existence may lead, on the whole, to a continued advance of organisation; yet they also lead in many cases to a retrogression, when such retrogression may aid in the preservation of any form under new conditions. They also lead to the persistence, with slight modifications, of numerous lowly organised forms which are suited to places which higher forms could not fully occupy, or to conditions under which they could not exist. Such are the ocean depths, the soil of the earth, the mud of rivers, deep caverns, subterranean waters, etc.; and it is in such places as these, as well as in some oceanic islands which competing higher forms have not been able to reach, that we find many curious relics of an earlier world, which, in the free air and sunlight and in the great continents, have long since been driven out or exterminated by higher types.

Summary of the first Five Chapters.

We have now passed in review, in more or less detail, the main facts on which the theory of "the origin of species by means of natural selection" is founded. In future chapters we shall have to deal mainly with the application of the theory to explain the varied and complex phenomena presented by the organic world; and, also, to discuss some of the theories put forth by modern writers, either as being more fundamental than that of Darwin or as supplementary to it. Before doing this, however, it will be well briefly to summarise the facts and arguments already set forth, because it is only by a clear comprehension of these that the full importance of the theory can be appreciated and its further applications understood.

The theory itself is exceedingly simple, and the facts on which it rests—though excessively numerous individually, and coextensive with the entire organic world—yet come under a few simple and easily understood classes. These facts are,—first, the enormous powers of increase in geometrical progression possessed by all organisms, and the inevitable struggle for existence among them; and, in the second place, the occurrence of much individual variation combined with the hereditary transmission of such variations. From these two great classes of facts, which are universal and indisputable, there necessarily arises, as Darwin termed it, the "preservation of favoured races in the struggle for life," the continuous action of which, under the ever-changing conditions both of the inorganic and organic universe, necessarily leads to the formation or development of new species.

But, although this general statement is complete and indisputable, yet to see its applications under all the complex conditions that actually occur in nature, it is necessary always to bear in mind the tremendous power and universality of the agencies at work. We must never for an instant lose sight of the fact of the enormously rapid increase of all organisms, which has been illustrated by actual cases, given in our second chapter, no less than by calculations of the results of unchecked increase for a few years. Then, never forgetting that the animal and plant population of any country is, on the whole, stationary, we must be always trying to realise the ever-recurring destruction of the enormous annual increase, and asking ourselves what determines, in each individual case, the death of the many, the survival of the few. We must think over all the causes of destruction to each organism,—to the seed, the young shoot, the growing plant, the full-grown tree, or shrub, or herb, and again the fruit and seed; and among animals, to the egg or new-born young, to the youthful, and to the adults. Then, we must always bear in mind that what goes on in the case of the individual or family group we may observe or think of, goes on also among the millions and scores of millions of individuals which are comprised in almost every species; and must get rid of the idea that *chance* determines which shall live and which die. For, although in many individual cases death may be due to chance rather than to any inferiority in those which die first, yet we cannot possibly believe that this can be the case on the large scale on which nature works. A plant, for instance, cannot be increased unless there are suitable vacant places its seeds can grow in, or stations where it can overcome other less vigorous and healthy plants. The seeds of all plants, by their varied modes of dispersal, may be said to be seeking out such places in which to grow; and we cannot doubt that, in the long run, those individuals whose seeds are the most numerous, have the greatest powers of dispersal, and the greatest vigour of growth, will leave more descendants than the individuals of the same species which are inferior in all these respects, although now and then some seed of an inferior individual may *chance* to be carried to a spot where it can grow and survive. The same rule will apply to every period of life and to every danger to which plants or animals are exposed. The best organised, or the most healthy, or the most active, or the best protected, or the most intelligent, will inevitably, in the long run, gain an advantage over those which are inferior in these qualities; that is, *the fittest will survive*, the fittest being, in each particular case, those which are superior in the special qualities on which safety depends. At one period of life, or to escape one kind of danger, concealment may be necessary; at another time, to escape another danger, swiftness; at another, intelligence or cunning; at another, the power to endure rain or cold or hunger; and those which possess all these faculties in the fullest perfection will generally survive.

Having fully grasped these facts in all their fulness and in their endless and complex results, we have next to consider the phenomena of variation, discussed in the third and fourth chapters; and it is here that perhaps the greatest difficulty will be felt in appreciating the full importance of the evidence as set forth. It has been so generally the practice to speak of variation as something exceptional and comparatively rare—as an abnormal deviation from the uniformity and stability of the characters of a species—and so few even among naturalists have ever compared, accurately, considerable numbers of individuals, that the conception of variability as a general characteristic of all dominant and widespread species, large in its amount and affecting, not a few, but considerable masses of the individuals which make up the species, will be to many entirely new. Equally important is the fact that the variability extends to every organ and every structure, external and internal; while perhaps most important of all is the independent variability of these several parts, each one varying without any constant or even usual dependence on, or correlation with, other parts. No doubt there is some such correlation in the differences that exist between species and species—more developed wings usually accompanying smaller feet and *vice versâ*—but this is, generally, a useful adaptation which has been brought about by natural selection, and does not apply to the individual variability which occurs within the species.

It is because these facts of variation are so important and so little understood, that they have been discussed in what will seem to some readers wearisome and unnecessary detail. Many naturalists, however, will hold that even more evidence is required; and more, to almost any amount, could easily have been given. The character and variety of that already adduced will, however, I trust, convince most readers that the facts are as stated; while they have been drawn from a sufficiently wide area to indicate a general principle throughout nature.

If, now, we fully realise these facts of variation, along with those of rapid multiplication and the struggle for existence, most of the difficulties in the way of comprehending how species have originated through natural selection will disappear. For whenever, through changes of climate, or of altitude, or of the nature of the soil, or of the area of the country, any species are exposed to new dangers, and have to maintain themselves and provide for the safety of their offspring under new and more arduous conditions, then, in the variability of all parts, organs, and structures, no less than of habits and intelligence, we have the means of producing modifications which will certainly bring the species into harmony with its new conditions. And if we remember that all such physical changes are slow and gradual in their operation, we shall see that the amount of variation which we know occurs in every new generation will be quite sufficient to enable modification and adaptation to go on at the same rate. Mr. Darwin was rather inclined to exaggerate the necessary slowness of the action of natural selection; but with the knowledge we now possess of the great amount and range of individual variation, there seems no difficulty in an amount of change, quite equivalent to that which usually distinguishes allied species, sometimes taking place in less than a century, should any rapid change of conditions necessitate an equally rapid adaptation. This may often have occurred, either to immigrants into a new land, or to residents whose country has been cut off by subsidence from a larger and more varied area over which they had formerly roamed. When no change of conditions occurs, species may remain unchanged for very long periods, and thus produce that appearance of stability of species which is even now often adduced as an argument against evolution by natural selection, but which is really quite in harmony with it.

On the principles, and by the light of the facts, now briefly summarised, we have been able, in the present chapter, to indicate how natural selection acts, how divergence of character is set up, how adaptation to conditions at various periods of life has been effected, how it is that low forms of life continue to exist, what kind of circumstances are most favourable to the formation of new species, and, lastly, to what extent the advance of organisation to higher types is produced by natural selection. We will now pass on to consider some of the more important objections and difficulties which have been advanced by eminent naturalists.

CHAPTER VI

DIFFICULTIES AND OBJECTIONS

Difficulty as to smallness of variations—As to the right variations occurring when required—The beginnings of important organs—The mammary glands—The eyes of flatfish—Origin of the eye—Useless or non-adaptive characters—Recent extension of the region of utility in plants—The same in animals—Uses of tails—Of the horns of deer—Of the scale-ornamentation of reptiles—Instability of non-adaptive characters—Delbœuf's law—No "specific" character proved to be useless—The swamping effects of intercrossing—Isolation as preventing intercrossing—Gulick on the effects of isolation—Cases in which isolation is ineffective.

In the present chapter I propose to discuss the more obvious and often repeated objections to Darwin's theory, and to show how far they affect its character as a true and sufficient explanation of the origin of species. The more recondite difficulties, affecting such fundamental questions as the causes and laws of variability, will be left for a future chapter, after we have become better acquainted with the applications of the theory to the more important adaptations and correlations of animal and plant life.

One of the earliest and most often repeated objections was, that it was difficult "to imagine a reason why variations tending in an infinitesimal degree in any special direction should be preserved," or to believe that the complex adaptation of living organisms could have been produced "by infinitesimal beginnings." Now this term "infinitesimal," used by a well-known early critic of the *Origin of Species*, was never made use of by Darwin himself, who spoke only of variations being "slight," and of the "small amount" of the variations that might be selected. Even in using these terms he undoubtedly afforded grounds for the objection above made, that such small and slight variations could be of no real use, and would not determine the survival of the individuals possessing them. We have seen, however, in our third chapter, that even Darwin's terms were hardly justified; and that the variability of many important species is of considerable amount, and may very often be properly described as large. As this is found to be the case both in animals and plants, and in all their chief groups and subdivisions, and also to apply to all the separate parts and organs that have been compared, we must take it as proved that the average *amount* of variability presents no difficulty whatever in the way of the action of natural selection. It may be here mentioned that, up to the time of the preparation of the last edition of *The Origin of Species*, Darwin had not seen the work of Mr. J. A. Allen of Harvard University (then only just published), which gave us the first body of accurate comparisons and measurements demonstrating this large amount of variability. Since then evidence of this nature has been accumulating, and we are, therefore, now in a far better position to appreciate the facilities for natural selection, in this respect, than was Mr. Darwin himself.

Another objection of a similar nature is, that the chances are immensely against the right variation or combination of variations occurring just when required; and further, that no variation can be perpetuated that is not accompanied by several concomitant variations of dependent parts—greater length of a wing in a bird, for example, would be of little use if unaccompanied by increased volume or contractility of the muscles which move it. This objection seemed a very strong one so long as it was supposed that variations occurred singly and at considerable intervals; but it ceases to have any weight now we know that they occur simultaneously in various parts of the organism, and also in a large proportion of the individuals which make up the species. A considerable number of individuals will, therefore, every year possess the required combination of characters; and it may also be considered probable that when the two characters are such that they always *act* together, there will be such a correlation between them that they will frequently *vary* together. But there is another consideration that seems to show that this coincident variation is not essential. All animals in a state of nature are kept, by the constant struggle for existence and the survival of the fittest, in such a state of perfect health and usually superabundant vigour, that in all ordinary circumstances they possess a surplus power in every important organ—a surplus only drawn upon in cases of the direst necessity when their very existence is at stake. It follows, therefore, that *any* additional power given to one of the component parts of an organ must be useful—an increase, for example, either in the wing muscles or in the form or length of the wing might give *some* increased powers of flight; and thus alternate variations—in one generation in the muscles, in another generation in the wing itself—might be as effective in permanently improving the powers of flight as coincident variations at longer intervals. On either supposition, however, this objection appears to have little weight if we take into consideration the large amount of coincident variability that has been shown to exist.

The Beginnings of Important Organs.

We now come to an objection which has perhaps been more frequently urged than any other, and which Darwin himself felt to have much weight—the first beginnings of important organs, such, for example, as wings, eyes, mammary glands, and numerous other structures. It is urged, that it is almost impossible to conceive how the first rudiments of these could have been of any use, and, if not of use they could not have been preserved and further developed by natural selection.

Now, the first remark to be made on objections of this nature is, that they are really outside the question of the origin of all existing species from allied species not very far removed from them, which is all that Darwin undertook to *prove* by means of his theory. Organs and structures such as those above mentioned all date back to a very remote past, when the world and its inhabitants were both very different from what they are now. To ask of a new theory that it shall reveal to us exactly what took place in remote geological epochs, and how it took place, is unreasonable. The most that should be asked is, that some probable or possible mode of origination should be pointed out in some at least of these difficult cases, and this Mr. Darwin has done. One or two of these may be briefly given here, but the whole series should be carefully read by any one who wishes to see how many curious facts and observations have been required in order to elucidate them; whence we may conclude that further knowledge will probably throw light on any difficulties that still remain.[1]

In the case of the mammary glands Mr. Darwin remarks that it is admitted that the ancestral mammals were allied to the marsupials. Now in the very earliest mammals, almost before they really deserved that name, the young may have been nourished by a fluid secreted by the interior surface of the marsupial sack, as is believed to be the case with the fish (Hippocampus) whose eggs are hatched within a somewhat similar sack. This being the case, those individuals which secreted a more nutritious fluid, and those whose young were able to obtain and swallow a more constant supply by suction, would be more likely to live and come to a healthy maturity, and would therefore be preserved by natural selection.

In another case which has been adduced as one of special difficulty, a more complete explanation is given. Soles, turbots, and other flatfish are, as is well known, unsymmetrical. They live and move on their sides, the under side being usually differently coloured from that which is kept uppermost. Now the eyes of these fish are curiously distorted in order that both eyes may be on the upper side, where alone they would be of any use. It was objected by Mr. Mivart that a sudden transformation of the eye from one side to the other was inconceivable, while, if the transit were gradual the first step could be of no use, since this would not remove the eye from the lower side. But, as Mr. Darwin shows by reference to the researches of Malm and others, the young of these fish are quite symmetrical, and during their growth exhibit to us the whole process of change. This begins by the fish (owing to the increasing depth of the body) being unable to maintain the vertical position, so that it falls on one side. It then twists the lower eye as much as possible towards the upper side; and, the whole bony structure of the head being at

[1] See *Origin of Species*, pp. 176-198.

this time soft and flexible, the constant repetition of this effort causes the eye gradually to move round the head till it comes to the upper side. Now if we suppose this process, which in the young is completed in a few days or weeks, to have been spread over thousands of generations during the development of these fish, those usually surviving whose eyes retained more and more of the position into which the young fish tried to twist them, the change becomes intelligible; though it still remains one of the most extraordinary cases of degeneration, by which symmetry—which is so universal a characteristic of the higher animals—is lost, in order that the creature may be adapted to a new mode of life, whereby it is enabled the better to escape danger and continue its existence.

The most difficult case of all, that of the eye—the thought of which even to the last, Mr. Darwin says, "gave him a cold shiver"—is nevertheless shown to be not unintelligible; granting of course the sensitiveness to light of some forms of nervous tissue. For he shows that there are, in several of the lower animals, rudiments of eyes, consisting merely of pigment cells covered with a translucent skin, which may possibly serve to distinguish light from darkness, but nothing more. Then we have an optic nerve and pigment cells; then we find a hollow filled with gelatinous substance of a convex form—the first rudiment of a lens. Many of the succeeding steps are lost, as would necessarily be the case, owing to the great advantage of each modification which gave increased distinctness of vision, the creatures possessing it inevitably surviving, while those below them became extinct. But we can well understand how, after the first step was taken, every variation tending to more complete vision would be preserved till we reached the perfect eye of birds and mammals. Even this, as we know, is not absolutely, but only relatively, perfect. Neither the chromatic nor the spherical aberration is absolutely corrected; while long- and short- sightedness, and the various diseases and imperfections to which the eye is liable, may be looked upon as relics of the imperfect condition from which the eye has been raised by variation and natural selection.

These few examples of difficulties as to the origin of remarkable or complex organs must suffice here; but the reader who wishes further information on the matter may study carefully the whole of the sixth and seventh chapters of the last edition of *The Origin of Species*, in which these and many other cases are discussed in considerable detail.

Useless or non-adaptive Characters.

Many naturalists seem to be of opinion that a considerable number of the characters which distinguish species are of no service whatever to their possessors, and therefore cannot have been produced or increased by natural selection. Professors Bronn and Broca have urged this objection on the continent. In America, Dr. Cope, the well-known palæontologist, has long since put forth the same objection, declaring that non-adaptive characters are as numerous as those which are adaptive; but he differs completely from most who hold the same general opinion in considering that they occur chiefly "in the characters of the classes, orders, families, and other higher groups;" and the objection, therefore, is quite distinct from that in which it is urged that "specific characters" are mostly useless. More recently, Professor G. J. Romanes has urged this difficulty in his paper on "Physiological Selection" (*Journ. Linn. Soc.*, vol. xix. pp. 338, 344). He says that the characters "which serve to distinguish allied species are frequently, if not usually, of a kind with which natural selection can have had nothing to do," being without any utilitarian significance. Again he speaks of "the enormous number," and further on of "the innumerable multitude" of specific peculiarities which are useless; and he finally declares that the question needs no further arguing, "because in the later editions of his works Mr. Darwin freely acknowledges that a large proportion of specific distinctions must be conceded to be useless to the species presenting them."

I have looked in vain in Mr. Darwin's works to find any such acknowledgment, and I think Mr. Romanes has not sufficiently distinguished between "useless characters" and "useless specific distinctions." On referring to all the passages indicated by him I find that, in regard to specific characters, Mr. Darwin is very cautious in admitting inutility. His most pronounced "admissions" on this question are the following: "But when, from the nature of the organism and of the conditions, modifications have been induced which are unimportant for the welfare of the species, they may be, and apparently often have been, transmitted in nearly the same state *to numerous, otherwise modified, descendants*" (*Origin*, p. 175). The words I have here italicised clearly show that such characters are usually not "specific," in the sense that they are such as distinguish species from each other, but are found in numerous allied species. Again: "Thus a large yet undefined extension may safely be given to the direct and indirect results of natural selection; but I now admit, after reading the essay of Nägeli on plants, and the remarks by various authors with respect to animals, more especially those recently made by Professor Broca, that in the earlier editions of my *Origin of Species* I perhaps attributed too much to the action of natural selection or the survival of the fittest. I have altered the fifth edition of the *Origin* so as to confine my remarks to adaptive changes of structure, *but I am convinced, from the light gained during even the last few years, that very many structures which now appear to us useless, will hereafter be proved to be useful, and will therefore come within the range of natural selection.* Nevertheless I did not formerly consider sufficiently the existence of structures which, *as far as we can at present judge*, are neither beneficial nor injurious; and this I believe to be one of the greatest oversights as yet detected in my work." Now it is to be remarked that neither in these passages nor in any of the other less distinct expressions of opinion on this question, does Darwin ever admit that "specific characters"—that is, the particular characters which serve to distinguish one species from another—are ever useless, much less that "a large proportion of them" are so, as Mr. Romanes makes him "freely acknowledge." On the other hand, in the passage which I have italicised he strongly expresses his view that much of what we suppose to be useless is due to our ignorance; and as I hold myself that, as regards many of the supposed useless characters, this is the true explanation, it may be well to give a brief sketch of the progress of knowledge in transferring characters from the one category to the other.

We have only to go back a single generation, and not even the most acute botanist could have suggested a reasonable use, for each species of plant, of the infinitely varied forms, sizes, and colours of the flowers, the shapes and arrangement of the leaves, and the numerous other external characters of the whole plant. But since Mr. Darwin showed that plants gained both in vigour and in fertility by being crossed with other individuals of the same species, and that this crossing was usually effected by insects which, in search of nectar or pollen, carried the pollen from one plant to the flowers of another plant, almost every detail is found to have a purpose and a use. The shape, the size, and the colour of the petals, even the streaks and spots with which they are adorned, the position in which they stand, the movements of the stamens and pistil at various times, especially at the period of, and just after, fertilisation, have been proved to be strictly adaptive in so many cases that botanists now believe that all the external characters of flowers either are or have been of use to the species.

It has also been shown, by Kerner and other botanists, that another set of characteristics have relation to the prevention of ants, slugs, and other animals from reaching the flowers, because these creatures would devour or injure them without effecting fertilisation. The spines, hairs, or sticky glands on the stem or flower-stalk, the curious hairs or processes shutting up the flower, or sometimes even the extreme smoothness and polish of the outside of the petals so that few insects can hang to the part, have been shown to be related to the possible intrusion of these "unbidden guests."[1] And, still more recently, attempts have been made by Grant Allen and Sir John Lubbock to account for the innumerable forms, textures, and groupings of leaves, by their relation to the needs of the plants themselves; and there can be little doubt that these attempts will be ultimately successful. Again, just as flowers have been adapted to secure fertilisation or cross-fertilisation, fruits have been developed to assist in the dispersal of seeds; and their forms, sizes, juices, and colours can be shown to be specially adapted to secure such dispersal by the agency of birds and mammals; while the same end is secured in other

[1] See Kerner's *Flowers and their Unbidden Guests* for numerous other structures and peculiarities of plants which are shown to be adaptive and useful.

cases by downy seeds to be wafted through the air, or by hooked or sticky seed-vessels to be carried away, attached to skin, wool, or feathers.

Here, then, we have an enormous extension of the region of utility in the vegetable kingdom, and one, moreover, which includes almost all the specific characters of plants. For the species of plants are usually characterised either by differences in the form, size, and colour of the flowers, or of the fruits; or, by peculiarities in the shape, size, dentation, or arrangement of the leaves; or by peculiarities in the spines, hairs, or down with which various parts of the plant are clothed. In the case of plants it must certainly be admitted that "specific" characters are pre-eminently adaptive; and though there may be some which are not so, yet all those referred to by Darwin as having been adduced by various botanists as useless, either pertain to genera or higher groups, or are found in some plants of a species only—that is, are individual variations not specific characters.

In the case of animals, the most recent wide extension of the sphere of utility has been in the matter of their colours and markings. It was of course always known that certain creatures gained protection by their resemblance to their normal surroundings, as in the case of white arctic animals, the yellow or brown tints of those living in deserts, and the green hues of many birds and insects surrounded by tropical vegetation. But of late years these cases have been greatly increased both in number and variety, especially in regard to those which closely imitate special objects among which they live; and there are other kinds of coloration which long appeared to have no use. Large numbers of animals, more especially insects, are gaudily coloured, either with vivid hues or with striking patterns, so as to be very easily seen. Now it has been found, that in almost all these cases the creatures possess some special quality which prevents their being attacked by the enemies of their kind whenever the peculiarity is known; and the brilliant or conspicuous colours or markings serve as a warning or signal flag against attack. Large numbers of insects thus coloured are nauseous and inedible; others, like wasps and bees, have stings; others are too hard to be eaten by small birds; while snakes with poisonous fangs usually have some characteristic either of rattle, hood, or unusual colour, which indicates that they had better be left alone.

But there is yet another form of coloration, which consists in special markings—bands, spots, or patches of white, or of bright colour, which vary in every species, and are often concealed when the creature is at rest but displayed when in motion,—as in the case of the bands and spots so frequent on the wings and tails of birds. Now these specific markings are believed, with good reason, to serve the purpose of enabling each species to be quickly recognised, even at a distance, by its fellows, especially the parents by their young and the two sexes by each other; and this recognition must often be an important factor in securing the safety of individuals, and therefore the wellbeing and continuance of the species. These interesting peculiarities will be more fully described in a future chapter, but they are briefly referred to here in order to show that the most common of all the characters by which species are distinguished from each other—their colours and markings—can be shown to be adaptive or utilitarian in their nature.

But besides colour there are almost always some structural characters which distinguish species from species, and, as regards many of these also, an adaptive character can be often discerned. In birds, for instance, we have differences in the size or shape of the bill or the feet, in the length of the wing or the tail, and in the proportions of the several feathers of which these organs are composed. All these differences in the organs on which the very existence of birds depends, which determine the character of flight, facility for running or climbing, for inhabiting chiefly the ground or trees, and the kind of food that can be most easily obtained for themselves and their offspring, must surely be in the highest degree utilitarian; although in each individual case we, in our ignorance of the minutiæ of their life-history, may be quite unable to see the use. In mammalia specific differences other than colour usually consist in the length or shape of the ears and tail, in the proportions of the limbs, or in the length and quality of the hair on different parts of the body. As regards the ears and tail, one of the objections by Professor

Bronn relates to this very point. He states that the length of these organs differ in the various species of hares and of mice, and he considers that this difference can be of no service whatever to their possessors. But to this objection Darwin replies, that it has been shown by Dr. Schöbl that the ears of mice "are supplied in an extraordinary manner with nerves, so that they no doubt serve as tactile organs." Hence, when we consider the life of mice, either nocturnal or seeking their food in dark and confined places, the length of the ears may be in each case adapted to the particular habits and surroundings of the species. Again, the tail, in the larger mammals, often serves the purpose of driving off flies and other insects from the body; and when we consider in how many parts of the world flies are injurious or even fatal to large mammals, we see that the peculiar characteristics of this organ may in each case have been adapted to its requirements in the particular area where the species was developed. The tail is also believed to have some use as a balancing organ, which assists an animal to turn easily and rapidly, much as our arms are used when running; while in whole groups it is a prehensile organ, and has become modified in accordance with the habits and needs of each species. In the case of mice it is thus used by the young. Darwin informs us that the late Professor Henslow kept some harvest-mice in confinement, and observed that they frequently curled their tails round the branches of a bush placed in the cage, and thus aided themselves in climbing; while Dr. Günther has actually seen a mouse suspend itself by the tail (*Origin*, p. 189).

Again, Mr. Lawson Tait has called attention to the use of the tail in the cat, squirrel, yak, and many other animals as a means of preserving the heat of the body during the nocturnal and the winter sleep. He says, that in cold weather animals with long or bushy tails will be found lying curled up, with their tails carefully laid over their feet like a rug, and with their noses buried in the fur of the tail, which is thus used exactly in the same way and for the same purpose as we use respirators.[1]

Another illustration is furnished by the horns of deer which, especially when very large, have been supposed to be

[1] *Nature*, vol. xx. p. 603.

a danger to the animal in passing rapidly through dense thickets. But Sir James Hector states, that the wapiti, in North America, throws back its head, thus placing the horns along the sides of the back, and is then enabled to rush through the thickest forest with great rapidity. The brow-antlers protect the face and eyes, while the widely spreading horns prevent injury to the neck or flanks. Thus an organ which was certainly developed as a sexual weapon, has been so guided and modified during its increase in size as to be of use in other ways. A similar use of the antlers of deer has been observed in India.[1]

The various classes of facts now referred to serve to show us that, in the case of the two higher groups—mammalia and birds—almost all the characters by which species are distinguished from each other are, or may be, adaptive. It is these two classes of animals which have been most studied and whose life-histories are supposed to be most fully known, yet even here the assertion of inutility, by an eminent naturalist, in the case of two important organs, has been sufficiently met by minute details either in the anatomy or in the habits of the groups referred to. Such a fact as this, together with the extensive series of characters already enumerated which have been of late years transferred from the "useless" to the "useful" class, should convince us, that the assertion of "inutility" in the case of any organ or peculiarity which is not a rudiment or a correlation, is not, and can never be, the statement of a fact, but merely an expression of our ignorance of its purpose or origin.[2]

[1] *Nature*, vol. xxxviii. p. 328.

[2] A very remarkable illustration of function in an apparently useless ornament is given by Semper. He says, "It is known that the skin of reptiles encloses the body with scales. These scales are distinguished by very various sculpturings, highly characteristic of the different species. Irrespective of their systematic significance they appear to be of no value in the life of the animal; indeed, they are viewed as ornamental without regard to the fact that they are microscopic and much too delicate to be visible to other animals of their own species. It might, therefore, seem hopeless to show the necessity for their existence on Darwinian principles, and to prove that they are physiologically active organs. Nevertheless, recent investigations on this point have furnished evidence that this is possible.

"It is known that many reptiles, and above all the snakes, cast off the whole skin at once, whereas human beings do so by degrees. If by any accident they are prevented doing so, they infallibly die, because the old

Instability of Non-adaptive Characters.

One very weighty objection to the theory that *specific* characters can ever be wholly useless (or wholly unconnected with useful organs by correlation of growth) appears to have been overlooked by those who have maintained the frequency of such characters, and that is, their almost necessary instability. Darwin has remarked on the extreme variability of secondary sexual characters—such as the horns, crests, plumes, etc., which are found in males only,—the reason being, that, although of some use, they are not of such direct and vital importance as those adaptive characters on which the wellbeing and very existence of the animals depend. But in the case of wholly useless structures,

skin has grown so tough and hard that it hinders the increase in volume which is inseparable from the growth of the animal. The casting of the skin is induced by the formation on the surface of the inner epidermis, of a layer of very fine and equally distributed hairs, which evidently serve the purpose of mechanically raising the old skin by their rigidity and position. These hairs then may be designated as *casting hairs.* That they are destined and calculated for this end is evident to me from the fact established by Dr. Braun, that the casting of the shells of the river cray-fish is induced in exactly the same manner by the formation of a coating of hairs which mechanically loosens the old skin or shell from the new. Now the researches of Braun and Cartier have shown that these casting hairs—which serve the same purpose in two groups of animals so far apart in the systematic scale—after the casting, are partly transformed into the concentric stripes, sharp spikes, ridges, or warts which ornament the outer edges of the skin-scales of reptiles or the carapace of crabs."[1] Professor Semper adds that this example, with many others that might be quoted, shows that we need not abandon the hope of explaining morphological characters on Darwinian principles, although their nature is often difficult to understand.

During a recent discussion of this question in the pages of *Nature*, Mr. St. George Mivart adduces several examples of what he deems useless specific characters. Among them are the aborted index finger of the lemurine Potto, and the thumbless hands of Colobus and Ateles, the "life-saving action" of either of which he thinks incredible. These cases suggest two remarks. In the first place, they involve *generic*, not *specific*, characters; and the three genera adduced are somewhat isolated, implying considerable antiquity and the extinction of many allied forms. This is important, because it affords ample time for great changes of conditions since the structures in question originated; and without a knowledge of these changes we can never safely assert that any detail of structure could not have been useful. In the second place, all three are cases of aborted or rudimentary organs; and these are admitted to be explained by non-use, leading to diminution of size, a further reduction being brought about by the action of the principle of economy

[1] *The Natural Conditions of Existence as they affect Animal Life*, p. 19.

which are not rudiments of once useful organs, we cannot see what there is to ensure any amount of constancy or stability. One of the cases on which Mr. Romanes lays great stress in his paper on "Physiological Selection" (*Journ. Linn. Soc.*, vol. xix. p. 384) is that of the fleshy appendages on the corners of the jaw of Normandy pigs and of some other breeds. But it is expressly stated that they are not constant; they appear "frequently," or "occasionally," they are "not strictly inherited, for they occur or fail in animals of the same litter;" and they are not always symmetrical, sometimes appearing on one side of the face alone. Now whatever may be the cause or explanation of these anomalous appendages they cannot be classed with "specific characters," the most essential features of which are, that they *are* symmetrical,

of growth. But, when so reduced, the rudiment might be inconvenient or even hurtful, and then natural selection would aid in its complete abortion; in other words, the abortion of the part would be *useful*, and would therefore be subject to the law of survival of the fittest. The genera Ateles and Colobus are two of the most purely arboreal types of monkeys, and it is not difficult to conceive that the constant use of the elongated fingers for climbing from tree to tree, and catching on to branches while making great leaps, might require all the nervous energy and muscular growth to be directed to the fingers, the small thumb remaining useless. The case of the Potto is more difficult, both because it is, presumably, a more ancient type, and its actual life-history and habits are completely unknown. These cases are, therefore, not at all to the point as proving that positive specific characters—not mere rudiments characterising whole genera—are in any case useless.

Mr. Mivart further objects to the alleged rigidity of the action of natural selection, because wounded or malformed animals have been found which had evidently lived a considerable time in their imperfect condition. But this simply proves that they were living under a temporarily favourable environment, and that the real struggle for existence, in their case, had not yet taken place. We must surely admit that, when the pinch came, and when perfectly formed stoats were dying for want of food, the one-footed animal, referred to by Mr. Mivart, would be among the first to succumb; and the same remark will apply to his abnormally toothed hares and rheumatic monkeys, which might, nevertheless, get on very well under favourable conditions. The struggle for existence, under which all animals and plants have been developed, is intermittent, and exceedingly irregular in its incidence and severity. It is most severe and fatal to the young; but when an animal has once reached maturity, and especially when it has gained experience by several years of an eventful existence, it may be able to maintain itself under conditions which would be fatal to a young and inexperienced creature of the same species. The examples adduced by Mr. Mivart do not, therefore, in any way impugn the hardness of nature as a taskmaster, or the extreme severity of the recurring struggle for existence.[1]

[1] See *Nature*, vol. xxxix. p. 127.

that they *are* inherited, and that they *are* constant. Admitting that this peculiar appendage is (as Mr. Romanes says rather confidently, "we happen to know it to be") wholly useless and meaningless, the fact would be rather an argument against specific characters being also meaningless, because the latter never have the characteristics which this particular variation possesses.

These useless or non-adaptive characters are, apparently, of the same nature as the "sports" that arise in our domestic productions, but which, as Mr. Darwin says, without the aid of selection would soon disappear; while some of them may be correlations with other characters which are or have been useful. Some of these correlations are very curious. Mr. Tegetmeier informed Mr. Darwin that the young of white, yellow, or dun-coloured pigeons are born almost naked, whereas other coloured pigeons are born well clothed with down. Now, if this difference occurred between wild species of different colours, it might be said that the nakedness of the young could not be of any use. But the colour with which it is correlated might, as has been shown, be useful in many ways. The skin and its various appendages, as horns, hoofs, hair, feathers, and teeth, are homologous parts, and are subject to very strange correlations of growth. In Paraguay, horses with curled hair occur, and these always have hoofs exactly like those of a mule, while the hair of the mane and tail is much shorter than usual. Now, if any one of these characters were useful, the others correlated with it might be themselves useless, but would still be tolerably constant because dependent on a useful organ. So the tusks and the bristles of the boar are correlated and vary in development together, and the former only may be useful, or both may be useful in unequal degrees.

The difficulty as to how individual differences or sports can become fixed and perpetuated, if altogether useless, is evaded by those who hold that such characters are exceedingly common. Mr. Romanes says that, upon his theory of physiological selection, "it is quite intelligible that when a varietal form is differentiated from its parent form by the bar of sterility, any little meaningless peculiarities of structure or of instinct *should at first be allowed to arise*, and that they should then *be allowed to perpetuate themselves* by heredity," until they are finally eliminated by disuse. But this is entirely begging the question. Do meaningless peculiarities, which we admit often arise as spontaneous variations, ever perpetuate themselves in all the individuals constituting a variety or race, without selection either human or natural? Such characters present themselves as unstable variations, and as such they remain, unless preserved and accumulated by selection; and they can therefore never become "specific" characters unless they are strictly correlated with some useful and important peculiarities.

As bearing upon this question we may refer to what is termed Delbœuf's law, which has been thus briefly stated by Mr. Murphy in his work on *Habit and Intelligence*, p. 241.

"If, in any species, a number of individuals, bearing a ratio not infinitely small to the entire number of births, are in every generation born with a particular variation which is neither beneficial nor injurious, and if it is not counteracted by reversion, then the proportion of the new variety to the original form will increase till it approaches indefinitely near to equality."

It is not impossible that some definite varieties, such as the melanic form of the jaguar and the bridled variety of the guillemot are due to this cause; but from their very nature such varieties are unstable, and are continually reproduced in varying proportions from the parent forms. They can, therefore, never constitute species unless the variation in question becomes beneficial, when it will be fixed by natural selection. Darwin, it is true, says—"There can be little doubt that the tendency to vary in the same manner has often been so strong that all the individuals of the same species have been similarly modified without the aid of any form of selection."[1] But no proof whatever is offered of this statement, and it is so entirely opposed to all we know of the facts of variation as given by Darwin himself, that the important word "all" is probably an oversight.

On the whole, then, I submit, not only has it not been proved that an "enormous number of specific peculiarities" are useless, and that, as a logical result, natural selection is "not a theory of the origin of species," but only of the origin

[1] *Origin of Species*, p. 72.

of adaptations which are usually common to many species, or, more commonly, to genera and families; but, I urge further, it has not even been proved that any truly "specific" characters—those which either singly or in combination distinguish each species from its nearest allies—are entirely unadaptive, useless, and meaningless; while a great body of facts on the one hand, and some weighty arguments on the other, alike prove that specific characters have been, and could only have been, developed and fixed by natural selection because of their utility. We may admit, that among the great number of variations and sports which continually arise many are altogether useless without being hurtful; but no cause or influence has been adduced adequate to render such characters fixed and constant throughout the vast number of individuals which constitute any of the more dominant species.[1]

The Swamping Effects of Intercrossing.

This supposed insuperable difficulty was first advanced in an article in the *North British Review* in 1867, and much attention has been attracted to it by the acknowledgment of Mr. Darwin that it proved to him that "single variations," or what are usually termed "sports," could very rarely, if ever, be perpetuated in a state of nature, as he had at first thought might occasionally be the case. But he had always considered that the chief part, and latterly the whole, of the materials with which natural selection works, was afforded by individual variations, or that amount of ever fluctuating variability which exists in all organisms and in all their parts. Other writers have urged the same objection, even as against individual variability, apparently in total ignorance of its amount and range; and quite recently Professor G. J. Romanes has adduced

[1] Darwin's latest expression of opinion on this question is interesting, since it shows that he was inclined to return to his earlier view of the general, or universal, utility of specific characters. In a letter to Semper (30th Nov. 1878) he writes: "As our knowledge advances, very slight differences, considered by systematists as of no importance in structure, are continually found to be functionally important; and I have been especially struck with this fact in the case of plants, to which my observations have, of late years, been confined. Therefore it seems to me rather rash to consider slight differences between representative species, for instance, those inhabiting the different islands of the same archipelago, as of no functional importance, and as not in any way due to natural selection" (*Life of Darwin*, vol. iii. p. 161).

it as one of the difficulties which can alone be overcome by his theory of physiological selection. He urges, that the same variation does not occur simultaneously in a number of individuals inhabiting the same area, and that it is mere assumption to say it does; while he admits that "if the assumption were granted there would be an end of the present difficulty; for if a sufficient number of individuals were thus simultaneously and similarly modified, there need be no longer any danger of the variety becoming swamped by intercrossing." I must again refer my readers to my third chapter for the proof that such simultaneous variability is not an assumption but a fact; but, even admitting this to be proved, the problem is not altogether solved, and there is so much misconception regarding variation, and the actual process of the origin of new species is so obscure, that some further discussion and elucidation of the subject are desirable.

In one of the preliminary chapters of Mr. Seebohm's recent work on the *Charadriidæ*, he discusses the differentiation of species; and he expresses a rather widespread view among naturalists when, speaking of the swamping effects of intercrossing, he adds: "This is unquestionably a very grave difficulty, to my mind an absolutely fatal one, to the theory of accidental variation." And in another passage he says: "The simultaneous appearance, and its repetition in successive generations, of a beneficial variation, in a large number of individuals in the same locality, cannot possibly be ascribed to chance." These remarks appear to me to exhibit an entire misconception of the facts of variation as they actually occur, and as they have been utilised by natural selection in the modification of species. I have already shown that every part of the organism, in common species, does vary to a very considerable amount, in a large number of individuals, and in the same locality; the only point that remains to be discussed is, whether any or most of these variations are "beneficial." But every one of these variations consists either in increase or diminution of size or power of the organ or faculty that varies; they can all be divided into a more effective and a less effective group—that is, into one that is more beneficial or less beneficial. If less size of body would be beneficial, then, as half the variations in size are above and half below the mean or existing standard of the species, there would be ample beneficial variations; if a darker colour or a longer beak or wing were required, there are always a considerable number of individuals darker and lighter in colour than the average, with longer or with shorter beaks and wings, and thus the beneficial variation must always be present. And so with every other part, organ, function, or habit; because, as variation, so far as we know, is and always must be in the two directions of excess and defect in relation to the mean amount, whichever kind of variation is wanted is always present in some degree, and thus the difficulty as to "beneficial" variations occurring, as if they were a special and rare class, falls to the ground. No doubt some organs may vary in three or perhaps more directions, as in the length, breadth, thickness, or curvature of the bill. But these may be taken as separate variations, each of which again occurs as "more" or "less"; and thus the "right" or "beneficial" or "useful" variation must always be present so long as any variation at all occurs; and it has not yet been proved that in any large or dominant species, or in any part, organ, or faculty of such species, there is no variation. And even were such a case found it would prove nothing, so long as in numerous other species variation was shown to exist; because we know that great numbers of species and groups throughout all geological time have died out, leaving no descendants; and the obvious and sufficient explanation of this fact is, that they did *not* vary enough at the time when variation was required to bring them into harmony with changed conditions. The objection as to the "right" or "beneficial" variation occurring when required, seems therefore to have no weight in view of the actual facts of variation.

Isolation to prevent Intercrossing.

Most writers on the subject consider the isolation of a portion of a species a very important factor in the formation of new species, while others maintain it to be absolutely essential. This latter view has arisen from an exaggerated opinion as to the power of intercrossing to keep down any variety or incipient species, and merge it in the parent stock. But it is evident that this can only occur with varieties which are not useful, or which, if useful, occur in very small numbers; and from this kind of variations it is clear that new species do not arise. Complete isolation, as in an oceanic island, will no doubt enable natural selection to act more rapidly, for several reasons. In the first place, the absence of competition will for some time allow the new immigrants to increase rapidly till they reach the limits of subsistence. They will then struggle among themselves, and by survival of the fittest will quickly become adapted to the new conditions of their environment. Organs which they formerly needed, to defend themselves against, or to escape from, enemies, being no longer required, would be encumbrances to be got rid of, while the power of appropriating and digesting new and varied food would rise in importance. Thus we may explain the origin of so many flightless and rather bulky birds in oceanic islands, as the dodo, the cassowary, and the extinct moas. Again, while this process was going on, the complete isolation would prevent its being checked by the immigration of new competitors or enemies, which would be very likely to occur in a continuous area; while, of course, any intercrossing with the original unmodified stock would be absolutely prevented. If, now, before this change has gone very far, the variety spreads into adjacent but rather distant islands, the somewhat different conditions in each may lead to the development of distinct forms constituting what are termed representative species; and these we find in the separate islands of the Galapagos, the West Indies, and other ancient groups of islands.

But such cases as these will only lead to the production of a few peculiar species, descended from the original settlers which happened to reach the islands; whereas, in wide areas, and in continents, we have variation and adaptation on a much larger scale; and, whenever important physical changes demand them, with even greater rapidity. The far greater complexity of the environment, together with the occurrence of variations in constitution and habits, will often allow of effective isolation, even here, producing all the results of actual physical isolation. As we have already explained, one of the most frequent modes in which natural selection acts is, by adapting some individuals of a species to a somewhat different mode of life, whereby they are able to seize upon unappropriated places in nature, and in so doing they become practically isolated from their parent form. Let us suppose, for example, that one portion of a species usually living in forests ranges into the open plains, and finding abundance of food remains there permanently. So long as the struggle for existence is not exceptionally severe, these two portions of the species may remain almost unchanged; but suppose some fresh enemies are attracted to the plains by the presence of these new immigrants, then variation and natural selection would lead to the preservation of those individuals best able to cope with the difficulty, and thus the open country form would become modified into a marked variety or into a distinct species; and there would evidently be little chance of this modification being checked by intercrossing with the parent form which remained in the forest.

Another mode of isolation is brought about by the variety—either owing to habits, climate, or constitutional change—breeding at a slightly different time from the parent species. This is known to produce complete isolation in the case of many varieties of plants. Yet another mode of isolation is brought about by changes of colour, and by the fact that in a wild state animals of similar colours prefer to keep together and refuse to pair with individuals of another colour. The probable reason and utility of this habit will be explained in another chapter, but the fact is well illustrated by the cattle which have run wild in the Falkland Islands. These are of several different colours, but each colour keeps in a separate herd, often restricted to one part of the island; and one of these varieties—the mouse-coloured—is said to breed a month earlier than the others; so that if this variety inhabited a larger area it might very soon be established as a distinct race or species.[1] Of course where the change of habits or of station is still greater, as when a terrestrial animal becomes sub-aquatic, or when aquatic animals come to live in tree-tops, as with the frogs and crustacea described at p. 118, the danger of intercrossing is reduced to a minimum.

Several writers, however, not content with the indirect effects of isolation here indicated, maintain that it is in itself a cause of modification, and ultimately of the origination of

[1] See *Variation of Animals and Plants*, vol. i. p. 86.

new species. This was the keynote of Mr. Vernon Wollaston's essay on "Variation of Species," published in 1856, and it is adopted by the Rev. J. G. Gulick in his paper on "Diversity of Evolution under one Set of External Conditions" (*Journ. Linn. Soc. Zool.*, vol. xi. p. 496). The idea seems to be that there is an inherent tendency to variation in certain divergent lines, and that when one portion of a species is isolated, even though under identical conditions, that tendency sets up a divergence which carries that portion farther and farther away from the original species. This view is held to be supported by the case of the land shells of the Sandwich Islands, which certainly present some very remarkable phenomena. In this comparatively small area there are about 300 species of land shells, almost all of which belong to one family (or sub-family), the Achatinellidæ, found nowhere else in the world. The interesting point is the extreme restriction of the species and varieties. The average range of each species is only five or six miles, while some are restricted to but one or two square miles, and only a very few range over a whole island. The forest region that extends over one of the mountain-ranges of the island of Oahu, is about forty miles in length and five or six miles in breadth; and this small territory furnishes about 175 species, represented by 700 or 800 varieties. Mr. Gulick states, that the vegetation of the different valleys on the same side of this range is much the same, yet each has a molluscan fauna differing in some degree from that of any other. "We frequently find a genus represented in several successive valleys by allied species, sometimes feeding on the same, sometimes on different plants. In every such case the valleys that are nearest to each other furnish the most nearly allied forms; and a full set of the varieties of each species presents a minute gradation of forms between the more divergent types found in the more widely separated localities." He urges, that these constant differences cannot be attributed to natural selection, because they occur in different valleys on the same side of the mountain, where food, climate, and enemies are the same; and also, because there is no greater difference in passing from the rainy to the dry side of the mountains than in passing from one valley to another on the same side an equal distance apart. In a very lengthy paper, presented to the Linnean Society last year, on "Divergent Evolution through Cumulative Segregation," Mr. Gulick endeavours to work out his views into a complete theory, the main point of which may perhaps be indicated by the following passage: "No two portions of a species possess exactly the same average character, and the initial differences are for ever reacting on the environment and on each other in such a way as to ensure increasing divergence in each successive generation as long as the individuals of the two groups are kept from intercrossing."[1]

It need hardly be said that the views of Mr. Darwin and myself are inconsistent with the notion that, if the environment were absolutely similar for the two isolated portions of the species, any such necessary and constant divergence would take place. It is an error to assume that what seem to us identical conditions are really identical to such small and delicate organisms as these land molluscs, of whose needs and difficulties at each successive stage of their existence, from the freshly-laid egg up to the adult animal, we are so profoundly ignorant. The exact proportions of the various species of plants, the numbers of each kind of insect or of bird, the peculiarities of more or less exposure to sunshine or to wind at certain critical epochs, and other slight differences which to us are absolutely immaterial and unrecognisable, may be of the highest significance to these humble creatures, and be quite sufficient to require some slight adjustments of size, form, or colour, which natural selection will bring about. All we know of the facts of variation leads us to believe that, without this action of natural selection, there would be produced over the whole area a series of inconstant varieties mingled together, not a distinct segregation of forms each confined to its own limited area.

Mr. Darwin has shown that, in the distribution and modification of species, the biological is of more importance than the physical environment, the struggle with other organisms being often more severe than that with the forces of nature. This is particularly evident in the case of plants, many of which, when protected from competition, thrive in a

[1] *Journal of the Linnean Society, Zoology*, vol. xx. p. 215.

soil, climate, and atmosphere widely different from those of their native habitat. Thus, many alpine plants only found near perpetual snow thrive well in our gardens at the level of the sea; as do the tritomas from the sultry plains of South Africa, the yuccas from the arid hills of Texas and Mexico, and the fuchsias from the damp and dreary shores of the Straits of Magellan. It has been well said that plants do not live where they like, but where they can; and the same remark will apply to the animal world. Horses and cattle run wild and thrive both in North and South America; rabbits, once confined to the south of Europe, have established themselves in our own country and in Australia; while the domestic fowl, a native of tropical India, thrives well in every part of the temperate zone.

If, then, we admit that when one portion of a species is separated from the rest, there will necessarily be a slight difference in the average characters of the two portions, it does not follow that this difference has much if any effect upon the characteristics that are developed by a long period of isolation. In the first place, the difference itself will necessarily be very slight unless there is an exceptional amount of variability in the species; and in the next place, if the average characters of the species are the expression of its exact adaptation to its whole environment, then, given a precisely similar environment, and the isolated portion will inevitably be brought back to the same average of characters. But, as a matter of fact, it is impossible that the environment of the isolated portion can be exactly like that of the bulk of the species. It cannot be so physically, since no two separated areas can be absolutely alike in climate and soil; and even if these are the same, the geographical features, size, contour, and relation to winds, seas, and rivers, would certainly differ. Biologically, the differences are sure to be considerable. The isolated portion of a species will almost always be in a much smaller area than that occupied by the species as a whole, hence it is at once in a different position as regards its own kind. The proportions of all the other species of animals and plants are also sure to differ in the two areas, and some species will almost always be absent in the smaller which are present in the larger country. These differences will act and react on the isolated portion of the species. The struggle for existence will differ in its severity and in its incidence from that which affects the bulk of the species. The absence of some one insect or other creature inimical to the young animal or plant may cause a vast difference in its conditions of existence, and may necessitate a modification of its external or internal characters in quite a different direction from that which happened to be present in the average of the individuals which were first isolated.

On the whole, then, we conclude that, while isolation is an important factor in effecting some modification of species, it is so, not on account of any effect produced, or influence exerted by isolation *per se*, but because it is always and necessarily accompanied by a change of environment, both physical and biological. Natural selection will then begin to act in adapting the isolated portion to its new conditions, and will do this the more quickly and the more effectually because of the isolation. We have, however, seen reason to believe that geographical or local isolation is by no means essential to the differentiation of species, because the same result is brought about by the incipient species acquiring different habits or frequenting a different station; and also by the fact that different varieties of the same species are known to prefer to pair with their like, and thus to bring about a physiological isolation of the most effective kind. This part of the subject will be again referred to when the very difficult problems presented by hybridity are discussed.[1]

Cases in which Isolation is Ineffective.

One objection to the views of those who, like Mr. Gulick, believe isolation itself to be a cause of modification of species deserves attention, namely, the entire absence of change where,

[1] In Mr. Gulick's last paper (*Journal of Linn. Soc. Zool.*, vol. xx. pp. 189-274) he discusses the various forms of isolation above referred to, under no less than thirty-eight different divisions and subdivisions, with an elaborate terminology, and he argues that these will frequently bring about divergent evolution without any change in the environment or any action of natural selection. The discussion of the problem here given will, I believe, sufficiently expose the fallacy of his contention; but his illustration of the varied and often recondite modes by which practical isolation may be brought about, may help to remove one of the popular difficulties in the way of the action of natural selection in the origination of species.

if this were a *vera causa*, we should expect to find it. In Ireland we have an excellent test case, for we know that it has been separated from Britain since the end of the glacial epoch, certainly many thousand years. Yet hardly one of its mammals, reptiles, or land molluscs has undergone the slightest change, even although there is certainly a distinct difference in the environment both inorganic and organic. That changes have not occurred through natural selection, is perhaps due to the less severe struggle for existence owing to the smaller number of competing species; but, if isolation itself were an efficient cause, acting continuously and cumulatively, it is incredible that a decided change should not have been produced in thousands of years. That no such change has occurred in this, and many other cases of isolation, seems to prove that it is not in itself a cause of modification.

There yet remain a number of difficulties and objections relating to the question of hybridity, which are so important as to require a separate chapter for their adequate discussion.

CHAPTER VII

ON THE INFERTILITY OF CROSSES BETWEEN DISTINCT SPECIES AND THE USUAL STERILITY OF THEIR HYBRID OFFSPRING

Statement of the problem—Extreme susceptibility of the reproductive functions—Reciprocal crosses—Individual differences in respect to cross-fertilisation—Dimorphism and trimorphism among plants—Cases of the fertility of hybrids and of the infertility of mongrels—The effects of close inter-breeding—Mr. Huth's objections—Fertile hybrids among animals—Fertility of hybrids among plants—Cases of sterility of mongrels—Parallelism between crossing and change of conditions—Remarks on the facts of hybridity—Sterility due to changed conditions and usually correlated with other characters—Correlation of colour with constitutional peculiarities—The isolation of varieties by selective association—The influence of natural selection upon sterility and fertility—Physiological selection—Summary and concluding remarks.

One of the greatest, or perhaps we may say the greatest, of all the difficulties in the way of accepting the theory of natural selection as a complete explanation of the origin of species, has been the remarkable difference between varieties and species in respect of fertility when crossed. Generally speaking, it may be said that the varieties of any one species, however different they may be in external appearance, are perfectly fertile when crossed, and their mongrel offspring are equally fertile when bred among themselves; while distinct species, on the other hand, however closely they may resemble each other externally, are usually infertile when crossed, and their hybrid offspring absolutely sterile. This used to be considered a fixed law of nature, constituting the absolute test and criterion of a *species* as distinct from a *variety;* and so long as it was believed that species were separate creations, or at all events had an origin quite distinct from that of varieties, this law could have no exceptions, because, if any two species had been found to be fertile when crossed and their hybrid offspring to be also fertile, this fact would have been held to prove them to be not *species* but *varieties*. On the other hand, if two varieties had been found to be infertile, or their mongrel offspring to be sterile, then it would have been said: These are not varieties but true species. Thus the old theory led to inevitable reasoning in a circle; and what might be only a rather common fact was elevated into a law which had no exceptions.

The elaborate and careful examination of the whole subject by Mr. Darwin, who has brought together a vast mass of evidence from the experience of agriculturists and horticulturists, as well as from scientific experimenters, has demonstrated that there is no such fixed law in nature as was formerly supposed. He shows us that crosses between some varieties are infertile or even sterile, while crosses between some species are quite fertile; and that there are besides a number of curious phenomena connected with the subject which render it impossible to believe that sterility is anything more than an incidental property of species, due to the extreme delicacy and susceptibility of the reproductive powers, and dependent on physiological causes we have not yet been able to trace. Nevertheless, the fact remains that most species which have hitherto been crossed produce sterile hybrids, as in the well-known case of the mule; while almost all domestic varieties, when crossed, produce offspring which are perfectly fertile among themselves. I will now endeavour to give such a sketch of the subject as may enable the reader to see something of the complexity of the problem, referring him to Mr. Darwin's works for fuller details.

Extreme Susceptibility of the Reproductive Functions.

One of the most interesting facts, as showing how susceptible to changed conditions or to slight constitutional changes are the reproductive powers of animals, is the very general difficulty of getting those which are kept in confinement to breed; and this is frequently the only bar to domesticating wild species. Thus, elephants, bears, foxes, and numbers of species of rodents, very rarely breed in confinement; while other species do so more or less freely. Hawks, vultures, and owls hardly ever breed in confinement; neither did the falcons kept for hawking ever breed. Of the numerous small seed-eating birds kept in aviaries, hardly any breed, neither do parrots. Gallinaceous birds usually breed freely in confinement, but some do not; and even the guans and curassows, kept tame by the South American Indians, never breed. This shows that change of climate has nothing to do with the phenomenon; and, in fact, the same species that refuse to breed in Europe do so, in almost every case, when tamed or confined in their native countries. This inability to reproduce is not due to ill-health, since many of these creatures are perfectly vigorous and live very long.

With our true domestic animals, on the other hand, fertility is perfect, and is very little affected by changed conditions. Thus, we see the common fowl, a native of tropical India, living and multiplying in almost every part of the world; and the same is the case with our cattle, sheep, and goats, our dogs and horses, and especially with domestic pigeons. It therefore seems probable, that this facility for breeding under changed conditions was an original property of the species which man has domesticated—a property which, more than any other, enabled him to domesticate them. Yet, even with these, there is evidence that great changes of conditions affect the fertility. In the hot valleys of the Andes sheep are less fertile; while geese taken to the high plateau of Bogota were at first almost sterile, but after some generations recovered their fertility. These and many other facts seem to show that, with the majority of animals, even a slight change of conditions may produce infertility or sterility; and also that after a time, when the animal has become thoroughly acclimatised, as it were, to the new conditions, the infertility is in some cases diminished or altogether ceases. It is stated by Bechstein that the canary was long infertile, and it is only of late years that good breeding birds have become common; but in this case no doubt selection has aided the change.

As showing that these phenomena depend on deep-seated causes and are of a very general nature, it is interesting to note that they occur also in the vegetable kingdom. Allowing for all the circumstances which are known to prevent the production of seed, such as too great luxuriance of foliage, too little or too much heat, or the absence of insects to cross-fertilise the flowers, Mr. Darwin shows that many species which grow and flower with us, apparently in perfect health, yet never produce seed. Other plants are affected by very slight changes of conditions, producing seed freely in one soil and not in another, though apparently growing equally well in both; while, in some cases, a difference of position even in the same garden produces a similar result.[1]

Reciprocal Crosses.

Another indication of the extreme delicacy of the adjustment between the sexes, which is necessary to produce fertility, is afforded by the behaviour of many species and varieties when reciprocally crossed. This will be best illustrated by a few of the examples furnished us by Mr. Darwin. The two distinct species of plants, Mirabilis jalapa and M. longiflora, can be easily crossed, and will produce healthy and fertile hybrids when the pollen of the latter is applied to the stigma of the former plant. But the same experimenter, Kölreuter, tried in vain, more than two hundred times during eight years, to cross them by applying the pollen of M. jalapa to the stigma of M. longiflora. In other cases two plants are so closely allied that some botanists class them as varieties (as with Matthiola annua and M. glabra), and yet there is the same great difference in the result when they are reciprocally crossed.

Individual Differences in respect to Cross-Fertilisation.

A still more remarkable illustration of the delicate balance of organisation needful for reproduction, is afforded by the individual differences of animals and plants, as regards both their power of intercrossing with other individuals or other species, and the fertility of the offspring thus produced. Among domestic animals, Darwin states that it is by no means rare to find certain males and females which will not breed

[1] Darwin's *Animals and Plants under Domestication*, vol. ii. pp. 163-170.

together, though both are known to be perfectly fertile with other males and females. Cases of this kind have occurred among horses, cattle, pigs, dogs, and pigeons; and the experiment has been tried so frequently that there can be no doubt of the fact. Professor G. J. Romanes states that he has a number of additional cases of this individual incompatibility, or of absolute sterility, between two individuals, each of which is perfectly fertile with other individuals.

During the numerous experiments that have been made on the hybridisation of plants similar peculiarities have been noticed, some individuals being capable, others incapable, of being crossed with a distinct species. The same individual peculiarities are found in varieties, species, and genera. Kölreuter crossed five varieties of the common tobacco (Nicotiana tabacum) with a distinct species, Nicotiana glutinosa, and they all yielded very sterile hybrids; but those raised from one variety were less sterile, in all the experiments, than the hybrids from the four other varieties. Again, most of the species of the genus Nicotiana have been crossed, and freely produce hybrids; but one species, N. acuminata, not particularly distinct from the others, could neither fertilise, nor be fertilised by, any of the eight other species experimented on. Among genera we find some—such as Hippeastrum, Crinum, Calceolaria, Dianthus—almost all the species of which will fertilise other species and produce hybrid offspring; while other allied genera, as Zephyranthes and Silene, notwithstanding the most persevering efforts, have not produced a single hybrid even between the most closely allied species.

Dimorphism and Trimorphism.

Peculiarities in the reproductive system affecting individuals of the same species reach their maximum in what are called heterostyled, or dimorphic and trimorphic flowers, the phenomena presented by which form one of the most remarkable of Mr. Darwin's many discoveries. Our common cowslip and primrose, as well as many other species of the genus Primula, have two kinds of flowers in about equal proportions. In one kind the stamens are short, being situated about the middle of the tube of the corolla, while the style is long, the globular stigma appearing just in the centre of the open flower. In the other kind the stamens are long, appearing in the centre or throat of the flower, while the style is short, the stigma being situated halfway down the tube at the same level as the stamens in the other form. These two forms have long been known to florists as the "pin-eyed" and the "thrum-eyed," but they are called by Darwin the long-styled and short-styled forms (see woodcut).

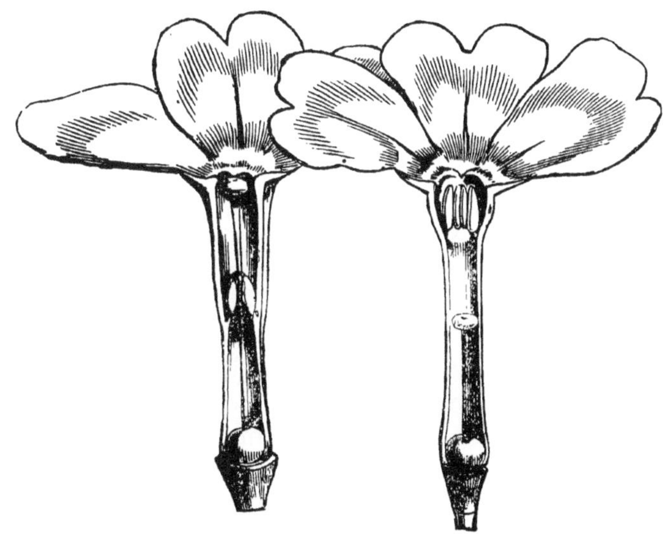

Long-styled form. Short-styled form.

FIG. 17.—Primula veris (Cowslip).

The meaning and use of these different forms was quite unknown till Darwin discovered, first, that cowslips and primroses are absolutely barren if insects are prevented from visiting them, and then, what is still more extraordinary, that each form is almost sterile when fertilised by its own pollen, and comparatively infertile when crossed with any other plant of its own form, but is perfectly fertile when the pollen of a long-styled is carried to the stigma of a short-styled plant, or *vice versâ*. It will be seen, by the figures, that the arrangement is such that a bee visiting the flowers will carry the pollen from the long anthers of the short-styled form to the stigma of the long-styled form, while it would never reach the stigma of another plant of the short-styled form.

But an insect visiting, first, a long-styled plant, would deposit the pollen on the stigma of another plant of the same kind if it were next visited; and this is probably the reason why the wild short-styled plants were found to be almost always most productive of seed, since they must be all fertilised by the other form, whereas the long-styled plants might often be fertilised by their own form. The whole arrangement, however, ensures cross-fertilisation; and this, as Mr. Darwin has shown by copious experiments, adds both to the vigour and fertility of almost all plants as well as animals.

Besides the primrose family, many other plants of several distinct natural orders present similar phenomena, one or two of the most curious of which must be referred to. The beautiful crimson flax (Linum grandiflorum) has also two forms, the styles only differing in length; and in this case Mr. Darwin found by numerous experiments, which have since been repeated and confirmed by other observers, that each form is absolutely sterile with pollen from another plant of its own form, but abundantly fertile when crossed with any plant of the other form. In this case the pollen of the two forms cannot be distinguished under the microscope (whereas that of the two forms of Primula differs in size and shape), yet it has the remarkable property of being absolutely powerless on the stigmas of half the plants of its own species. The crosses between the opposite forms, which are fertile, are termed by Mr. Darwin "legitimate," and those between similar forms, which are sterile, "illegitimate"; and he remarks that we have here, within the limits of the same species, a degree of sterility which rarely occurs except between plants or animals not only of different *species* but of different *genera*.

But there is another set of plants, the trimorphic, in which the styles and stamens have each three forms—long, medium, and short, and in these it is possible to have eighteen different crosses. By an elaborate series of experiments it was shown that the six legitimate unions—that is, when a plant was fertilised by pollen from stamens of length corresponding to that of its style in the two other forms—were all abundantly fertile; while the twelve illegitimate unions, when a plant was fertilised by pollen from stamens of a different length from its own style, in any of the three forms, were either comparatively or wholly sterile.[1]

We have here a wonderful amount of constitutional difference of the reproductive organs within a single species, greater than usually occurs within the numerous distinct species of a genus or group of genera; and all this diversity appears to have arisen for a purpose which has been obtained by many other, and apparently simpler, changes of structure or of function, in other plants. This seems to show us, in the first place, that variations in the mutual relations of the reproductive organs of different individuals must be as frequent as structural variations have been shown to be; and, also, that sterility in itself can be no test of specific distinctness. But this point will be better considered when we have further illustrated and discussed the complex phenomena of hybridity.

Cases of the Fertility of Hybrids, and of the Infertility of Mongrels.

I now propose to adduce a few cases in which it has been proved, by experiment, that hybrids between two distinct species are fertile *inter se;* and then to consider why it is that such cases are so few in number.

The common domestic goose (Anser ferus) and the Chinese goose (A. cygnoides) are very distinct species, so distinct that some naturalists have placed them in different genera; yet they have bred together, and Mr. Eyton raised from a pair of these hybrids a brood of eight. This fact was confirmed by Mr. Darwin himself, who raised several fine birds from a pair of hybrids which were sent him.[2] In India, according to Mr. Blyth and Captain Hutton, whole flocks of these hybrid geese are kept in various parts of the country where neither of the pure parent species exists, and as they are kept for profit they must certainly be fully fertile.

Another equally striking case is that of the Indian humped and the common cattle, species which differ osteologically, and also in habits, form, voice, and constitution, so that they are by no means closely allied; yet Mr. Darwin assures us that he

[1] For a full account of these interesting facts and of the various problems to which they give rise, the reader must consult Darwin's volume on *The Different Forms of Flowers in Plants of the same Species*, chaps. i.-iv.

[2] See *Nature*, vol. xxi. p. 207.

has received decisive evidence that the hybrids between these are perfectly fertile *inter se.*

Dogs have been frequently crossed with wolves and with jackals, and their hybrid offspring have been found to be fertile *inter se* to the third or fourth generation, and then usually to show some signs of sterility or of deterioration. The wolf and dog may be originally the same species, but the jackal is certainly distinct; and the appearance of infertility or of weakness is probably due to the fact that, in almost all these experiments, the offspring of a single pair—themselves usually from the same litter—were bred in-and-in, and this alone sometimes produces the most deleterious effects. Thus, Mr. Low in his great work on the *Domesticated Animals of Great Britain*, says: "If we shall breed a pair of dogs from the same litter, and unite again the offspring of this pair, we shall produce at once a feeble race of creatures; and the process being repeated for one or two generations more, the family will die out, or be incapable of propagating their race. A gentleman of Scotland made the experiment on a large scale with certain foxhounds, and he found that the race actually became monstrous and perished utterly." The same writer tells us that hogs have been made the subject of similar experiments: "After a few generations the victims manifest the change induced in the system. They become of diminished size; the bristles are changed into hairs; the limbs become feeble and short; the litters diminish in frequency, and in the number of the young produced; the mother becomes unable to nourish them, and, if the experiment be carried as far as the case will allow, the feeble, and frequently monstrous offspring, will be incapable of being reared up, and the miserable race will utterly perish."[1]

These precise statements, by one of the greatest authorities on our domesticated animals, are sufficient to show that the fact of infertility or degeneracy appearing in the offspring of hybrids after a few generations need not be imputed to the fact of the first parents being distinct species, since exactly the same phenomena appear when individuals of the same species are bred under similar adverse conditions. But in almost all the experiments that have hitherto been made in crossing distinct species, no care has been taken to avoid close inter-

[1] Low's *Domesticated Animals of Great Britain*, Introduction, p. lxiv.

breeding by securing several hybrids from quite distinct stocks to start with, and by having two or more sets of experiments carried on at once, so that crosses between the hybrids produced may be occasionally made. Till this is done no experiments, such as those hitherto tried, can be held to prove that hybrids are in all cases infertile *inter se*.

It has, however, been denied by Mr. A. H. Huth, in his interesting work on *The Marriage of Near Kin*, that any amount of breeding in-and-in is in itself hurtful; and he quotes the evidence of numerous breeders whose choicest stocks have always been so bred, as well as cases like the Porto Santo rabbits, the goats of Juan Fernandez, and other cases in which animals allowed to run wild have increased prodigiously and continued in perfect health and vigour, although all derived from a single pair. But in all these cases there has been rigid selection by which the weak or the infertile have been eliminated, and with such selection there is no doubt that the ill effects of close interbreeding can be prevented for a long time; but this by no means proves that no ill effects are produced. Mr. Huth himself quotes M. Allié, M. Aubé, Stephens, Giblett, Sir John Sebright, Youatt, Druce, Lord Weston, and other eminent breeders, as finding from experience that close interbreeding *does* produce bad effects; and it cannot be supposed that there would be such a consensus of opinion on this point if the evil were altogether imaginary. Mr. Huth argues, that the evil results which do occur do not depend on the close interbreeding itself, but on the tendency it has to perpetuate any constitutional weakness or other hereditary taints; and he attempts to prove this by the argument that "if crosses act by virtue of being a cross, and not by virtue of removing an hereditary taint, then the greater the difference between the two animals crossed the more beneficial will that act be." He then shows that, the wider the difference the less is the benefit, and concludes that a cross, as such, has no beneficial effect. A parallel argument would be, that change of air, as from inland to the sea-coast, or from a low to an elevated site, is not beneficial in itself, because, if so, a change to the tropics or to the polar regions should be more beneficial. In both these cases it may well be that no benefit would accrue to a person in perfect health; but then there is no such thing as "perfect health" in man, and probably no such thing as absolute freedom from constitutional taint in animals. The experiments of Mr. Darwin, showing the great and immediate good effects of a cross between distinct strains in plants, cannot be explained away; neither can the innumerable arrangements to secure cross-fertilisation by insects, the real use and purport of which will be discussed in our eleventh chapter. On the whole, then, the evidence at our command proves that, whatever may be its ultimate cause, close inter-breeding *does* usually produce bad results; and it is only by the most rigid selection, whether natural or artificial, that the danger can be altogether obviated.

Fertile Hybrids among Animals.

One or two more cases of fertile hybrids may be given before we pass on to the corresponding experiments in plants. Professor Alfred Newton received from a friend a pair of hybrid ducks, bred from a common duck (Anas boschas), and a pintail (Dafila acuta). From these he obtained four ducklings, but these latter, when grown up, proved infertile, and did not breed again. In this case we have the results of close inter-breeding, with too great a difference between the original species, combining to produce infertility, yet the fact of a hybrid from such a pair producing healthy offspring is itself noteworthy.

Still more extraordinary is the following statement of Mr. Low: "It has been long known to shepherds, though questioned by naturalists, that the progeny of the cross between the sheep and goat is fertile. Breeds of this mixed race are numerous in the north of Europe."[1] Nothing appears to be known of such hybrids either in Scandinavia or in Italy; but Professor Giglioli of Florence has kindly given me some useful references to works in which they are described. The following extract from his letter is very interesting: "I need not tell you that there being such hybrids is now generally accepted as a fact. Buffon (*Supplements*, tom. iii. p. 7, 1756) obtained one such hybrid in 1751 and eight in 1752. Sanson (*La Culture*, vol. vi. p. 372, 1865) mentions a case observed in the Vosges, France. Geoff. St. Hilaire (*Hist. Nat. Gén. des reg. org.*, vol. iii. p.

[1] Low's *Domesticated Animals*, p. 28.

163) was the first to mention, I believe, that in different parts of South America the ram is more usually crossed with the she-goat than the sheep with the he-goat. The well-known 'pellones' of Chile are produced by the second and third generation of such hybrids (Gay, 'Hist. de Chile,' vol. i. p. 466, *Agriculture*, 1862). Hybrids bred from goat and sheep are called 'chabin' in French, and 'cabruno' in Spanish. In Chile such hybrids are called 'carneros lanudos'; their breeding *inter se* appears to be not always successful, and often the original cross has to be recommenced to obtain the proportion of three-eighths of he-goat and five-eighths of sheep, or of three-eighths of ram and five-eighths of she-goat; such being the reputed best hybrids."

With these numerous facts recorded by competent observers we can hardly doubt that races of hybrids between these very distinct species have been produced, and that such hybrids are fairly fertile *inter se;* and the analogous facts already given lead us to believe that whatever amount of infertility may at first exist could be eliminated by careful selection, if the crossed races were bred in large numbers and over a considerable area of country. This case is especially valuable, as showing how careful we should be in assuming the infertility of hybrids when experiments have been made with the progeny of a single pair, and have been continued only for one or two generations.

Among insects one case only appears to have been recorded. The hybrids of two moths (Bombyx cynthia and B. arrindia) were proved in Paris, according to M. Quatrefages, to be fertile *inter se* for eight generations.

Fertility of Hybrids among Plants.

Among plants the cases of fertile hybrids are more numerous, owing, in part, to the large scale on which they are grown by gardeners and nurserymen, and to the greater facility with which experiments can be made. Darwin tells us that Kölreuter found ten cases in which two plants considered by botanists to be distinct species were quite fertile together, and he therefore ranked them all as varieties of each other. In some cases these were grown for six to ten successive generations, but after a time the fertility decreased, as we saw to be the case in animals, and presumably from the same cause, too close interbreeding.

Dean Herbert, who carried on experiments with great care and skill for many years, found numerous cases of hybrids which were perfectly fertile *inter se.* Crinum capense, fertilised by three other species—C. pedunculatum, C. canaliculatum, or C. defixum—all very distinct from it, produced perfectly fertile hybrids; while other species less different in appearance were quite sterile with the same C. capense.

All the species of the genus Hippeastrum produce hybrid offspring which are invariably fertile. Lobelia syphylitica and L. fulgens, two very distinct species, have produced a hybrid which has been named Lobelia speciosa, and which reproduces itself abundantly. Many of the beautiful pelargoniums of our greenhouses are hybrids, such as P. ignescens from a cross between P. citrinodorum and P. fulgidum, which is quite fertile, and has become the parent of innumerable varieties of beautiful plants. All the varied species of Calceolaria, however different in appearance, intermix with the greatest readiness, and the hybrids are all more or less fertile. But the most remarkable case is that of two species of Petunia, of which Dean Herbert says: "It is very remarkable that, although there is a great difference in the form of the flower, especially of the tube, of P. nyctanigenæflora and P. phœnicea the mules between them are not only fertile, but I have found them seed much more freely with me than either parent. . . . . From a pod of the above-mentioned mule, to which no pollen but its own had access, I had a large batch of seedlings in which there was no variability or difference from itself; and it is evident that the mule planted by itself, in a congenial climate, would reproduce itself as a species; at least as much deserving to be so considered as the various Calceolarias of different districts of South America."[1]

Darwin was informed by Mr. C. Noble that he raises stocks for grafting from a hybrid between Rhododendron ponticum and R. catawbiense, and that this hybrid seeds as freely as it is possible to imagine. He adds that horticulturists raise large beds of the same hybrid, and such alone are fairly treated; for, by insect agency, the several individuals are freely

[1] *Amaryllidaceæ*, by the Hon. and Rev. William Herbert, p. 379.

crossed with each other, and the injurious influence of close interbreeding is thus prevented. Had hybrids, when fairly treated, always gone on decreasing in fertility in each successive generation, as Gärtner believed to be the case, the fact would have been notorious to nurserymen.[1]

Cases of Sterility of Mongrels.

The reverse phenomenon to the fertility of hybrids, the sterility of mongrels or of the crosses between *varieties* of the same species, is a comparatively rare one, yet some undoubted cases have occurred. Gärtner, who believed in the absolute distinctness of species and varieties, had two varieties of maize—one dwarf with yellow seeds, the other taller with red seeds; yet they never naturally crossed, and, when fertilised artificially, only a single head produced any seeds, and this one only five grains. Yet these few seeds were fertile; so that in this case the first cross was almost sterile, though the hybrid when at length produced was fertile. In like manner, dissimilarly coloured varieties of Verbascum or mullein have been found by two distinct observers to be comparatively infertile. The two pimpernels (Anagallis arvensis and A. cœrulea), classed by most botanists as varieties of one species, have been found, after repeated trials, to be perfectly sterile when crossed.

No cases of this kind are recorded among animals; but this is not to be wondered at, when we consider how very few experiments have been made with natural varieties; while there is good reason for believing that domestic varieties are exceptionally fertile, partly because one of the conditions of domestication was fertility under changed conditions, and also because long continued domestication is believed to have the effect of increasing fertility and eliminating whatever sterility may exist. This is shown by the fact that, in many cases, domestic animals are descended from two or more distinct species. This is almost certainly the case with the dog, and probably with the hog, the ox, and the sheep; yet the various breeds are now all perfectly fertile, although we have every reason to suppose that there would be some degree of infertility if the several aboriginal species were crossed together for the first time.

[1] *Origin of Species*, p. 239.

Parallelism between Crossing and Change of Conditions.

In the whole series of these phenomena, from the beneficial effects of the crossing of different stocks and the evil effects of close interbreeding, up to the partial or complete sterility induced by crosses between species belonging to different genera, we have, as Mr. Darwin points out, a curious parallelism with the effects produced by change of physical conditions. It is well known that slight changes in the conditions of life are beneficial to all living things. Plants, if constantly grown in one soil and locality from their own seeds, are greatly benefited by the importation of seed from some other locality. The same thing happens with animals; and the benefit we ourselves experience from "change of air" is an illustration of the same phenomenon. But the amount of the change which is beneficial has its limits, and then a greater amount is injurious. A change to a climate a few degrees warmer or colder may be good, while a change to the tropics or to the arctic regions might be injurious.

Thus we see that, both slight changes of conditions and a slight amount of crossing, are beneficial; while extreme changes, and crosses between individuals too far removed in structure or constitution, are injurious. And there is not only a parallelism but an actual connection between the two classes of facts, for, as we have already shown, many species of animals and plants are rendered infertile, or altogether sterile, by the change from their natural conditions which occurs in confinement or in cultivation; while, on the other hand, the increased vigour or fertility which is invariably produced by a judicious cross may be also effected by a judicious change of climate and surroundings. We shall see in a subsequent chapter, that this interchangeability of the beneficial effects of crossing and of new conditions, serves to explain some very puzzling phenomena in the forms and economy of flowers.

Remarks on the Facts of Hybridity.

The facts that have now been adduced, though not very numerous, are sufficiently conclusive to prove that the old belief, of the universal sterility of hybrids and fertility of mongrels, is incorrect. The doctrine that such a universal law existed was never more than a plausible generalisation, founded on a few inconclusive facts derived from domesticated animals and cultivated plants. The facts were, and still are, inconclusive for several reasons. They are founded, primarily, on what occurs among animals in domestication; and it has been shown that domestication both tends to increase fertility, and was itself rendered possible by the fertility of those particular species being little affected by changed conditions. The exceptional fertility of all the varieties of domesticated animals does not prove that a similar fertility exists among natural varieties. In the next place, the generalisation is founded on too remote crosses, as in the case of the horse and the ass, the two most distinct and widely separated species of the genus Equus, so distinct indeed that they have been held by some naturalists to form distinct genera. Crosses between the two species of zebra, or even between the zebra and the quagga, or the quagga and the ass, might have led to a very different result. Again, in pre-Darwinian times it was so universally the practice to argue in a circle, and declare that the fertility of the offspring of a cross proved the identity of species of the parents, that experiments in hybridity were usually made between very remote species and even between species of different genera, to avoid the possibility of the reply: "They are both really the same species;" and the sterility of the hybrid offspring of such remote crosses of course served to strengthen the popular belief.

Now that we have arrived at a different standpoint, and look upon a species, not as a distinct entity due to special creation, but as an assemblage of individuals which have become somewhat modified in structure, form, and constitution so as to adapt them to slightly different conditions of life; which can be differentiated from other allied assemblages; which reproduce their like, and which usually breed together—we require a fresh set of experiments calculated to determine the matter of fact,—whether such species crossed with their near allies do always produce offspring which are more or less sterile *inter se.* Ample materials for such experiments exist, in the numerous "representative species" inhabiting distinct areas on a continent or different islands of a group; or even in those found in the same area but frequenting somewhat different stations.

To carry out these experiments with any satisfactory result, it will be necessary to avoid the evil effects of confinement and of too close interbreeding. If birds are experimented with, they should be allowed as much liberty as possible, a plot of ground with trees and bushes being enclosed with wire netting overhead so as to form a large open aviary. The species experimented with should be obtained in considerable numbers, and by two separate persons, each making the opposite reciprocal cross, as explained at p. 155. In the second generation these two stocks might be themselves crossed to prevent the evil effects of too close interbreeding. By such experiments, carefully carried out with different groups of animals and plants, we should obtain a body of facts of a character now sadly wanting, and without which it is hopeless to expect to arrive at a complete solution of this difficult problem. There are, however, some other aspects of the question that need to be considered, and some theoretical views which require to be carefully examined, having done which we shall be in a condition to state the general conclusions to which the facts and reasonings at our command seem to point.

Sterility due to changed Conditions and usually correlated with other Characters, especially with Colour.

The evidence already adduced as to the extreme susceptibility of the reproductive system, and the curious irregularity with which infertility or sterility appears in the crosses between some varieties or species while quite absent in those between others, seem to indicate that sterility is a characteristic which has a constant tendency to appear, either by itself or in correlation with other characters. It is known to be especially liable to occur under changed conditions of life; and, as such change is usually the starting-point and cause of the development of new species, we have already found a reason why it should so often appear when species become fully differentiated.

In almost all the cases of infertility or sterility between varieties or species, we have some external differences with which it is correlated; and though these differences are sometimes slight, and the amount of the infertility is not always, or even usually, proportionate to the external difference between the two forms crossed, we must believe that there is some connection between the two classes of facts. This is especially the case as regards colour; and Mr. Darwin has collected a body of facts which go far to prove that colour, instead of being an altogether trifling and unimportant character, as was supposed by the older naturalists, is really one of great significance, since it is undoubtedly often correlated with important constitutional differences. Now colour is one of the characters that most usually distinguishes closely allied species; and when we hear that the most closely allied species of plants are infertile together, while those more remote are fertile, the meaning usually is that the former differ chiefly in the *colour* of their flowers, while the latter differ in the form of the flowers or foliage, in habit, or in other structural characters.

It is therefore a most curious and suggestive fact, that in all the recorded cases, in which a decided infertility occurs between varieties of the same species, those varieties are distinguished by a difference of colour. The infertile varieties of Verbascum were white and yellow flowered respectively; the infertile varieties of maize were red and yellow seeded; while the infertile pimpernels were the red and the blue flowered varieties. So, the differently coloured varieties of hollyhocks, though grown close together, each reproduce their own colour from seed, showing that they are not capable of freely intercrossing. Yet Mr. Darwin assures us that the agency of bees is necessary to carry the pollen from one plant to another, because in each flower the pollen is shed before the stigma is ready to receive it. We have here, therefore, either almost complete sterility between varieties of different colours, or a prepotent effect of pollen from a flower of the same colour, bringing about the same result.

Similar phenomena have not been recorded among animals; but this is not to be wondered at when we consider that most of our pure and valued domestic breeds are characterised by definite colours which constitute one of their distinctive marks, and they are, therefore, seldom crossed with these of another colour; and even when they are so crossed, no notice would be taken of any slight diminution of fertility, since this is liable to occur from many causes. We have also reason to believe that fertility has been increased by long domestication, in addition to the fact of the original stocks being exceptionally fertile; and no experiments have been made on the differently coloured varieties of wild animals. There are, however, a number of very curious facts showing that colour in animals, as in plants, is often correlated with constitutional differences of a remarkable kind, and as these have a close relation to the subject we are discussing, a brief summary of them will be here given.

Correlation of Colour with Constitutional Peculiarities.

The correlation of a white colour and blue eyes in male cats with deafness, and of the tortoise-shell marking with the female sex of the same animal, are two well-known but most extraordinary cases. Equally remarkable is the fact, communicated to Darwin by Mr. Tegetmeier, that white, yellow, pale blue, or dun pigeons, of all breeds, have the young birds born naked, while in all other colours they are well covered with down. Here we have a case in which colour seems of more physiological importance than all the varied structural differences between the varieties and breeds of pigeons. In Virginia there is a plant called the paint-root (Lachnanthes tinctoria), which, when eaten by pigs, colours their bones pink, and causes the hoofs of all but the black varieties to drop off; so that black pigs only can be kept in the district.[1] Buckwheat in flower is also said to be injurious to white pigs but not to black. In the Tarentino, black sheep are not injured by eating the Hypericum crispum—a species of St. John's-wort—which kills white sheep. White terriers suffer most from distemper; white chickens from the gapes. White-haired horses or cattle are subject to cutaneous diseases from which the dark coloured are free; while, both in Thuringia and the West Indies, it has been noticed that white or pale coloured cattle are much more troubled by flies than are those which are brown or black. The same law even extends

[1] *Origin of Species*, sixth edition, p. 9.

to insects, for it is found that silkworms which produce white cocoons resist the fungus disease much better than do those which produce yellow cocoons.[1] Among plants, we have in North America green and yellow-fruited plums not affected by a disease that attacked the purple-fruited varieties. Yellow-fleshed peaches suffer more from disease than white-fleshed kinds. In Mauritius, white sugar-canes were attacked by a disease from which the red canes were free. White onions and verbenas are most liable to mildew; and red-flowered hyacinths were more injured by the cold during a severe winter in Holland than any other kinds.[2]

These curious and inexplicable correlations of colour with constitutional peculiarities, both in animals and plants, render it probable that the correlation of colour with infertility, which has been detected in several cases in plants, may also extend to animals in a state of nature; and if so, the fact is of the highest importance as throwing light on the origin of the infertility of many allied species. This will be better understood after considering the facts which will be now described.

The Isolation of Varieties by Selective Association.

In the last chapter I have shown that the importance of geographical isolation for the formation of new species by natural selection has been greatly exaggerated, because the

[1] In the *Medico-Chirurgical Transactions*, vol. liii. (1870), Dr. Ogle has adduced some curious physiological facts bearing on the presence or absence of white colours in the higher animals. He states that a dark pigment in the olfactory region of the nostrils is essential to perfect smell, and that this pigment is rarely deficient except when the whole animal is pure white, and the creature is then almost without smell or taste. He observes that there is no proof that, in any of the cases given above, the black animals actually eat the poisonous root or plant; and that the facts are readily understood if the senses of smell and taste are dependent on a pigment which is absent in the white animals, who therefore eat what those gifted with normal senses avoid. This explanation however hardly seems to cover the facts. We cannot suppose that almost all the sheep in the world (which are mostly white) are without smell or taste. The cutaneous disease on the white patches of hair on horses, the special liability of white terriers to distemper, of white chickens to the gapes, and of silkworms which produce yellow silk to the fungus, are not explained by it. The analogous facts in plants also indicate a real constitutional relation with colour, not an affection of the sense of smell and taste only.

[2] For all these facts, see *Animals and Plants under Domestication*, vol. ii. pp. 335-338.

very change of conditions, which is the initial power in starting such new forms, leads also to a local or stational segregation of the forms acted upon. But there is also a very powerful cause of isolation in the mental nature—the likes and dislikes—of animals; and to this is probably due the fact of the comparative rarity of hybrids in a state of nature. The differently coloured herds of cattle in the Falkland Islands, each of which keeps separate, have been already mentioned; and it may be added, that the white variety seem to have already developed a physiological peculiarity in breeding three months earlier than the others. Similar facts occur, however, among our domestic animals and are well known to breeders. Professor Low, one of the greatest authorities on our domesticated animals, says: "The female of the dog, when not under restraint, makes selection of her mate, the mastiff selecting the mastiff, the terrier the terrier, and so on." And again: "The Merino sheep and Heath sheep of Scotland, if two flocks are mixed together, each will breed with its own variety." Mr. Darwin has collected many facts illustrating this point. One of the chief pigeon-fanciers in England informed him that, if free to choose, each breed would prefer pairing with its own kind. Among the wild horses in Paraguay those of the same colour and size associate together; while in Circassia there are three races of horses which have received special names, and which, when living a free life, almost always refuse to mingle and cross, and will even attack one another. On one of the Faröe Islands, not more than half a mile in diameter, the half-wild native black sheep do not readily mix with imported white sheep. In the Forest of Dean, and in the New Forest, the dark and pale coloured herds of fallow deer have never been known to mingle; and even the curious Ancon sheep of quite modern origin have been observed to keep together, separating themselves from the rest of the flock when put into enclosures with other sheep. The same rule applies to birds, for Darwin was informed by the Rev. W. D. Fox that his flocks of white and Chinese geese kept distinct.[1]

This constant preference of animals for their like, even in the case of slightly different varieties of the same species, is evidently

[1] *Animals and Plants under Domestication*, vol. ii. pp. 102, 103.

a fact of great importance in considering the origin of species by natural selection, since it shows us that, so soon as a slight differentiation of form or colour has been effected, isolation will at once arise by the selective association of the animals themselves; and thus the great stumbling-block of "the swamping effects of intercrossing," which has been so prominently brought forward by many naturalists, will be completely obviated.

If now we combine with this fact the correlation of colour with important constitutional peculiarities, and, in some cases, with infertility; and consider, further, the curious parallelism that has been shown to exist between the effects of changed conditions and the intercrossing of varieties in producing either an increase or a decrease of fertility, we shall have obtained, at all events, a starting-point for the production of that infertility which is so characteristic a feature of distinct species when intercrossed. All we need, now, is some means of increasing or accumulating this initial tendency; and to a discussion of this problem we will therefore address ourselves.

The Influence of Natural Selection upon Sterility and Fertility.

It will occur to many persons that, as the infertility or sterility of incipient species would be useful to them when occupying the same or adjacent areas, by neutralising the effects of intercrossing, this infertility might have been increased by the action of natural selection; and this will be thought the more probable if we admit, as we have seen reason to do, that variations in fertility occur, perhaps as frequently as other variations. Mr. Darwin tells us that, at one time, this appeared to him probable, but he found the problem to be one of extreme complexity; and he was also influenced against the view by many considerations which seemed to render such an origin of the sterility or infertility of species when intercrossed very improbable. The fact that species which occupy distinct areas, and which nowhere come in contact with each other, are often sterile when crossed, is one of the difficulties; but this may perhaps be overcome by the consideration that, though now isolated, they may, and often must, have been in contact at their origination. More important is the objection that natural selection could not possibly have produced the difference that often occurs between reciprocal crosses, one of these being sometimes fertile, while the other is sterile. The extremely different amounts of infertility or sterility between different species of the same genus, the infertility often bearing no proportion to the difference between the species crossed, is also an important objection. But none of these objections would have much weight if it could be clearly shown that natural selection *is* able to increase the infertility variations of incipient species, as it is certainly able to increase and develop all useful variations of form, structure, instincts, or habits. Ample causes of infertility have been shown to exist, in the nature of the organism and the laws of correlation; the agency of natural selection is only needed to accumulate the effects produced by these causes, and to render their final results more uniform and more in accordance with the facts that exist.

About twenty years ago I had much correspondence and discussion with Mr. Darwin on this question. I then believed that I was able to demonstrate the action of natural selection in accumulating infertility; but I could not convince him, owing to the extreme complexity of the process under the conditions which he thought most probable. I have recently returned to the question; and, with the fuller knowledge of the facts of variation we now possess, I think it may be shown that natural selection *is*, in some probable cases at all events, able to accumulate variations in infertility between incipient species.

The simplest case to consider, will be that in which two forms or varieties of a species, occupying an extensive area, are in process of adaptation to somewhat different modes of life within the same area. If these two forms freely intercross with each other, and produce mongrel offspring which are quite fertile *inter se*, then the further differentiation of the forms into two distinct species will be retarded, or perhaps entirely prevented; for the offspring of the crossed unions will be, perhaps, more vigorous on account of the cross, although less perfectly adapted to the conditions of existence than either of the pure breeds; and this would certainly establish a powerful antagonistic influence to the further differentiation of the two forms.

Now, let us suppose that a partial sterility of the hybrids between the two forms arises, in correlation with the different modes of life and the slight external or internal peculiarities that exist between them, both of which we have seen to be real causes of infertility. The result will be that, even if the hybrids between the two forms are still freely produced, these hybrids will not themselves increase so rapidly as the two pure forms; and as these latter are, by the terms of the problem, better suited to their conditions of life than are the hybrids between them, they will not only increase more rapidly, but will also tend to supplant the hybrids altogether whenever the struggle for existence becomes exceptionally severe. Thus, the more complete the sterility of the hybrids the more rapidly will they die out and leave the two parent forms pure. Hence it will follow that, if there is greater infertility between the two forms in one part of the area than the other, these forms will be kept more pure wherever this greater infertility prevails, will therefore have an advantage at each recurring period of severe struggle for existence, and will thus ultimately supplant the less infertile or completely fertile forms that may exist in other portions of the area. It thus appears that, in such a case as here supposed, natural selection would preserve those portions of the two breeds which were most infertile with each other, or whose hybrid offspring were most infertile; and would, therefore, if variations in fertility continued to arise, tend to increase that infertility. It must particularly be noted that this effect would result, not by the preservation of the infertile variations on account of their infertility, but by the inferiority of the hybrid offspring, both as being fewer in numbers, less able to continue their race, and less adapted to the conditions of existence than either of the pure forms. It is this inferiority of the hybrid offspring that is the essential point; and as the number of these hybrids will be permanently less where the infertility is greatest, therefore those portions of the two forms in which infertility is greatest will have the advantage, and will ultimately survive in the struggle for existence.

The differentiation of the two forms into distinct species, with the increase of infertility between them, would be greatly assisted by two other important factors in the problem. It has already been shown that, with each modification of form and habits, and especially with modifications of colour, there arises a disinclination of the two forms to pair together; and this would produce an amount of isolation which would greatly assist the specialisation of the forms in adaptation to their different conditions of life. Again, evidence has been adduced that change of conditions or of mode of life is a potent cause of disturbance of the reproductive system, and, consequently, of infertility. We may therefore assume that, as the two forms adopted more and more different modes of life, and perhaps acquired also decided peculiarities of form and coloration, the infertility between them would increase or become more general; and as we have seen that every such increase of infertility would give that portion of the species in which it arose an advantage over the remaining portions in which the two varieties were more fertile together, all this induced infertility would maintain itself, and still further increase the general infertility between the two forms of the species.

It follows, then, that specialisation to separate conditions of life, differentiation of external characters, disinclination to cross-unions, and the infertility of the hybrid produce of these unions, would all proceed *pari passu*, and would ultimately lead to the production of two distinct forms having all the characteristics, physiological as well as structural, of true species.

In the case now discussed it has been supposed, that some amount of general infertility might arise in correlation with the different modes of life of two varieties or incipient species. A considerable body of facts already adduced renders it probable that this *is* the mode in which any widespread infertility would arise; and, if so, it has been shown that, by the influence of natural selection and the known laws which affect varieties, the infertility would be gradually increased. But, if we suppose the infertility to arise sporadically within the two forms, and to affect only a small proportion of the individuals in any area, it will be difficult, if not impossible, to show that such infertility would have any tendency to increase, or would produce any but a prejudicial effect. If, for example, five per cent of each form thus varied so as to be infertile with the other form, the result would be hardly perceptible, because the individuals which formed cross-unions and produced hybrids would constitute a very small portion of the whole species; and the hybrid offspring, being at a disadvantage in the struggle for existence and being themselves infertile, would soon die out, while the much more numerous fertile portion of the two forms would increase rapidly, and furnish a sufficient number of pure-bred offspring of each form to take the place of the somewhat inferior hybrids between them whenever the struggle for existence became severe. We must suppose that the normal fertile forms would transmit their fertility to their progeny, and the few infertile forms their infertility; but the latter would necessarily lose half their proper increase by the sterility of their hybrid offspring whenever they crossed with the other form, and when they bred with their own form the tendency to sterility would die out except in the very minute proportion of the five per cent (one-twentieth) that chance would lead to pair together. Under these circumstances the incipient sterility between the two forms would rapidly be eliminated, and could never rise much above the numbers which were produced by sporadic variation each year.

It was, probably, by a consideration of some such case as this that Mr. Darwin came to the conclusion that infertility arising between incipient species could not be increased by natural selection; and this is the more likely, as he was always disposed to minimise both the frequency and the amount even of structural variations.

We have yet to notice another mode of action of natural selection in favouring and perpetuating any infertility that may arise between two incipient species. If several distinct species are undergoing modification at the same time and in the same area, to adapt them to some new conditions that have arisen there, then any species in which the structural or colour differences that have arisen between it and its varieties or close allies were correlated with infertility of the crosses between them, would have an advantage over the corresponding varieties of other species in which there was no such physiological peculiarity. Thus, incipient species which were infertile together would have an advantage over other incipient species which were fertile, and, whenever the struggle for existence became severe, would prevail over them and take their place. Such infertility, being correlated with constitutional or structural differences, would probably, as already suggested, go on increasing as these differences increased; and thus, by the time the new species became fully differentiated from its parent form (or brother variety) the infertility might have become as well marked as we usually find it to be between distinct species.

This discussion has led us to some conclusions of the greatest importance as bearing on the difficult problem of the cause of the sterility of the hybrids between distinct species. Accepting, as highly probable, the fact of variations in fertility occurring in correlation with variations in habits, colour, or structure, we see, that so long as such variations occurred only sporadically, and affected but a small proportion of the individuals in any area, the infertility could not be increased by natural selection, but would tend to die out almost as fast as it was produced. If, however, it was so closely correlated with physical variations or diverse modes of life as to affect, even in a small degree, a considerable proportion of the individuals of the two forms in definite areas, it would be preserved by natural selection, and the portion of the varying species thus affected would increase at the expense of those portions which were more fertile when crossed. Each further variation towards infertility between the two forms would be again preserved, and thus the incipient infertility of the hybrid offspring might be increased till it became so great as almost to amount to sterility. Yet further, we have seen that if several competing species in the same area were being simultaneously modified, those between whose varieties infertility arose would have an advantage over those whose varieties remained fertile *inter se*, and would ultimately supplant them.

The preceding argument, it will be seen, depends entirely upon the assumption that some amount of infertility characterises the distinct varieties which are in process of differentiation into species; and it may be objected that of such infertility there is no proof. This is admitted; but it is urged that facts have been adduced which render such infertility probable, at least in some cases, and this is all that is required. It is by no means necessary that *all* varieties should exhibit incipient infertility, but only some varieties; for we know that, of the innumerable varieties that occur but few become developed into distinct species, and it may be that the absence of infertility, to obviate the effects of intercrossing, is one of the usual causes of their failure. All I have attempted to show is, that *when* incipient infertility does occur in correlation with other varietal differences, that infertility can be, and in fact must be, increased by natural selection; and this, it appears to me, is a decided step in advance in the solution of the problem.[1]

[1] As this argument is a rather difficult one to follow, while its theoretical importance is very great, I add here the following briefer exposition of it, in a series of propositions; being, with a few verbal alterations, a copy of what I wrote on the subject about twenty years back. Some readers may find this easier to follow than the fuller discussion in the text:—

Can Sterility of Hybrids have been Produced by Natural Selection?

1. Let there be a species which has varied into *two forms* each adapted to certain existing conditions better than the parent form, which they soon supplant.

2. If these *two forms*, which are supposed to coexist in the same district, do not intercross, natural selection will accumulate all favourable variations till they become well suited to their conditions of life, and form two slightly differing species.

3. But if these *two forms* freely intercross with each other, and produce hybrids, which are also quite fertile *inter se*, then the formation of the two distinct races or species will be retarded, or perhaps entirely prevented; for the offspring of the crossed unions will be *more vigorous* owing to the cross, although *less adapted* to their conditions of life than either of the pure breeds.

4. Now, let a partial sterility of the hybrids of some considerable proportion of these two forms arise; and, as this would probably be due to some special conditions of life, we may fairly suppose it to arise in some definite portion of the area occupied by the two forms.

5. The result will be that, in that area, the hybrids (although continually produced by first crosses almost as freely as before) will not themselves increase so rapidly as the two pure forms; and as the two pure forms are, by the terms of the problem, better suited to their several conditions of life than the hybrids, they will inevitably increase more rapidly, and will continually tend to supplant the hybrids altogether at every recurrent severe struggle for existence.

6. We may fairly suppose, also, that as soon as any sterility appears some disinclination to *cross unions* will appear, and this will further tend to the diminution of the production of hybrids.

Physiological Selection.

Another form of infertility has been suggested by Professor G. J. Romanes as having aided in bringing about the characteristic infertility or sterility of hybrids. It is founded on the fact, already noticed, that certain individuals of some species possess what may be termed selective sterility—that is, while fertile with some individuals of the species they are sterile with others, and this altogether independently of any differences of form, colour, or structure. The phenomenon, in the only form in which it has been observed, is that of "infertility or absolute sterility between two individuals, each of which is perfectly fertile with all other individuals;" but Mr. Romanes thinks that "it would not be nearly so remarkable, or physiologically improbable, that such incompatibility should run through a whole race or strain."[1] Admitting that this may be

7. In the other part of the area, however, where hybridism occurs with perfect freedom, hybrids of various degrees may increase till they equal or even exceed in number the pure species—that is, the incipient species will be liable to be swamped by intercrossing.

8. The first result, then, of a partial sterility of crosses appearing in one part of the area occupied by the two forms, will be—that the great majority of the individuals will there consist of the two pure forms only, while in the remaining part these will be in a minority,—which is the same as saying that the new *physiological variety* of the two forms will be better suited to the conditions of existence than the remaining portion which has not varied physiologically.

9. But when the struggle for existence becomes severe, that variety which is best adapted to the conditions of existence always supplants that which is imperfectly adapted; therefore, *by natural selection* the *varieties* which are *sterile* when crossed will become established as the only ones.

10. Now let variations in the *amount of sterility* and in the *disinclination to crossed unions* continue to occur—also in certain parts of the area: exactly the same result must recur, and the progeny of this new physiological variety will in time occupy the whole area.

11. There is yet another consideration that would facilitate the process. It seems probable that the *sterility variations* would, to some extent, concur with, and perhaps depend upon, the *specific variations;* so that, just in proportion as the *two forms* diverged and became better adapted to the conditions of existence, they would become more sterile when intercrossed. If this were the case, then natural selection would act with double strength; and those which were better adapted to survive both structurally and physiologically would certainly do so.

[1] Cases of this kind are referred to at p. 155. It must, however, be noted, that such sterility in first crosses appears to be equally rare between different species of the same genus as between individuals of the same species. Mules and other hybrids are freely produced between very distinct species, but are

so, though we have at present no evidence whatever in support of it, it remains to be considered whether such physiological varieties could maintain themselves, or whether, as in the cases of sporadic infertility already discussed, they would necessarily die out unless correlated with useful characters. Mr. Romanes thinks that they would persist, and urges that "whenever this one kind of variation occurs *it cannot escape the preserving agency* of physiological selection. Hence, even if it be granted that the variation which affects the reproductive system in this particular way is a variation of comparatively rare occurrence, still, as *it must always be preserved* whenever it does occur, its influence in the manufacture of specific types *must be cumulative*." The very positive statements which I have italicised would lead most readers to believe that the alleged fact had been demonstrated by a careful working out of the process in some definite supposed cases. This, however, has nowhere been done in Mr. Romanes' paper; and as it is *the* vital theoretical point on which any possible value of the new theory rests, and as it appears so opposed to the self-destructive effects of simple infertility, which we have already demonstrated when it occurs between the intermingled portion of two varieties, it must be carefully examined. In doing so, I will suppose that the required variation is not of "rare occurrence," but of considerable amount, and that it appears afresh each year to about the same extent, thus giving the theory every possible advantage.

Let us then suppose that a given species consists of 100,000 individuals of each sex, with only the usual amount of fluctuating external variability. Let a physiological variation arise, so that 10 per cent of the whole number—10,000 individuals of each sex—while remaining fertile *inter se* become quite sterile with the remaining 90,000. This peculiarity is not correlated with any external differences of

themselves infertile or quite sterile; and it is this infertility or sterility of the hybrids that is the characteristic—and was once thought to be the criterion—of species, not the sterility of their first crosses. Hence we should not expect to find any constant infertility in the first crosses between the distinct strains or varieties that formed the starting-point of new species, but only a slight amount of infertility in their mongrel offspring. It follows, that Mr. Romanes' theory of *Physiological Selection*—which assumes sterility or infertility between first crosses as the fundamental fact in the origin of species—does not accord with the general phenomena of hybridism in nature.

form or colour, or with inherent peculiarities of likes or dislikes leading to any choice as to the pairing of the two sets of individuals. We have now to inquire, What would be the result?

Taking, first, the 10,000 pairs of the physiological or abnormal variety, we find that each male of these might pair with any one of the whole 100,000 of the opposite sex. If, therefore, there was nothing to limit their choice to particular individuals of either variety, the probabilities are that 9000 of them would pair with the opposite variety, and only 1000 with their own variety—that is, that 9000 would form sterile unions, and only *one* thousand would form fertile unions.

Taking, next, the 90,000 normal individuals of either sex, we find, that each male of these has also a choice of 100,000 to pair with. The probabilities are, therefore, that nine-tenths of them—that is, 81,000—would pair with their normal fellows, while 9000 would pair with the opposite abnormal variety forming the above-mentioned sterile unions.

Now, as the number of individuals forming a species remains constant, generally speaking, from year to year, we shall have next year also 100,000 pairs, of which the two physiological varieties will be in the proportion of eighty-one to one, or 98,780 pairs of the normal variety to 1220[1] of the abnormal, that being the proportion of the fertile unions of each. In this year we shall find, by the same rule of probabilities, that only 15 males of the abnormal variety will pair with their like and be fertile, the remaining 1205 forming sterile unions with some of the normal variety. The following year the total 100,000 pairs will consist of 99,984 of the normal, and only 16 of the abnormal variety; and the probabilities, of course, are, that the whole of these latter will pair with some of the enormous preponderance of normal individuals, and, their unions being sterile, the physiological variety will become extinct in the third year.

If now in the second and each succeeding year a similar proportion as at first (10 per cent) of the physiological variety is produced afresh from the ranks of the normal variety, the same rate of diminution will go on, and it will be found that,

[1] The exact number is 1219·51, but the fractions are omitted for clearness.

on the most favourable estimate, the physiological variety can never exceed 12,000 to the 88,000 of the normal form of the species, as shown by the following table:—

1st Year.	10,000 of physiological variety to 90,000 of normal variety.						
2d ,,	1,220 +	10,000 again produced.					
3d ,,	16 +	1,220 +	10,000	do.			= 11,236
4th ,,	0 +	16 +	1,220 +	10,000	do.		= 11,236
5th ,,		0 +	16 +	1,220 +	10,000		= 11,236

and so on for any number of generations.

In the preceding discussion we have given the theory the advantage of the large proportion of 10 per cent of this very exceptional variety arising in its midst year by year, and we have seen that, even under these favourable conditions, it is unable to increase its numbers much above its starting-point, and that it remains wholly dependent on the continued renewal of the variety for its existence beyond a few years. It appears, then, that this form of inter-specific sterility cannot be increased by natural or any other known form of selection, but that it contains within itself its own principle of destruction. If it is proposed to get over the difficulty by postulating a larger percentage of the variety annually arising within the species, we shall not affect the law of decrease until we approach equality in the numbers of the two varieties. But with any such increase of the physiological variety the species itself would inevitably suffer by the large proportion of sterile unions in its midst, and would thus be at a great disadvantage in competition with other species which were fertile throughout. Thus, natural selection will always tend to weed out any species with too great a tendency to sterility among its own members, and will therefore prevent such sterility from becoming the general characteristic of varying species, which this theory demands should be the case.

On the whole, then, it appears clear that no form of infertility or sterility between the individuals of a species, can be increased by natural selection unless correlated with some useful variation, while all infertility not so correlated has a constant tendency to effect its own elimination. But the opposite property, fertility, is of vital importance to every species, and gives the offspring of the individuals which possess it, in consequence of their superior numbers, a greater chance of survival in the battle of life. It is, therefore, directly under the control of natural selection, which acts both by the self-preservation of fertile and the self-destruction of infertile stocks—except always where correlated as above, when they become useful, and therefore subject to be increased by natural selection.

Summary and Concluding Remarks on Hybridity.

The facts which are of the greatest importance to a comprehension of this very difficult subject are those which show the extreme susceptibility of the reproductive system both in plants and animals. We have seen how both these classes of organisms may be rendered infertile, by a change of conditions which does not affect their general health, by captivity, or by too close interbreeding. We have seen, also, that infertility is frequently correlated with a difference of colour, or with other characters; that it is not proportionate to divergence of structure; that it varies in reciprocal crosses between pairs of the same species; while in the cases of dimorphic and trimorphic plants the different crosses between the same pair of individuals may be fertile or sterile at the same time. It appears as if fertility depended on such a delicate adjustment of the male and female elements to each other, that, unless constantly kept up by the preservation of the most fertile individuals, sterility is always liable to arise. This preservation always occurs within the limits of each species, both because fertility is of the highest importance to the continuance of the race, and also because sterility (and to a less extent infertility) is self-destructive as well as injurious to the species.

So long therefore as a species remains undivided, and in occupation of a continuous area, its fertility is kept up by natural selection; but the moment it becomes separated, either by geographical or selective isolation, or by diversity of station or of habits, then, while each portion must be kept fertile *inter se*, there is nothing to prevent infertility arising between the two separated portions. As the two portions will necessarily exist under somewhat different conditions of life, and will usually have acquired some diversity of form and colour—both which circumstances we know to be either the cause of infertility or to be correlated with it,—the fact of some degree of infertility usually appearing between closely allied but locally or physiologically segregated species is exactly what we should expect.

The reason why varieties do not usually exhibit a similar amount of infertility is not difficult to explain. The popular conclusions on this matter have been drawn chiefly from what occurs among domestic animals, and we have seen that the very first essential to their becoming domesticated was that they should continue fertile under changed conditions of life. During the slow process of the formation of new varieties by conscious or unconscious selection, fertility has always been an essential character, and has thus been invariably preserved or increased; while there is some evidence to show that domestication itself tends to increase fertility.

Among plants, wild species and varieties have been more frequently experimented on than among animals, and we accordingly find numerous cases in which distinct species of plants are perfectly fertile when crossed, their hybrid offspring being also fertile *inter se*. We also find some few examples of the converse fact—varieties of the same species which when crossed are infertile or even sterile.

The idea that either infertility or geographical isolation is absolutely essential to the formation of new species, in order to prevent the swamping effects of intercrossing, has been shown to be unsound, because the varieties or incipient species will, in most cases, be sufficiently isolated by having adopted different habits or by frequenting different stations; while selective association, which is known to be general among distinct varieties or breeds of the same species, will produce an effective isolation even when the two forms occupy the same area.

From the various considerations now adverted to, Mr. Darwin arrived at the conclusion that the sterility or infertility of species with each other, whether manifested in the difficulty of obtaining first crosses between them or in the sterility of the hybrids thus obtained, is not a constant or necessary result of specific difference, but is incidental on unknown peculiarities of the reproductive system. These peculiarities constantly tend to arise under changed conditions owing to the extreme susceptibility of that system, and they are usually correlated with variations of form or of colour. Hence, as fixed differences of form and colour, slowly gained by natural selection in adaptation to changed conditions, are what essentially characterise distinct species, some amount of infertility between species is the usual result.

Here the problem was left by Mr. Darwin; but we have shown that its solution may be carried a step further. If we accept the association of some degree of infertility, however slight, as a not unfrequent accompaniment of the external differences which always arise in a state of nature between varieties and incipient species, it has been shown that natural selection *has* power to increase that infertility just as it has power to increase other favourable variations. Such an increase of infertility will be beneficial, whenever new species arise in the same area with the parent form; and we thus see how, out of the fluctuating and very unequal amounts of infertility correlated with physical variations, there may have arisen that larger and more constant amount which appears usually to characterise well-marked species.

The great body of facts of which a condensed account has been given in the present chapter, although from an experimental point of view very insufficient, all point to the general conclusion we have now reached, and afford us a not unsatisfactory solution of the great problem of hybridism in relation to the origin of species by means of natural selection. Further experimental research is needed in order to complete the elucidation of the subject; but until these additional facts are forthcoming no new theory seems required for the explanation of the phenomena.

CHAPTER VIII

THE ORIGIN AND USES OF COLOUR IN ANIMALS

The Darwinian theory threw new light on organic colour—The problem to be solved—The constancy of animal colour indicates utility—Colour and environment—Arctic animals white—Exceptions prove the rule—Desert, forest, nocturnal, and oceanic animals—General theories of animal colour—Variable protective colouring—Mr. Poulton's experiments—Special or local colour adaptations—Imitation of particular objects—How they have been produced—Special protective colouring of butterflies—Protective resemblance among marine animals—Protection by terrifying enemies—Alluring coloration—The coloration of birds' eggs—Colour as a means of recognition—Summary of the preceding exposition—Influence of locality or of climate on colour—Concluding remarks.

AMONG the numerous applications of the Darwinian theory in the interpretation of the complex phenomena presented by the organic world, none have been more successful, or are more interesting, than those which deal with the colours of animals and plants. To the older school of naturalists colour was a trivial character, eminently unstable and untrustworthy in the determination of species; and it appeared to have, in most cases, no use or meaning to the objects which displayed it. The bright and often gorgeous coloration of insect, bird, or flower, was either looked upon as having been created for the enjoyment of mankind, or as due to unknown and perhaps undiscoverable laws of nature.

But the researches of Mr. Darwin totally changed our point of view in this matter. He showed, clearly, that some of the colours of animals are useful, some hurtful to them; and he believed that many of the most brilliant colours were developed by sexual choice; while his great general principle, that all the fixed characters of organic beings have been developed under the action of the law of utility, led to the inevitable conclusion that so remarkable and conspicuous a character as colour, which so often constitutes the most obvious distinction of species from species, or group from group, must also have arisen from survival of the fittest, and must, therefore, in most cases have some relation to the wellbeing of its possessors. Continuous observation and research, carried on by multitudes of observers during the last thirty years, have shown this to be the case; but the problem is found to be far more complex than was at first supposed. The modes in which colour is of use to different classes of organisms is very varied, and have probably not yet been all discovered; while the infinite variety and marvellous beauty of some of its developments are such as to render it hopeless to arrive at a complete and satisfactory explanation of every individual case. So much, however, has been achieved, so many curious facts have been explained, and so much light has been thrown on some of the most obscure phenomena of nature, that the subject deserves a prominent place in any account of the Darwinian theory.

The Problem to be Solved.

Before dealing with the various modifications of colour in the animal world it is necessary to say a few words on colour in general, on its prevalence in nature, and how it is that the colours of animals and plants require any special explanation. What we term colour is a subjective phenomenon, due to the constitution of our mind and nervous system; while, objectively, it consists of light-vibrations of different wave-lengths emitted by, or reflected from, various objects. Every visible object must be coloured, because to be visible it must send rays of light to our eye. The kind of light it sends is modified by the molecular constitution or the surface texture of the object. Pigments absorb certain rays and reflect the remainder, and this reflected portion has to our eyes a definite colour, according to the portion of the rays constituting white light which are absorbed. Interference colours are produced either by thin films or by very fine striæ on the surfaces of bodies, which cause rays of certain wave-lengths to neutralise each other, leaving the remainder to produce the effects of colour. Such are the colours of soap-bubbles, or of steel or glass on which extremely fine lines have been ruled; and these colours often produce the effect of metallic lustre, and are the cause of most of the metallic hues of birds and insects.

As colour thus depends on molecular or chemical constitution or on the minute surface texture of bodies, and, as the matter of which organic beings are composed consists of chemical compounds of great complexity and extreme instability, and is also subject to innumerable changes during growth and development, we might naturally expect the phenomena of colour to be more varied here than in less complex and more stable compounds. Yet even in the inorganic world we find abundant and varied colours; in the earth and in the water; in metals, gems, and minerals; in the sky and in the ocean; in sunset clouds and in the many-tinted rainbow. Here we can have no question of *use* to the coloured object, and almost as little perhaps in the vivid red of blood, in the brilliant colours of red snow and other low algæ and fungi, or even in the universal mantle of green which clothes so large a portion of the earth's surface. The presence of some colour, or even of many brilliant colours, in animals and plants would require no other explanation than does that of the sky or the ocean, of the ruby or the emerald —that is, it would require a purely physical explanation only. It is the wonderful individuality of the colours of animals and plants that attracts our attention—the fact that the colours are localised in definite patterns, sometimes in accordance with structural characters, sometimes altogether independent of them; while often differing in the most striking and fantastic manner in allied species. We are thus compelled to look upon colour not merely as a physical but also as a biological characteristic, which has been differentiated and specialised by natural selection, and must, therefore, find its explanation in the principle of adaptation or utility.

The Constancy of Animal Colour indicates Utility.

That the colours and markings of animals have been acquired under the fundamental law of utility is indicated by a general fact which has received very little attention. As a rule, colour and marking are constant in each species of wild animal, while, in almost every domesticated animal, there arises great variability. We see this in our horses and cattle, our dogs and cats, our pigeons and poultry. Now, the essential difference between the conditions of life of domesticated and wild animals is, that the former are protected by man, while the latter have to protect themselves. The extreme variations in colour that immediately arise under domestication indicate a tendency to vary in this way, and the occasional occurrence of white or piebald or other exceptionally coloured individuals of many species in a state of nature, shows that this tendency exists there also; and, as these exceptionally coloured individuals rarely or never increase, there must be some constant power at work to keep it in check. This power can only be natural selection or the survival of the fittest, which again implies that some colours are useful, some injurious, in each particular case. With this principle as our guide, let us see how far we can account both for the general and special colours of the animal world.

Colour and Environment.

The fact that first strikes us in our examination of the colours of animals as a whole, is the close relation that exists between these colours and the general environment. Thus, white prevails among arctic animals; yellow or brown in desert species; while green is only a common colour in tropical evergreen forests. If we consider these cases somewhat carefully we shall find, that they afford us excellent materials for forming a judgment on the various theories that have been suggested to account for the colours of the animal world.

In the arctic regions there are a number of animals which are wholly white all the year round, or which only turn white in winter. Among the former are the polar bear and the American polar hare, the snowy owl and the Greenland falcon; among the latter the arctic fox, the arctic hare, the ermine, and the ptarmigan. Those which are permanently white remain among the snow nearly all the year round, while those which change their colour inhabit regions which are free from snow in summer. The obvious explanation of this style of coloration is, that it is protective, serving to conceal the herbivorous species from their enemies, and enabling carnivorous animals to approach their prey unperceived. Two other explanations have, however, been suggested. One is, that the prevalent white of the arctic regions has a direct effect in producing the white colour in animals, either by some photographic or chemical action on the skin or by a reflex action through vision. The other is, that the white colour is chiefly beneficial as a means of checking radiation and so preserving animal heat during the severity of an arctic winter. The first is part of the general theory that colour is the effect of coloured light on the objects—a pure hypothesis which has, I believe, no facts whatever to support it. The second suggestion is also an hypothesis merely, since it has not been proved by experiment that a white colour, *per se*, independently of the fur or feathers which is so coloured, has any effect whatever in checking the radiation of low-grade heat like that of the animal body. But both alike are sufficiently disproved by the interesting exceptions to the rule of white coloration in the arctic regions, which exceptions are, nevertheless, quite in harmony with the theory of protection.

Whenever we find arctic animals which, from whatever cause, do not require protection by the white colour, then neither the cold nor the snow-glare has any effect upon their coloration. The sable retains its rich brown fur throughout the Siberian winter; but it frequents trees at that season and not only feeds partially on fruits or seeds, but is able to catch birds among the branches of the fir-trees, with the bark of which its colour assimilates. Then we have that thoroughly arctic animal, the musk-sheep, which is brown and conspicuous; but this animal is gregarious, and its safety depends on its association in small herds. It is, therefore, of more importance for it to be able to recognise its kind at a distance than to be concealed from its enemies, against which it can well protect itself so long as it keeps together in a compact body. But the most striking example is that of the common raven, which is a true arctic bird, and is found even in mid-winter as far north as any known bird or mammal. Yet it always retains its black coat, and the reason, from our point of view, is obvious. The raven is a powerful bird and fears no enemy, while, being a carrion-feeder, it has no need for concealment in order to approach its prey. The colour of the raven and of the musk-sheep are, therefore, both inconsistent with any other theory than that the white colour of arctic animals has been acquired for concealment, and to that theory both afford a strong support. Here we have a striking example of the exception proving the rule.

In the desert regions of the earth we find an even more general accordance of colour with surroundings. The lion, the camel, and all the desert antelopes have more or less the colour of the sand or rock among which they live. The Egyptian cat and the Pampas cat are sandy or earth coloured. The Australian kangaroos are of similar tints, and the original colour of the wild horse is supposed to have been sandy or clay coloured. Birds are equally well protected by assimilative hues; the larks, quails, goatsuckers, and grouse which abound in the North African and Asiatic deserts are all tinted or mottled so as closely to resemble the average colour of the soil in the districts they inhabit. Canon Tristram, who knows these regions and their natural history so well, says, in an often quoted passage: "In the desert, where neither trees, brushwood, nor even undulations of the surface afford the slightest protection to its foes, a modification of colour which shall be assimilated to that of the surrounding country is absolutely necessary. Hence, without exception, the upper plumage of every bird, whether lark, chat, sylvain, or sand-grouse, and also the fur of all the smaller mammals, and the skin of all the snakes and lizards, is of one uniform isabelline or sand colour."

Passing on to the tropical regions, it is among their evergreen forests alone that we find whole groups of birds whose ground colour is green. Parrots are very generally green, and in the East we have an extensive group of green fruit-eating pigeons; while the barbets, bee-eaters, turacos, leaf-thrushes (Phyllornis), white-eyes (Zosterops), and many other groups, have so much green in their plumage as to tend greatly to their concealment among the dense foliage. There can be no doubt that these colours have been acquired as a protection, when we see that in all the temperate regions, where the leaves are deciduous, the ground colour of the great majority of birds, especially on the upper surface, is a rusty brown of various shades, well corresponding with the bark, withered leaves, ferns, and bare thickets among which they live in autumn and winter, and especially in early spring when so many of them build their nests.

Nocturnal animals supply another illustration of the same rule, in the dusky colours of mice, rats, bats, and moles, and in the soft mottled plumage of owls and goatsuckers which, while almost equally inconspicuous in the twilight, are such as to favour their concealment in the daytime.

An additional illustration of general assimilation of colour to the surroundings of animals, is furnished by the inhabitants of the deep oceans. Professor Moseley of the Challenger Expedition, in his British Association lecture on this subject, says: "Most characteristic of pelagic animals is the almost crystalline transparency of their bodies. So perfect is this transparency that very many of them are rendered almost entirely invisible when floating in the water, while some, even when caught and held up in a glass globe, are hardly to be seen. The skin, nerves, muscles, and other organs are absolutely hyaline and transparent, but the liver and digestive tract often remain opaque and of a yellow or brown colour, and exactly resemble when seen in the water small pieces of floating seaweed." Such marine organisms, however, as are of larger size, and either occasionally or habitually float on the surface, are beautifully tinged with blue above, thus harmonising with the colour of the sea as seen by hovering birds; while they are white below, and are thus invisible against the wave-foam and clouds as seen by enemies beneath the surface. Such are the tints of the beautiful nudibranchiate mollusc, Glaucus atlanticus, and many others.

General Theories of Animal Colour.

We are now in a position to test the general theories, or, to speak more correctly, the popular notions, as to the origin of animal coloration, before proceeding to apply the principle of utility to the explanation of some among the many extraordinary manifestations of colour in the animal world. The most generally received theory undoubtedly is, that brilliancy and variety of colour are due to the direct action of light and heat; a theory no doubt derived from the abundance of bright-coloured birds, insects, and flowers which are brought from tropical regions. There are, however, two strong arguments against this theory. We have already seen how generally bright coloration is wanting in desert animals, yet here heat and light are both at a maximum, and if these alone were the agents in the production of colour, desert animals should be the most brilliant. Again, all naturalists who have lived in tropical regions know that the proportion of bright to dull coloured species is little if any greater there than in the temperate zone, while there are many tropical groups in which bright colours are almost entirely unknown. No part of the world presents so many brilliant birds as South America, yet there are extensive families, containing many hundreds of species, which are as plainly coloured as our average temperate birds. Such are the families of the bush-shrikes and ant-thrushes (Formicariidæ), the tyrant-shrikes (Tyrannidæ), the American creepers (Dendrocolaptidæ), together with a large proportion of the wood-warblers (Mniotiltidæ), the finches, the wrens, and some other groups. In the eastern hemisphere, also, we have the babbling-thrushes (Timaliidæ), the cuckoo-shrikes (Campephagidæ), the honey-suckers (Meliphagidæ), and several other smaller groups which are certainly not coloured above the average standard of temperate birds.

Again, there are many families of birds which spread over the whole world, temperate and tropical, and among these the tropical species rarely present any exceptional brilliancy of colour. Such are the thrushes, goatsuckers, hawks, plovers, and ducks; and in the last-named group it is the temperate and arctic zones that afford the most brilliant coloration.

The same general facts are found to prevail among insects. Although tropical insects present some of the most gorgeous coloration in the whole realm of nature, yet there are thousands and tens of thousands of species which are as dull coloured as any in our cloudy land. The extensive family of the carnivorous ground-beetles (Carabidæ) attains its greatest brilliancy in the temperate zone; while by far the larger proportion of the great families of the longicorns and the weevils, are of obscure colours even in the tropics. In butterflies, there is undoubtedly a larger proportion of brilliant colour in the tropics; but if we compare families which are almost equally developed over the globe—as the Pieridæ or whites and yellows, and the Satyridæ or ringlets—we shall find no great disproportion in colour between those of temperate and tropical regions.

The various facts which have now briefly been noticed are sufficient to indicate that the light and heat of the sun are not the direct causes of the colours of animals, although they may favour the production of colour when, as in tropical regions, the persistent high temperature favours the development of the maximum of life. We will now consider the next suggestion, that light reflected from surrounding coloured objects tends to produce corresponding colours in the animal world.

This theory is founded on a number of very curious facts which prove, that such a change does sometimes occur and is directly dependent on the colours of surrounding objects; but these facts are comparatively rare and exceptional in their nature, and the same theory will certainly not apply to the infinitely varied colours of the higher animals, many of which are exposed to a constantly varying amount of light and colour during their active existence. A brief sketch of these dependent changes of colour may, however, be advantageously given here.

Variable Protective Colouring.

There are two distinct kinds of change of colour in animals due to the colouring of the environment. In one case the change is caused by reflex action set up by the animal *seeing* the colour to be imitated, and the change produced can be altered or repeated as the animal changes its position. In the other case the change occurs but once, and is probably not due to any conscious or sense action, but to some direct influence on the surface tissues while the creature is undergoing a moult or change to the pupa form.

The most striking example of the first class is that of the chameleon, which changes to white, brown, yellowish, or green, according to the colour of the object on which it rests. This change is brought about by means of two layers of pigment cells, deeply seated in the skin, and of bluish and yellowish colours. By suitable muscles these cells can be forced upwards so as to modify the colour of the skin, which, when they are not brought into action, is a dirty white. These animals are excessively sluggish and defenceless, and the power of changing their colour to that of their immediate surroundings is no doubt of great service to them. Many of the flatfish are also capable of changing their colour according to the colour of the bottom they rest on; and frogs have a similar power to a limited extent. Some crustacea also change colour, and the power is much developed in the Chameleon shrimp (Mysis Chamæleon) which is gray when on sand, but brown or green when among brown or green seaweed. It has been proved by experiment that when this animal is blinded the change does not occur. In all these cases, therefore, we have some form of reflex or sense action by which the change is produced, probably by means of pigment cells beneath the skin as in the chameleon.

The second class consists of certain larvæ, and pupæ, which undergo changes of colour when exposed to differently coloured surroundings. This subject has been carefully investigated by Mr. E. B. Poulton, who has communicated the results of his experiments to the Royal Society.[1] It had been noticed that some species of larvæ which fed on several different plants had colours more or less corresponding to the particular plant the individual fed on. Numerous cases are given in Professor Meldola's article on "Variable Protective Colouring" (*Proc. Zool. Soc.*, 1873, p. 153), and while the general green coloration was attributed to the presence of chlorophyll beneath the skin, the particular change in correspondence to each food-plant was attributed to a special function which had been developed by natural selection. Later on, in a note to his translation of Weissmann's *Theory of Descent*, Professor Meldola seemed disposed to think that the variations of colour of some of the species might be phytophagic—that is, due to the direct action of the differently coloured leaves on which the insect fed. Mr. Poulton's experiments have thrown much light on this question, since he has conclusively proved that, in the case of the sphinx caterpillar of Smerinthus ocellatus, the change of colour is not due to the food but to the coloured light reflected from the leaves.

[1] *Proceedings of the Royal Society*, No. 243, 1886; *Transactions of the Royal Society*, vol. clxxviii. B. pp. 311-441.

This was shown by feeding two sets of larvæ on the same plant but exposed to differently coloured surroundings, obtained by sewing the leaves together, so that in one case only the dark upper surface, in the other the whitish under surface was exposed to view. The result in each case was a corresponding change of colour in the larvæ, confirming the experiments on different individuals of the same batch of larvæ which had been supplied with different food-plants or exposed to a different coloured light.

An even more interesting series of experiments was made on the colours of pupæ, which in many cases were known to be affected by the material on which they underwent their transformations. The late Mr. T. W. Wood proved, in 1867, that the pupæ of the common cabbage butterflies (Pieris brassicæ and P. rapæ) were either light, or dark, or green, according to the coloured boxes they were kept in, or the colours of the fences, walls, etc., against which they were suspended. Mrs. Barber in South Africa found that the pupæ of Papilio Nireus underwent a similar change, being deep green when attached to orange leaves of the same tint, pale yellowish-green when on a branch of the bottle-brush tree whose half-dried leaves were of this colour, and yellowish when attached to the wooden frame of a box. A few other observers noted similar phenomena, but nothing more was done till Mr. Poulton's elaborate series of experiments with the larvæ of several of our common butterflies were the means of clearing up several important points. He showed that the action of the coloured light did not affect the pupa itself but the larva, and that only for a limited period of time. After a caterpillar has done feeding it wanders about seeking a suitable place to undergo its transformation. When this is found it rests quietly for a day or two, spinning the web from which it is to suspend itself; and it is during this period of quiescence, and perhaps also the first hour or two after its suspension, that the action of the surrounding coloured surfaces determines, to a considerable extent, the colour of the pupa. By the application of various surrounding colours during this period, Mr. Poulton was able to modify the colour of the pupa of the common tortoise-shell butterfly from nearly black to pale, or to a brilliant golden; and that of Pieris rapæ from dusky through pinkish to pale green. It is interesting to note, that the colours produced were in all cases such only as assimilated with the surroundings usually occupied by the species, and also, that colours which did not occur in such surroundings, as dark red or blue, only produced the same effects as dusky or black.

Careful experiments were made to ascertain whether the effect was produced through the sight of the caterpillar. The ocelli were covered with black varnish, but neither this, nor cutting off the spines of the tortoise-shell larva to ascertain whether they might be sense-organs, produced any effect on the resulting colour. Mr. Poulton concludes, therefore, that the colour-action probably occurs over the whole surface of the body, setting up physiological processes which result in the corresponding colour-change of the pupa. Such changes are, however, by no means universal, or even common, in protectively coloured pupæ, since in Papilio machaon and some others which have been experimented on, both in this country and abroad, no change can be produced on the pupa by any amount of exposure to differently coloured surroundings. It is a curious point that,. with the small tortoise-shell larva, exposure to light from gilded surfaces produced pupæ with a brilliant golden lustre; and the explanation is supposed to be that mica abounded in the original habitat of the species, and that the pupæ thus obtained protection when suspended against micaceous rock. Looking, however, at the wide range of the species and the comparatively limited area in which micaceous rocks occur, this seems a rather improbable explanation, and the occurrence of this metallic appearance is still a difficulty. It does not, however, commonly occur in this country in a natural state.

The two classes of variable colouring here discussed are evidently exceptional, and can have little if any relation to the colours of those more active creatures which are continually changing their position with regard to surrounding objects, and whose colours and markings are nearly constant throughout the life of the individual, and (with the exception of sexual differences) in all the individuals of the species. We will now briefly pass in review the various characteristics and uses of the colours which more generally prevail in nature;

and having already discussed those protective colours which serve to harmonise animals with their general environment, we have to consider only those cases in which the colour resemblance is more local or special in its character.

Special or Local Colour Adaptations.

This form of colour adaptation is generally manifested by markings rather than by colour alone, and is extremely prevalent both among insects and vertebrates, so that we shall be able to notice only a few illustrative cases. Among our native birds we have the snipe and woodcock, whose markings and tints strikingly accord with the dead marsh vegetation among which they live; the ptarmigan in its summer dress is mottled and tinted exactly like the lichens which cover the stones of the higher mountains; while young unfledged plovers are spotted so as exactly to resemble the beach pebbles among which they crouch for protection, as beautifully exhibited in one of the cases of British birds in the Natural History Museum at South Kensington.

In mammalia, we notice the frequency of rounded spots on forest or tree haunting animals of large size, as the forest deer and the forest cats; while those that frequent reedy or grassy places are striped vertically, as the marsh antelopes and the tiger. I had long been of opinion that the brilliant yellow and black stripes of the tiger were adaptive, but have only recently obtained proof that it is so. An experienced tiger-hunter, Major Walford, states in a letter, that the haunts of the tiger are invariably full of the long grass, dry and pale yellow for at least nine months of the year, which covers the ground wherever there is water in the rainy season, and he adds: "I once, while following up a wounded tiger, failed for at least a minute to see him under a tree in grass at a distance of about twenty yards—jungle open—but the natives saw him, and I eventually made him out well enough to shoot him, but even then I could not see at what part of him I was aiming. There can be no doubt whatever that the colour of both the tiger and the panther renders them almost invisible, especially in a strong blaze of light, when among grass, and one does not seem to notice stripes or spots till they are dead." It is the black shadows of the vegetation that assimilate with the black stripes of the tiger; and, in like manner, the spotty shadows of leaves in the forest so harmonise with the spots of ocelots, jaguars, tiger-cats, and spotted deer as to afford them a very perfect concealment.

In some cases the concealment is effected by colours and markings which are so striking and peculiar that no one who had not seen the creature in its native haunts would imagine them to be protective. An example of this is afforded by the banded fruit pigeon of Timor, whose pure white head and neck, black wings and back, yellow belly, and deeply-curved black band across the breast, render it a very handsome and conspicuous bird. Yet this is what Mr. H. O. Forbes says of it: "On the trees the white-headed fruit pigeon (Ptilopus cinctus) sate motionless during the heat of the day in numbers, on well-exposed branches; but it was with the utmost difficulty that I or my sharp-eyed native servant could ever detect them, even in trees where we knew they were sitting."[1] The trees referred to are species of Eucalyptus which abound in Timor. They have whitish or yellowish bark and very open foliage, and it is the intense sunlight casting black curved shadows of one branch upon another, with the white and yellow bark and deep blue sky seen through openings of the foliage, that produces the peculiar combination of colours and shadows to which the colours and markings of this bird have become so closely assimilated.

Even such brilliant and gorgeously coloured birds as the sun-birds of Africa are, according to an excellent observer, often protectively coloured. Mrs. M. E. Barber remarks that "A casual observer would scarcely imagine that the highly varnished and magnificently coloured plumage of the various species of Noctarinea could be of service to them, yet this is undoubtedly the case. The most unguarded moments of the lives of these birds are those that are spent amongst the flowers, and it is then that they are less wary than at any other time. The different species of aloes, which blossom in succession, form the principal sources of their winter supplies of food; and a legion of other gay flowering plants in spring and summer, the aloe blossoms especially, are all brilliantly coloured, and they harmonise admirably with the gay plumage

[1] *A Naturalist's Wanderings in the Eastern Archipelago*, p. 460.

of the different species of sun-birds. Even the keen eye of a hawk will fail to detect them, so closely do they resemble the flowers they frequent. The sun-birds are fully aware of this fact, for no sooner have they relinquished the flowers than they become exceedingly wary and rapid in flight, darting arrow-like through the air and seldom remaining in exposed situations. The black sun-bird (Nectarinea amethystina) is never absent from that magnificent forest-tree, the 'Kaffir Boom' (Erythrina caffra); all day long the cheerful notes of these birds may be heard amongst its spreading branches, yet the general aspect of the tree, which consists of a huge mass of scarlet and purple-black blossoms without a single green leaf, blends and harmonises with the colours of the black sun-bird to such an extent that a dozen of them may be feeding amongst its blossoms without being conspicuous, or even visible."[1]

Some other cases will still further illustrate how the colours of even very conspicuous animals may be adapted to their peculiar haunts.

The late Mr. Swinhoe says of the Kerivoula picta, which he observed in Formosa: "The body of this bat was of an orange colour, but the wings were painted with orange-yellow and black. It was caught suspended, head downwards, on a cluster of the fruit of the longan tree (Nephelium longanum). Now this tree is an evergreen, and all the year round some portion of its foliage is undergoing decay, the particular leaves being, in such a stage, partially orange and black. This bat can, therefore, at all seasons suspend from its branches and elude its enemies by its resemblance to the leaves of the tree."[2]

Even more curious is the case of the sloths—defenceless animals which feed upon leaves, and hang from the branches of trees with their back downwards. Most of the species have a curious buff-coloured spot on the back, rounded or oval in shape and often with a darker border, which seems placed there on purpose to make them conspicuous; and this was a great puzzle to naturalists, because the long coarse gray or greenish hair was evidently like tree-moss and therefore protective. But an old writer, Baron von Slack, in his *Voyage*

[1] *Trans. Phil. Soc.* (? *of S. Africa*), 1878, part iv, p. 27.
[2] *Proc. Zool. Soc.*, 1862 p. 357.

to Surinam (1810), had already explained the matter. He says: "The colour and even the shape of the hair are much like withered moss, and serve to hide the animal in the trees, but particularly when it has that orange-coloured spot between the shoulders and lies close to the tree; it looks then exactly like a piece of branch where the rest has been broken off, by which the hunters are often deceived." Even such a huge animal as the giraffe is said to be perfectly concealed by its colour and form when standing among the dead and broken trees that so often occur on the outskirts of the thickets where it feeds. The large blotch-like spots on the skin and the strange shape of the head and horns, like broken branches, so tend to its concealment that even the keen-eyed natives have been known to mistake trees for giraffes or giraffes for trees.

Innumerable examples of this kind of protective colouring occur among insects; beetles mottled like the bark of trees or resembling the sand or rock or moss on which they live, with green caterpillars of the exact general tints of the foliage they feed on; but there are also many cases of detailed imitation of particular objects by insects that must be briefly described.[1]

Protective Imitation of Particular Objects.

The insects which present this kind of imitation most perfectly are the Phasmidæ, or stick and leaf insects. The well-

[1] With reference to this general resemblance of insects to their environment the following remarks by Mr. Poulton are very instructive. He says: "Holding the larva of Sphinx ligustri in one hand and a twig of its food-plant in the other, the wonder we feel is, not at the resemblance but at the difference; we are surprised at the difficulty experienced in detecting so conspicuous an object. And yet the protection is very real, for the larvæ will be passed over by those who are not accustomed to their appearance, although the searcher may be told of the presence of a large caterpillar. An experienced entomologist may also fail to find the larvæ till after a considerable search. This is general protective resemblance, and it depends upon a general harmony between the appearance of the organism and its whole environment. It is impossible to understand the force of this protection for any larva, without seeing it on its food-plant and in an entirely normal condition. The artistic effect of green foliage is more complex than we often imagine; numberless modifications are wrought by varied lights and shadows upon colours which are in themselves far from uniform. In the larva of Papilio machaon the protection is very real when the larva is on the food-plant, and can hardly be appreciated at all when the two are apart." Numerous other examples are given in the chapter on "Mimicry and other Protective Resemblances among Animals," in my *Contributions to the Theory of Natural Selection.*

known leaf-insects of Ceylon and of Java, species of Phyllium, are so wonderfully coloured and veined, with leafy expansions on the legs and thorax, that not one person in ten can see them when resting on the food-plant close beneath their eyes. Others resemble pieces of stick with all the minutiæ of knots and branches, formed by the insects' legs, which are stuck out rigidly and unsymmetrically. I have often been unable to distinguish between one of these insects and a real piece of stick, till I satisfied myself by touching it and found it to be alive. One species, which was brought me in Borneo, was covered with delicate semitransparent green foliations, exactly resembling the hepaticæ which cover pieces of rotten stick in the damp forests. Others resemble dead leaves in all their varieties of colour and form; and to show how perfect is the protection obtained and how important it is to the possessors of it, the following incident, observed by Mr. Belt in Nicaragua, is most instructive. Describing the armies of foraging ants in the forest which devour every insect they can catch, he says: "I was much surprised with the behaviour of a green leaf-like locust. This insect stood immovably among a host of ants, many of which ran over its legs without ever discovering there was food within their reach. So fixed was its instinctive knowledge that its safety depended on its immovability, that it allowed me to pick it up and replace it among the ants without making a single effort to escape. This species closely resembles a green leaf."[1]

Caterpillars also exhibit a considerable amount of detailed resemblance to the plants on which they live. Grass-feeders are striped longitudinally, while those on ordinary leaves are always striped obliquely. Some very beautiful protective resemblances are shown among the caterpillars figured in Smith and Abbott's *Lepidopterous Insects of Georgia*, a work published in the early part of the century, before any theories of protection were started. The plates in this work are most beautifully executed from drawings made by Mr. Abbott, representing the insects, in every case, on the plants which they frequented, and no reference is made in the descriptions to the remarkable protective details which appear upon the plates. We have, first, the larva of Sphinx fuciformis feeding

[1] *The Naturalist in Nicaragua*, p. 19.

on a plant with linear grass-like leaves and small blue flowers; and we find the insect of the same green as the leaves, striped longitudinally in accordance with the linear leaves, and with the head blue corresponding both in size and colour with the flowers. Another species (Sphinx tersa) is represented feeding on a plant with small red flowers situated in the axils of the leaves; and the larva has a row of seven red spots, unequal in size, and corresponding very closely with the colour and size of the flowers. Two other figures of sphinx larvæ are very curious. That of Sphinx pampinatrix feeds on a wild vine (Vitis indivisa), having green tendrils, and in this species the curved horn on the tail is green, and closely imitates in its curve the tip of the tendril. But in another species (Sphinx cranta), which feeds on the fox-grape (Vitis vulpina), the horn is very long and red, corresponding with the long red-tipped tendrils of the plant. Both these larvæ are green with oblique stripes, to harmonise with the veined leaves of the vines; but a figure is also given of the last-named species after it has done feeding, when it is of a decided brown colour and has entirely lost its horn. This is because it then descends to the ground to bury itself, and the green colour and red horn would be conspicuous and dangerous; it therefore loses both at the last moult. Such a change of colour occurs in many species of caterpillars. Sometimes the change is seasonal; and, in those which hibernate with us, the colour of some species, which is brownish in autumn in adaptation to the fading foliage, becomes green in spring to harmonise with the newly-opened leaves at that season.[1]

Some of the most curious examples of minute imitation are afforded by the caterpillars of the geometer moths, which are always brown or reddish, and resemble in form little twigs of the plant on which they feed. They have the habit, when at rest, of standing out obliquely from the branch, to which they hold on by their hind pair of prolegs or claspers, and remain motionless for hours. Speaking of these protective resemblances Mr. Jenner Weir says: "After being thirty years an entomologist I was deceived myself, and took out my pruning scissors to cut from a plum tree a spur which I thought I had overlooked. This turned out to be the larva

[1] R. Meldola, in *Proc. Zool. Soc.*, 1873, p. 155.

of a geometer two inches long. I showed it to several members of my family, and defined a space of four inches in which it was to be seen, but none of them could perceive that it was a caterpillar."[1]

One more example of a protected caterpillar must be given. Mr. A. Everett, writing from Sarawak, Borneo, says: "I had a caterpillar brought me, which, being mixed by my boy with some other things, I took to be a bit of moss with two exquisite pinky-white seed-capsules; but I soon saw that it moved, and examining it more closely found out its real character: it is covered with hair, with two little pink spots on the upper surface, the general hue being more green. Its motions are very slow, and when eating the head is withdrawn beneath a fleshy mobile hood, so that the action of feeding does not produce any movement externally. It was found in the limestone hills at Busan, the situation of all others where mosses are most plentiful and delicate, and where they partially clothe most of the protruding masses of rock."

How these Imitations have been Produced.

To many persons it will seem impossible that such beautiful and detailed resemblances as those now described—and these are only samples of thousands that occur in all parts of the world—can have been brought about by the preservation of accidental useful variations. But this will not seem so surprising if we keep in mind the facts set forth in our earlier chapters—the rapid multiplication, the severe struggle for existence, and the constant variability of these and all other organisms. And, further, we must remember that these delicate adjustments are the result of a process which has been going on for millions of years, and that we now see the small percentage of successes among the myriads of failures. From the very first appearance of insects and their various kinds of enemies the need of protection arose, and was usually most easily met by modifications of colour. Hence, we may be sure that the earliest leaf-eating insects acquired a green colour as one of the necessities of their existence; and, as the species became modified and specialised,

[1] *Nature*, vol. iii. p. 166.

those feeding on particular species of plants would rapidly acquire the peculiar tints and markings best adapted to conceal them upon those plants. Then, every little variation that, once in a hundred years perhaps, led to the preservation of some larva which was thereby rather better concealed than its fellows, would form the starting-point of a further development, leading ultimately to that perfection of imitation in details which now astonishes us. The researches of Dr. Weissmann illustrate this progressive adaptation. The very young larvæ of several species are green or yellowish without any markings; they then, in subsequent moults, obtain certain markings, some of which are often lost again before the larva is fully grown. The early stages of those species which, like elephant hawk-moths (Chærocampa), have the anterior segments elongated and retractile, with large eye-like spots to imitate the head of a vertebrate, are at first like those of non-retractile species, the anterior segments being as large as the rest. After the first moult they become smaller, comparatively; but it is only after the second moult that the ocelli begin to appear, and these are not fully defined till after the third moult. This progressive development of the individual—the ontogeny—gives us a clue to the ancestral development of the whole race—the phylogeny; and we are enabled to picture to ourselves the very slow and gradual steps by which the existing perfect adaptation has been brought about. In many larvæ great variability still exists, and in some there are two or more distinctly-coloured forms —usually a dark and a light or a brown and a green form. The larva of the humming-bird hawk-moth (Macroglossa stellatarum) varies in this manner, and Dr. Weissmann raised five varieties from a batch of eggs from one moth. It feeds on species of bedstraw (Galium verum and G. mollugo), and as the green forms are less abundant than the brown, it has probably undergone some recent change of food-plant or of habits which renders brown the more protective colour.

Special Protective Colouring of Butterflies.

We will now consider a few cases of special protective colouring in the perfect butterfly or moth. Mr. Mansel Weale states that in South Africa there is a great prevalence of white and silvery foliage or bark, sometimes of dazzling brilliancy, and that many insects and their larvæ have brilliant silvery tints which are protective, among them being three species of butterflies whose undersides are silvery, and which are thus effectually protected when at rest.[1] A common African butterfly (Aterica meleagris) always settles on the ground with closed wings, which so closely resemble the soil of the district that it can with difficulty be seen, and the colour varies with the soil in different localities. Thus specimens from Senegambia were dull brown, the soil being reddish sand and iron-clay; those from Calabar and Cameroons were light brown with numerous small white spots, the soil of those countries being light brown clay with small quartz pebbles; while in other localities where the colours of the soil were more varied the colours of the butterfly varied also. Here we have variation in a single species which has become specialised in certain areas to harmonise with the colour of the soil.[2]

Many butterflies, in all parts of the world, resemble dead leaves on their under side, but those in which this form of protection is carried to the greatest perfection are the species of the Eastern genus Kallima. In India K. inachis, and in the larger Malay islands K. paralekta, are very common. They are rather large and showy butterflies, orange and bluish on the upper side, with a very rapid flight, and frequenting dry forests. Their habit is to settle always where there is some dead or decaying foliage, and the shape and colour of the wings (on the under surface), together with the attitude of the insect, is such as to produce an absolutely perfect imitation of a dead leaf. This is effected by the butterfly always settling on a twig, with the short tail of the hind wings just touching it and forming the leaf-stalk. From this a dark curved line runs across to the elongated tip of the upper wings, imitating the midrib, on both sides of which are oblique lines, formed partly by the nervures and partly by markings, which give the effect of the usual veining of a leaf. The head and antennæ fit exactly between the closed upper wings so as not to interfere with the outline,

[1] *Trans. Ent. Soc. Lond.*, 1878, p. 185.
[2] *Ibid.* (*Proceedings*, p. xlii.)

which has just that amount of irregular curvature that is seen in dry and withered leaves. The colour is very remarkable for its extreme amount of variability, from deep reddish-brown to olive or pale yellow, hardly two specimens being exactly alike, but all coming within the range of colour of leaves in various stages of decay. Still more curious is the fact that the paler wings, which imitate leaves most decayed, are usually covered with small black dots, often gathered into circular groups, and so exactly resembling the minute fungi on decaying leaves that it is hard at first to believe that the insects themselves are not attacked by some such fungus. The concealment produced by this wonderful imitation is most complete, and in Sumatra I have often seen one enter a bush and then disappear like magic. Once I was so fortunate as to see the exact spot on which the insect settled; but even then I lost sight of it for some time, and only after a persistent search discovered that it was close before my eyes.[1] Here we have a kind of imitation, which is very common in a less developed form, carried to extreme perfection, with the result that the species is very abundant over a considerable area of country.

Protective Resemblance among Marine Animals.

Among marine animals this form of protection is very common. Professor Moseley tells us that all the inhabitants of the Gulf-weed are most remarkably coloured, for purposes of protection and concealment, exactly like the weed itself. "The shrimps and crabs which swarm in the weed are of exactly the same shade of yellow as the weed, and have white markings upon their bodies to represent the patches of Membranipora. The small fish, Antennarius, is in the same way weed-colour with white spots. Even a Planarian worm, which lives in the weed, is similarly yellow-coloured, and also a mollusc, Scyllæa pelagica." The same writer tells us that "a number of little crabs found clinging to the floats of the blue-shelled mollusc, Ianthina, were all coloured of a corresponding blue for concealment."[2]

[1] Wallace's *Malay Archipelago*, vol. i. p. 204 (fifth edition, p. 130), with figure.

[2] Moseley's *Notes by a Naturalist on the Challenger.*

Professor E. S. Morse of Salem, Mass., found that most of the New England marine mollusca were protectively coloured; instancing among others a little red chiton on rocks clothed with red calcareous algæ, and Crepidula plana, living within the apertures of the shells of larger species of Gasteropods and of a pure white colour corresponding to its habitat, while allied species living on seaweed or on the outside of dark shells were dark brown.[1] A still more interesting case has been recorded by Mr. George Brady. He says: "Amongst the Nullipore which matted together the laminaria roots in the Firth of Clyde were living numerous small starfishes (Ophiocoma bellis) which, except when their writhing movements betrayed them, were quite undistinguishable from the calcareous branches of the alga; their rigid angularly twisted rays had all the appearance of the coralline, and exactly assimilated to its dark purple colour, so that though I held in my hand a root in which were half a dozen of the starfishes, I was really unable to detect them until revealed by their movements."[2]

These few examples are sufficient to show that the principle of protective coloration extends to the ocean as well as over the earth; and if we consider how completely ignorant we are of the habits and surroundings of most marine animals, it may well happen that many of the colours of tropical fishes, which seem to us so strange and so conspicuous, are really protective, owing to the number of equally strange and brilliant forms of corals, sea-anemones, sponges, and seaweeds among which they live.

Protection by Terrifying Enemies.

A considerable number of quite defenceless insects obtain protection from some of their enemies by having acquired a resemblance to dangerous animals, or by some threatening or unusual appearance. This is obtained either by a modification of shape, of habits, of colour, or of all combined. The simplest form of this protection is the aggressive attitude of the caterpillars of the Sphingidæ, the forepart of the body

[1] *Proceedings of the Boston Soc. of Nat. Hist.*, vol. xiv. 1871.
[2] *Nature*, 1870, p. 376.

being erected so as to produce a rude resemblance to the figure of a sphinx, hence the name of the family. The protection is carried further by those species which retract the first three segments and have large ocelli on each side of the fourth segment, thus giving to the caterpillar, when the forepart of its body is elevated, the appearance of a snake in a threatening attitude.

The blood-red forked tentacle, thrown out of the neck of the larvæ of the genus Papilio when alarmed, is, no doubt, a protection against the attacks of ichneumons, and may, perhaps, also frighten small birds; and the habit of turning up the tail possessed by the harmless rove-beetles (Staphylinidæ), giving the idea that they can sting, has, probably, a similar use. Even an unusual angular form, like a crooked twig or inorganic substance, may be protective; as Mr. Poulton thinks is the case with the curious caterpillar of Notodonta ziczac, which, by means of a few slight protuberances on its body, is able to assume an angular and very unorganic-looking appearance. But perhaps the most perfect example of this kind of protection is exhibited by the large caterpillar of the Royal Persimmon moth (Bombyx regia), a native of the southern states of North America, and known there as the "Hickory-horned devil." It is a large green caterpillar, often six inches long, ornamented with an immense crown of orange-red tubercles, which, if disturbed, it erects and shakes from side to side in a very alarming manner. In its native country the negroes believe it to be as deadly as a rattlesnake, whereas it is perfectly innocuous. The green colour of the body suggests that its ancestors were once protectively coloured; but, growing too large to be effectually concealed, it acquired the habit of shaking its head about in order to frighten away its enemies, and ultimately developed the crown of tentacles as an addition to its terrifying powers. This species is beautifully figured in Abbott and Smith's *Lepidopterous Insects of Georgia.*

Alluring Coloration.

Besides those numerous insects which obtain protection through their resemblance to the natural objects among which they live, there are some whose disguise is not used for concealment, but as a direct means of securing their prey by attracting them within the enemy's reach. Only a few cases of this kind of coloration have yet been observed, chiefly among spiders and mantidæ; but, no doubt, if attention were given to the subject in tropical countries, many more would be discovered. Mr. H. O. Forbes has described a most interesting example of this kind of simulation in Java. While pursuing a large butterfly through the jungle, he was stopped by a dense bush, on a leaf of which he observed one of the skipper butterflies sitting on a bird's dropping. "I had often," he says, "observed small Blues at rest on similar spots on the ground, and have wondered what such a refined and beautiful family as the Lycænidæ could find to enjoy, in food apparently so incongruous for a butterfly. I approached with gentle steps, but ready net, to see if possible how the present species was engaged. It permitted me to get quite close, and even to seize it between my fingers; to my surprise, however, part of the body remained behind, adhering as I thought to the excreta. I looked closely, and finally touched with my finger the excreta to find if it were glutinous. To my delighted astonishment I found that my eyes had been most perfectly deceived, and that what seemed to be the excreta was a most artfully coloured spider, lying on its back with its feet crossed over and closely adpressed to the body." Mr. Forbes then goes on to describe the exact appearance of such excreta, and how the various parts of the spider are coloured to produce the imitation, even to the liquid portion which usually runs a little down the leaf. This is exactly imitated by a portion of the thin web which the spider first spins to secure himself firmly to the leaf; thus producing, as Mr. Forbes remarks, a living bait for butterflies and other insects so artfully contrived as to deceive a pair of human eyes, even when intently examining it.[1]

A native species of spider (Thomisus citreus) exhibits a somewhat similar alluring protection by its close resemblance to buds of the wayfaring tree, Viburnum lantana. It is pure creamy-white, the abdomen exactly resembling in shape and colour the unopened buds of the flowers among which it takes

[1] *A Naturalist's Wanderings in the Eastern Archipelago*, p. 63.

its station; and it has been seen to capture flies which came to the flowers.

But the most curious and beautiful case of alluring protection is that of a wingless Mantis in India, which is so formed and coloured as to resemble a pink orchis or some other fantastic flower. The whole insect is of a bright pink colour, the large and oval abdomen looking like the labellum of an orchid. On each side, the two posterior legs have immensely dilated and flattened thighs which represent the petals of a flower, while the neck and forelegs imitate the upper sepal and column of an orchid. The insect rests motionless, in this symmetrical attitude, among bright green foliage, being of course very conspicuous, but so exactly resembling a flower that butterflies and other insects settle upon it and are instantly captured. It is a living trap, baited in the most alluring manner to catch the unwary flower-haunting insects.[1]

The Coloration of Birds' Eggs.

The colours of birds' eggs have long been a difficulty on the theory of adaptive coloration, because, in so many cases it has not been easy to see what can be the use of the particular colours, which are often so bright and conspicuous that they seem intended to attract attention rather than to be concealed. A more careful consideration of the subject in all its bearings shows, however, that here too, in a great number of cases, we have examples of protective coloration. When, therefore, we cannot see the meaning of the colour, we may suppose that it has been protective in some ancestral form, and, not being hurtful, has persisted under changed conditions which rendered the protection needless.

We may divide all eggs, for our present purpose, into two

[1] A beautiful drawing of this rare insect, Hymenopus bicornis (in the nymph or active pupa state), was kindly sent me by Mr. Wood-Mason, Curator of the Indian Museum at Calcutta. A species, very similar to it, inhabits Java, where it is said to resemble a pink orchid. Other Mantidæ, of the genus Gongylus, have the anterior part of the thorax dilated and coloured either white, pink, or purple; and they so closely resemble flowers that, according to Mr. Wood-Mason, one of them, having a bright violet-blue prothoracic shield, was found in Pegu by a botanist, and was for a moment mistaken by him for a flower. See *Proc. Ent. Soc. Lond.*, 1878, p. liii.

great divisions; those which are white or nearly so, and those which are distinctly coloured or spotted. Egg-shells being composed mainly of carbonate of lime, we may assume that the primitive colour of birds' eggs was white, a colour that prevails now among the other egg-bearing vertebrates—lizards, crocodiles, turtles, and snakes; and we might, therefore, expect that this colour would continue where its presence had no disadvantages. Now, as a matter of fact, we find that in all the groups of birds which lay their eggs in concealed places, whether in holes of trees or in the ground, or in domed or covered nests, the eggs are either pure white or of very pale uniform coloration. Such is the case with kingfishers, bee-eaters, penguins, and puffins, which nest in holes in the ground; with the great parrot family, the woodpeckers, the rollers, hoopoes, trogons, owls, and some others, which build in holes in trees or other concealed places; while martins, wrens, willow-warblers, and Australian finches, build domed or covered nests, and usually have white eggs.

There are, however, many other birds which lay their white eggs in open nests; and these afford some very interesting examples of the varied modes by which concealment may be obtained. All the duck tribe, the grebes, and the pheasants belong to this class; but these birds all have the habit of covering their eggs with dead leaves or other material whenever they leave the nest, so as effectually to conceal them. Other birds, as the short-eared owl, the goatsucker, the partridge, and some of the Australian ground pigeons, lay their white or pale eggs on the bare soil; but in these cases the birds themselves are protectively coloured, so that, when sitting, they are almost invisible; and they have the habit of sitting close and almost continuously, thus effectually concealing their eggs.

Pigeons and doves offer a very curious case of the protection of exposed eggs. They usually build very slight and loose nests of sticks and twigs, so open that light can be seen through them from below, while they are generally well concealed by foliage above. Their eggs are white and shining; yet it is a difficult matter to discover, from beneath, whether there are eggs in the nest or not, while they are well hidden by the thick foliage above. The Australian podargi—

huge goatsuckers—build very similar nests, and their white eggs are protected in the same manner. Some large and powerful birds, as the swans, herons, pelicans, cormorants, and storks, lay white eggs in open nests; but they keep careful watch over them, and are able to drive away intruders. On the whole, then, we see that, while white eggs are conspicuous, and therefore especially liable to attack by egg-eating animals, they are concealed from observation in many and various ways. We may, therefore, assume that, in cases where there seems to be no such concealment, we are too ignorant of the whole of the conditions to form a correct judgment.

We now come to the large class of coloured or richly spotted eggs, and here we have a more difficult task, though many of them decidedly exhibit protective tints or markings. There are two birds which nest on sandy shores—the lesser tern and the ringed plover,—and both lay sand-coloured eggs, the former spotted so as to harmonise with coarse shingle, the latter minutely speckled like fine sand, which are the kinds of ground the two birds choose respectively for their nests. "The common sandpipers' eggs assimilate so closely with the tints around them as to make their discovery a matter of no small difficulty, as every oologist can testify who has searched for them. The pewits' eggs, dark in ground colour and boldly marked, are in strict harmony with the sober tints of moor and fallow, and on this circumstance alone their concealment and safety depend. The divers' eggs furnish another example of protective colour; they are generally laid close to the water's edge, amongst drift and shingle, where their dark tints and black spots conceal them by harmonising closely with surrounding objects. The snipes and the great army of sandpipers furnish innumerable instances of protectively coloured eggs. In all the instances given the sitting-bird invariably leaves the eggs uncovered when it quits them, and consequently their safety depends solely on the colours which adorn them."[1] The wonderful range of colour and marking in the eggs of the guillemot may be imputed to the inaccessible rocks on which

[1] C. Dixon, in Seebohm's *History of British Birds*, vol. ii. Introduction, p. xxvi. Many of the other examples here cited are taken from the same valuable work.

it breeds, giving it complete protection from enemies. Thus the pale or bluish ground colour of the eggs of its allies, the auks and puffins, has become intensified and blotched and spotted in the most marvellous variety of patterns, owing to there being no selective agency to prevent individual variation having full sway.

The common black coot (Fulica atra) has eggs which are coloured in a specially protective manner. Dr. William Marshall writes, that it only breeds in certain localities where a large water reed (Phragmites arundinacea) abounds. The eggs of the coot are stained and spotted with black on a yellowish-gray ground, and the dead leaves of the reed are of the same colour, and are stained black by small parasitic fungi of the Uredo family; and these leaves form the bed on which the eggs are laid. The eggs and the leaves agree so closely in colour and markings that it is a difficult thing to distinguish the eggs at any distance. It is to be noted that the coot never covers up its eggs, as its ally the moor-hen usually does.

The beautiful blue or greenish eggs of the hedge-sparrow, the song-thrush, the blackbird, and the lesser redpole seem at first sight especially calculated to attract attention, but it is very doubtful whether they are really so conspicuous when seen at a little distance among their usual surroundings. For the nests of these birds are either in evergreens, as holly or ivy, or surrounded by the delicate green tints of our early spring vegetation, and may thus harmonise very well with the colours around them. The great majority of the eggs of our smaller birds are so spotted or streaked with brown or black on variously tinted grounds that, when lying in the shadow of the nest and surrounded by the many colours and tints of bark and moss, of purple buds and tender green or yellow foliage, with all the complex glittering lights and mottled shades produced among these by the spring sunshine and by sparkling raindrops, they must have a quite different aspect from that which they possess when we observe them torn from their natural surroundings. We have here, probably, a similar case of general protective harmony to that of the green caterpillars with beautiful white or purple bands and spots, which, though gaudily conspicuous when seen alone, become practically invisible among the complex lights and shadows of the foliage they feed upon.

In the case of the cuckoo, which lays its eggs in the nests of a variety of other birds, the eggs themselves are subject to considerable variations of colour, the most common type, however, resembling those of the pipits, wagtails, or warblers, in whose nests they are most frequently laid. It also often lays in the nest of the hedge-sparrow, whose bright blue eggs are usually not at all nearly matched, although they are sometimes said to be so on the Continent. It is the opinion of many ornithologists that each female cuckoo lays the same coloured eggs, and that it usually chooses a nest the owners of which lay somewhat similar eggs, though this is by no means universally the case. Although birds which have cuckoos' eggs imposed upon them do not seem to neglect them on account of any difference of colour, yet they probably do so occasionally; and if, as seems probable, each bird's eggs are to some extent protected by their harmony of colour with their surroundings, the presence of a larger and very differently coloured egg in the nest might be dangerous, and lead to the destruction of the whole set. Those cuckoos, therefore, which most frequently placed their eggs among the kinds which they resembled, would in the long run leave most progeny, and thus the very frequent accord in colour might have been brought about.

Some writers have suggested that the varied colours of birds' eggs are primarily due to the effect of surrounding coloured objects on the female bird during the period preceding incubation; and have expended much ingenuity in suggesting the objects that may have caused the eggs of one bird to be blue, another brown, and another pink.[1] But no evidence has been presented to prove that any effects whatever are produced by this cause, while there seems no difficulty in accounting for the facts by individual variability and the action of natural selection. The changes that occur in the conditions of existence of birds must sometimes render the concealment less perfect than it may once have been; and when any danger arises from this cause, it may be met either

[1] See A. H. S. Lucas, in *Proceedings of Royal Society of Victoria*, 1887, p. 56.

by some change in the colour of the eggs, or in the structure or position of the nest, or by the increased care which the parents bestow upon the eggs. In this way the various divergences which now so often puzzle us may have arisen.

Colour as a Means of Recognition.

If we consider the habits and life-histories of those animals which are more or less gregarious, comprising a large proportion of the herbivora, some carnivora, and a considerable number of all orders of birds, we shall see that a means of ready recognition of its own kind, at a distance or during rapid motion, in the dusk of twilight or in partial cover, must be of the greatest advantage and often lead to the preservation of life. Animals of this kind will not usually receive a stranger into their midst. While they keep together they are generally safe from attack, but a solitary straggler becomes an easy prey to the enemy; it is, therefore, of the highest importance that, in such a case, the wanderer should have every facility for discovering its companions with certainty at any distance within the range of vision.

Some means of easy recognition must be of vital importance to the young and inexperienced of each flock, and it also enables the sexes to recognise their kind and thus avoid the evils of infertile crosses; and I am inclined to believe that its necessity has had a more widespread influence in determining the diversities of animal coloration than any other cause whatever. To it may probably be imputed the singular fact that, whereas bilateral symmetry of coloration is very frequently lost among domesticated animals, it almost universally prevails in a state of nature; for if the two sides of an animal were unlike, and the diversity of coloration among domestic animals occurred in a wild state, easy recognition would be impossible among numerous closely allied forms.[1]

[1] Professor Wm. H. Brewer of Yale College has shown that the white marks or the spots of domesticated animals are rarely symmetrical, but have a tendency to appear more frequently on the left side. This is the case with horses, cattle, dogs, and swine. Among wild animals the skunk varies considerably in the amount of white on the body, and this too was found to be usually greatest on the left side. A close examination of numerous striped or spotted species, as tigers, leopards, jaguars, zebras, etc., showed that the bilateral symmetry was not exact, although the general effect of the two sides

The wonderful diversity of colour and of marking that prevails, especially in birds and insects, may be due to the fact that one of the first needs of a new species would be, to keep separate from its nearest allies, and this could be most readily done by some easily seen external mark of difference. A few illustrations will serve to show how this principle acts in nature.

My attention was first called to the subject by a remark of Mr. Darwin's that, though, "the hare on her form is a familiar instance of concealment through colour, yet the principle partly fails in a closely allied species, the rabbit; for when running to its burrow it is made conspicuous to the sportsman, and no doubt to all beasts of prey, by its upturned white tail."[1] But a little consideration of the habits of the animal will show that the white upturned tail is of the greatest value, and is really, as it has been termed by a writer in *The Field*, a "signal flag of danger." For the rabbit is usually a crepuscular animal, feeding soon after sunset or on moonlight nights. When disturbed or alarmed it makes for its burrow, and the white upturned tails of those in front serve as guides and signals to those more remote from home, to the young and the feeble; and thus each following the one or two before it, all are able with the least possible delay to regain a place of comparative safety. The apparent danger, therefore, becomes a most important means of security.

The same general principle enables us to understand the singular, and often conspicuous, markings on so many gregarious herbivora which are yet, on the whole, protectively coloured. Thus, the American prong-buck has a white patch behind and a black muzzle. The Tartarian antelope, the Ovis poli of High Asia, the Java wild ox, several species of deer, and a large number of antelopes have a similar conspicuous white patch behind, which, in contrast to the dusky body, must enable them to be seen and followed from a distance by their fellows. Where there are many species of nearly the same general size and form inhabiting the same region—as with the antelopes

was the same. This is precisely what we should expect if the symmetry is not the result of a general law of the organisation, but has been, in part at least, produced and preserved for the useful purpose of recognition by the animal's fellows of the same species, and especially by the sexes and the young. See *Proc. of the Am. Ass. for Advancement of Science*, vol. xxx. p. 246.

[1] *Descent of Man*, p. 542.

of Africa—we find many distinctive markings of a similar kind. The gazelles have variously striped and banded faces, besides white patches behind and on the flanks, as shown in the woodcut. The spring-bok has a white patch on the face and one on the sides, with a curiously distinctive white stripe above the tail, which is nearly concealed when the animal is at rest by a fold of skin but comes into full view when it is in motion, being thus quite analogous to the

FIG. 18.—Gazella sœmmerringi.

upturned white tail of the rabbit. In the pallah the white rump-mark is bordered with black, and the peculiar shape of the horns distinguishes it when seen from the front. The sable-antelope, the gems-bok, the oryx, the hart-beest, the bonte-bok, and the addax have each peculiar white markings; and they are besides characterised by horns so remarkably different in each species and so conspicuous, that it seems probable that the peculiarities in length, twist, and curvature have been differentiated for the purpose of recognition, rather than for any speciality of defence in species whose general habits are so similar.

It is interesting to note that these markings for recognition are very slightly developed in the antelopes of the woods and marshes. Thus, the grys-bok is nearly uniform in colour, except the long black-tipped ears; and it frequents the wooded mountains. The duyker-bok and the rhoode-bok are wary bush-haunters, and have no marks but the small white patch behind. The wood-haunting bosch-bok goes in pairs, and has hardly any distinctive marks on its dusky chestnut coat, but the male alone is horned. The large and handsome koodoo frequents brushwood, and its vertical white stripes are no doubt protective, while its magnificent spiral horns afford easy recognition. The eland, which is an inhabitant of the open country, is uniformly coloured, being sufficiently recognisable by its large size and distinctive form; but the Derbyan eland is a forest animal, and has a protectively striped coat. In like manner, the fine Speke's antelope, which lives entirely in the swamps and among reeds, has pale vertical stripes on the sides (protective), with white markings on face and breast for recognition. An inspection of the figures of antelopes and other animals in Wood's *Natural History*, or in other illustrated works, will give a better idea of the peculiarities of recognition markings than any amount of description.

Other examples of such coloration are to be seen in the dusky tints of the musk-sheep and the reindeer, to whom recognition at a distance on the snowy plains is of more importance than concealment from their few enemies. The conspicuous stripes and bands of the zebra and the quagga are probably due to the same cause, as may be the singular crests and face-marks of several of the monkeys and lemurs.[1]

[1] It may be thought that such extremely conspicuous markings as those of the zebra would be a great danger in a country abounding with lions, leopards, and other beasts of prey; but it is not so. Zebras usually go in bands, and are so swift and wary that they are in little danger during the day. It is in the evening, or on moonlight nights, when they go to drink, that they are chiefly exposed to attack; and Mr. Francis Galton, who has studied these animals in their native haunts, assures me, that in twilight they are not at all conspicuous, the stripes of white and black so merging together into a gray tint that it is very difficult to see them at a little distance. We have here an admirable illustration of how a glaringly conspicuous style of marking for recognition may be so arranged as to become also protective at the time when protection is most needed; and we may also learn how impossible it is for us to decide on the inutility of any kind of coloration without a careful study of the habits of the species in its native country.

C. forbesi. Charadrius bifrontatus. C. tricollaris.

Fig. 19.—Recognition marks of three African plovers.

Among birds, these recognition marks are especially numerous and suggestive. Species which inhabit open districts are usually protectively coloured; but they generally possess some distinctive markings for the purpose of being easily recognised by their kind, both when at rest and during flight. Such are, the white bands or patches on the breast or belly of many birds, but more especially the head and neck markings in the form of white or black caps, collars, eye-marks or frontal patches, examples of which are seen in the three species of African plovers figured on page 221.

Recognition marks during flight are very important for all birds which congregate in flocks or which migrate together; and it is essential that, while being as conspicuous as possible, the marks shall not interfere with the general protective tints of the species when at rest. Hence they usually consist of well-contrasted markings on the wings and tail, which are concealed during repose but become fully visible when the bird takes flight. Such markings are well seen in our four British species of shrikes, each having quite different white marks on the expanded wings and on the tail feathers; and the same is the case with our three species of Saxicola—the stone-chat, whin-chat, and wheat-ear—which are thus easily recognisable on the wing, especially when seen from above, as they would be by stragglers looking out for their companions. The figures opposite, of the wings of two African species of stone-curlew which are sometimes found in the same districts, well illustrates these specific recognition marks. Though not very greatly different to our eyes, they are no doubt amply so to the sharp vision of the birds themselves.

Besides the white patches on the primaries here shown, the secondary feathers are, in some cases, so coloured as to afford very distinctive markings during flight, as seen in the central secondary quills of two African coursers (Fig. 21).

Most characteristic of all, however, are the varied markings of the outer tail-feathers, whose purpose is so well shown by their being almost always covered during repose by the two middle feathers, which are themselves quite unmarked and protectively tinted like the rest of the upper surface of the body. The figures of the expanded tails of two species of East Asiatic snipe, whose geographical ranges overlap each other,

Fig. 20.—Œdicnemus vermiculatus (above). Œ. senegalensis (below).

will serve to illustrate this difference; which is frequently much greater and modified in an endless variety of ways (Fig. 22).

Numbers of species of pigeons, hawks, finches, warblers, ducks, and innumerable other birds possess this class of markings; and they correspond so exactly in general character with

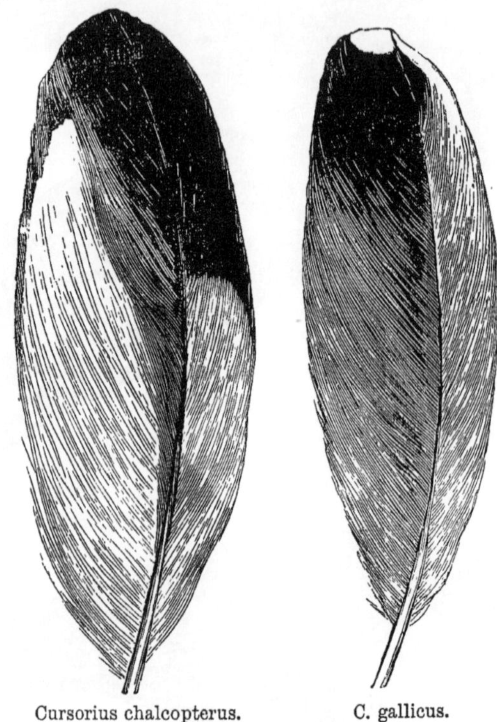

Cursorius chalcopterus. C. gallicus.

FIG. 21.—Secondary quills.

those of the mammalia, already described, that we cannot doubt they serve a similar purpose.[1]

Those birds which are inhabitants of tropical forests, and which need recognition marks that shall be at all times visible among the dense foliage, and not solely or chiefly during flight, have usually small but brilliant patches of colour

[1] The principle of colouring for recognition was, I believe, first stated in my article on "The Colours of Animals and Plants" in Macmillan's *Magazine*, and more fully in my volume on *Tropical Nature*. Subsequently Mrs. Barber gave a few examples under the head of "Indicative or Banner Colours," but she applied it to the distinctive colours of the males of birds, which I explain on another principle, though this may assist.

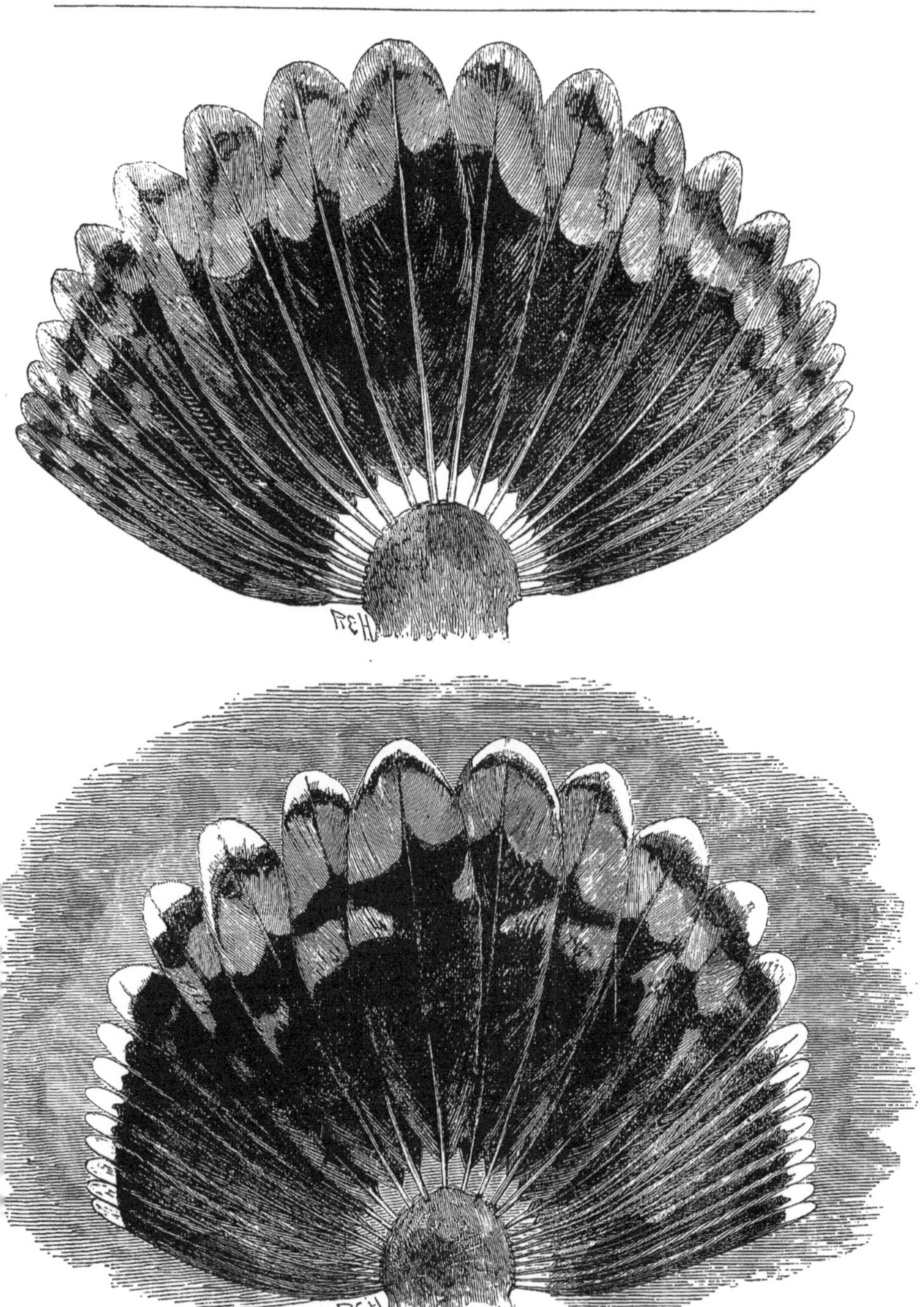

Fig. 22.—Scolopax megala (upper). S. stenura (lower).

on the head or neck, often not interfering with the generally protective character of their plumage. Such are the bright patches of blue, red, or yellow, by which the usually green Eastern barbets are distinguished; and similar bright patches of colour characterise the separate species of small green fruit-doves. To this necessity for specialisation in colour, by which each bird may easily recognise its kind, is probably due that marvellous variety in the peculiar beauties of some groups of birds. The Duke of Argyll, speaking of the humming birds, made the objection that "A crest of topaz is no better in the struggle for existence than a crest of sapphire. A frill ending in spangles of the emerald is no better in the battle of life than a frill ending in spangles of the ruby. A tail is not affected for the purposes of flight, whether its marginal or its central feathers are decorated with white;" and he goes on to urge that mere beauty and variety for their own sake are the only causes of these differences. But, on the principles here suggested, the divergence itself is useful, and must have been produced *pari passu* with the structural differences on which the differentiation of species depends; and thus we have explained the curious fact that prominent differences of colour often distinguish species otherwise very closely allied to each other.

Among insects, the principle of distinctive coloration for recognition has probably been at work in the production of the wonderful diversity of colour and marking we find everywhere, more especially among the butterflies and moths; and here its chief function may have been to secure the pairing together of individuals of the same species. In some of the moths this has been secured by a peculiar odour, which attracts the males to the females from a distance; but there is no evidence that this is universal or even general, and among butterflies, especially, the characteristic colour and marking, aided by size and form, afford the most probable means of recognition. That this is so is shown by the fact that "the common white butterfly often flies down to a bit of paper on the ground, no doubt mistaking it for one of its own species;" while, according to Mr. Collingwood, in the Malay Archipelago, "a dead butterfly pinned upon a conspicuous twig will often arrest an insect of the same species in its headlong flight, and bring it down within easy reach of the net, especially if it be of the opposite sex."[1] In a great number of insects, no doubt, form, motions, stridulating sounds, or peculiar odours, serve to distinguish allied species from each other, and this must be especially the case with nocturnal insects, or with those whose colours are nearly uniform and are determined by the need of protection; but by far the larger number of day-flying and active insects exhibit varieties of colour and marking, forming the most obvious distinction between allied species, and which have, therefore, in all probability been acquired in the process of differentiation for the purpose of checking the intercrossing of closely allied forms.[2]

Whether this principle extends to any of the less highly organised animals is doubtful, though it may perhaps have affected the higher mollusca. But in marine animals it seems probable that the colours, however beautiful, varied, and brilliant they may often be, are in most cases protective, assimilating them to the various bright-coloured seaweeds, or to some other animals which it is advantageous for them to imitate.[3]

Summary of the Preceding Exposition.

Before proceeding to discuss some of the more recondite phenomena of animal coloration, it will be well to consider for a moment the extent of the ground we have already covered. Protective coloration, in some of its varied forms, has not improbably modified the appearance of one-half of the animals living on the globe. The white of arctic animals, the yellowish tints of the desert forms, the dusky hues of crepuscular and nocturnal species, the transparent or bluish tints of oceanic creatures, represent a vast host in themselves; but we have an equally numerous body whose tints are adapted to tropical foliage, to the bark of trees, or to the soil

[1] Quoted by Darwin in *Descent of Man*, p. 317.

[2] In the *American Naturalist* of March 1888, Mr. J. E. Todd has an article on "Directive Coloration in Animals," in which he recognises many of the cases here referred to, and suggests a few others, though I think he includes many forms of coloration—as "paleness of belly and inner side of legs"—which do not belong to this class.

[3] For numerous examples of this protective colouring of marine animals see Moseley's *Voyage of the Challenger*, and Dr. E. S. Morse in *Proc. of Bost. Soc. of Nat. Hist.*, vol. xiv. 1871.

or dead leaves on or among which they habitually live. Then we have the innumerable special adaptations to the tints and forms of leaves, or twigs, or flowers; to bark or moss; to rock or pebble; by which such vast numbers of the insect tribes obtain protection; and we have seen that these various forms of coloration are equally prevalent in the waters of the seas and oceans, and are thus coextensive with the domain of life upon the earth. The comparatively small numbers which possess "terrifying" or "alluring" coloration may be classed under the general head of the protectively coloured.

But under the next head—colour for recognition—we have a totally distinct category, to some extent antagonistic or complementary to the last, since its essential principle is visibility rather than concealment. Yet it has been shown, I think, that this mode of coloration is almost equally important, since it not only aids in the preservation of existing species and in the perpetuation of pure races, but was, perhaps, in its earlier stages, a not unimportant factor in their development. To it we owe most of the variety and much of the beauty in the colours of animals; it has caused at once bilateral symmetry and general permanence of type; and its range of action has been perhaps equally extensive with that of coloration for concealment.

Influence of Locality or of Climate on Colour.

Certain relations between locality and coloration have long been noticed. Mr. Gould observed that birds from inland or continental localities were more brightly coloured than those living near the sea-coast or on islands, and he supposed that the more brilliant atmosphere of the inland stations was the explanation of the phenomenon.[1] Many American naturalists have observed similar facts, and they assert that the intensity of the colours of birds and mammals increases from north to south, and also with the increase of humidity. This change is imputed by Mr. J. A. Allen to the direct action of the environment. He says: "In respect to the correlation of intensity of colour in animals with the degree of humidity, it would perhaps be more in accordance with cause and effect to express the law of correlation as a *decrease* of intensity of colour with

[1] See *Origin of Species*, p. 107.

a *decrease* of humidity, the paleness evidently resulting from exposure and the blanching effect of intense sunlight, and a dry, often intensely heated atmosphere. With the decrease of the aqueous precipitation the forest growth and the protection afforded by arborescent vegetation gradually also decreases, as of course does also the protection afforded by clouds, the excessively humid regions being also regions of extreme cloudiness, while the dry regions are comparatively cloudless districts."[1] Almost identical changes occur in birds, and are imputed by Mr. Allen to similar causes.

It will be seen that Mr. Gould and Mr. Allen impute opposite effects to the same cause, brilliancy or intensity of colour being due to a brilliant atmosphere according to the former, while paleness of colour is imputed by the latter to a too brilliant sun. According to the principles which have been established by the consideration of arctic, desert, and forest animals respectively, we shall be led to conclude that there has been no direct action in this case, but that the effects observed are due to the greater or less need of protection. The pale colour that is prevalent in arid districts is in harmony with the general tints of the surface; while the brighter tints or more intense coloration, both southward and in humid districts, are sufficiently explained by the greater shelter due to a more luxuriant vegetation and a shorter winter. The advocates of the theory that intensity of light directly affects the colours of organisms, are led into perpetual inconsistencies. At one time the brilliant colours of tropical birds and insects are imputed to the intensity of a tropical sun, while the same intensity of sunlight is now said to have a "bleaching" effect. The comparatively dull and sober hues of our northern fauna were once supposed to be the result of our cloudy skies; but now we are told that cloudy skies and a humid atmosphere intensify colour.

In my *Tropical Nature* (pp. 257-264) I have called attention to what is perhaps the most curious and decided relation of colour to locality which has yet been observed—the prevalence of white markings in the butterflies and birds of islands.

[1] The "Geographical Variation of North American Squirrels," *Proc. Bost. Soc. of Nat. Hist.*, 1874, p. 284; and *Mammals and Winter Birds of Florida*, pp. 233-241.

So many cases are adduced from so many different islands, both in the eastern and western hemisphere, that it is impossible to doubt the existence of some common cause; and it seems probable to me now, after a fuller consideration of the whole subject of colour, that here too we have one of the almost innumerable results of the principle of protective coloration. White is, as a rule, an uncommon colour in animals, but probably only because it is so conspicuous. Whenever it becomes protective, as in the case of arctic animals and aquatic birds, it appears freely enough; while we know that white varieties of many species occur occasionally in the wild state, and that, under domestication, white or parti-coloured breeds are freely produced. Now in all the islands in which exceptionally white-marked birds and butterflies have been observed, we find two features which would tend to render the conspicuous white markings less injurious—a luxuriant tropical vegetation, and a decided scarcity of rapacious mammals and birds. White colours, therefore, would not be eliminated by natural selection; but variations in this direction would bear their part in producing the recognition marks which are everywhere essential, and which, in these islands, need not be so small or so inconspicuous as elsewhere.

Concluding Remarks.

On a review of the whole subject, then, we must conclude that there is no evidence of the individual or prevalent colours of organisms being directly determined by the amount of light, or heat, or moisture, to which they are exposed; while, on the other hand, the two great principles of the need of concealment from enemies or from their prey, and of recognition by their own kind, are so wide-reaching in their application that they appear at first sight to cover almost the whole ground of animal coloration. But, although they are indeed wonderfully general and have as yet been very imperfectly studied, we are acquainted with other modes of coloration which have a different origin. These chiefly appertain to the very singular class of warning colours, from which arise the yet more extraordinary phenomena of mimicry; and they open up so curious a field of inquiry and present so many interesting problems, that a chapter must be devoted to them. Yet another chapter will be required by the subject of sexual differentiation of colour and ornament, as to the origin and meaning of which I have arrived at different conclusions from Mr. Darwin. These various forms of coloration having been discussed and illustrated, we shall be in a position to attempt a brief sketch of the fundamental laws which have determined the general coloration of the animal world.

CHAPTER IX

WARNING COLORATION AND MIMICRY

The skunk as an example of warning coloration—Warning colours among insects—Butterflies—Caterpillars—Mimicry—How mimicry has been produced—Heliconidæ—Perfection of the imitation—Other cases of mimicry among Lepidoptera—Mimicry among protected groups—Its explanation—Extension of the principle—Mimicry in other orders of insects—Mimicry among the vertebrata—Snakes—The rattlesnake and the cobra—Mimicry among birds—Objections to the theory of mimicry—Concluding remarks on warning colours and mimicry.

We have now to deal with a class of colours which are the very opposite of those we have hitherto considered, since, instead of serving to conceal the animals that possess them or as recognition marks to their associates, they are developed for the express purpose of rendering the species conspicuous. The reason of this is that the animals in question are either the possessors of some deadly weapons, as stings or poison fangs, or they are uneatable, and are thus so disagreeable to the usual enemies of their kind that they are never attacked when their peculiar powers or properties are known. It is, therefore, important that they should not be mistaken for defenceless or eatable species of the same class or order, since in that case they might suffer injury, or even death, before their enemies discovered the danger or the uselessness of the attack. They require some signal or danger-flag which shall serve as a warning to would-be enemies not to attack them, and they have usually obtained this in the form of conspicuous or brilliant coloration, very distinct from the protective tints of the defenceless animals allied to them.

The Skunk as illustrating Warning Coloration.

While staying a few days, in July 1887, at the Summit Hotel on the Central Pacific Railway, I strolled out one evening after dinner, and on the road, not fifty yards from the house, I saw a pretty little white and black animal with a bushy tail coming towards me. As it came on at a slow pace and without any fear, although it evidently saw me, I thought at first that it must be some tame creature, when it suddenly occurred to me that it was a skunk. It came on till within five or six yards of me, then quietly climbed over a dwarf wall and disappeared under a small outhouse, in search of chickens, as the landlord afterwards told me. This animal possesses, as is well known, a most offensive secretion, which it has the power of ejecting over its enemies, and which effectually protects it from attack. The odour of this substance is so penetrating that it taints, and renders useless, everything it touches, or in its vicinity. Provisions near it become uneatable, and clothes saturated with it will retain the smell for several weeks, even though they are repeatedly washed and dried. A drop of the liquid in the eyes will cause blindness, and Indians are said not unfrequently to lose their sight from this cause. Owing to this remarkable power of offence the skunk is rarely attacked by other animals, and its black and white fur, and the bushy white tail carried erect when disturbed, form the danger-signals by which it is easily distinguished in the twilight or moonlight from unprotected animals. Its consciousness that it needs only to be seen to be avoided gives it that slowness of motion and fearlessness of aspect which are, as we shall see, characteristic of most creatures so protected.

Warning Colours among Insects.

It is among insects that warning colours are best developed, and most abundant. We all know how well marked and conspicuous are the colours and forms of the stinging wasps and bees, no one of which in any part of the world is known to be protectively coloured like the majority of defenceless insects. Most of the great tribe of Malacoderms among beetles are distasteful to insect-eating animals. Our red and black Telephoridæ, commonly called "soldiers and sailors," were found, by Mr. Jenner Weir, to be refused by small birds. These and the allied Lampyridæ (the fire-flies and glow-worms) in Nicaragua, were rejected by Mr. Belt's tame monkey and by his fowls, though most other insects were greedily eaten by them. The Coccinellidæ or lady-birds are another uneatable group, and their conspicuous and singularly spotted bodies serve to distinguish them at a glance from all other beetles.

These uneatable insects are probably more numerous than is supposed, although we already know immense numbers that are so protected. The most remarkable are the three families of butterflies—Heliconidæ, Danaidæ, and Acræidæ—comprising more than a thousand species, and characteristic respectively of the three great tropical regions—South America, Southern Asia, and Africa. All these butterflies have peculiarities which serve to distinguish them from every other group in their respective regions. They all have ample but rather weak wings, and fly slowly; they are always very abundant; and they all have conspicuous colours or markings, so distinct from those of other families that, in conjunction with their peculiar outline and mode of flight, they can usually be recognised at a glance. Other distinctive features are, that their colours are always nearly the same on the under surface of their wings as on the upper; they never try to conceal themselves, but rest on the upper surfaces of leaves or flowers; and, lastly, they all have juices which exhale a powerful scent, so that when one kills them by pinching the body, the liquid that exudes stains the fingers yellow, and leaves an odour that can only be removed by repeated washings.

Now, there is much direct evidence to show that this odour, though not very offensive to us, is so to most insect-eating creatures. Mr. Bates observed that, when set out to dry, specimens of Heliconidæ were less subject to the attacks of vermin; while both he and I noticed that they were not attacked by insect-eating birds or dragonflies, and that their wings were not found in the forest paths among the numerous wings of other butterflies whose bodies had been devoured. Mr. Belt once observed a pair of birds capturing insects for their young; and although the Heliconidæ swarmed in the vicinity, and from their slow flight could have been easily caught, not one was ever pursued, although other butterflies did not escape. His tame monkey also, which would greedily munch up other butterflies, would never eat the Heliconidæ. It would sometimes smell them, but always rolled them up in its hand and then dropped them.

We have also some corresponding evidence as to the distastefulness of the Eastern Danaidæ. The Hon. Mr. Justice Newton, who assiduously collected and took notes upon the Lepidoptera of Bombay, informed Mr. Butler of the British Museum that the large and swift-flying butterfly Charaxes psaphon, was continually persecuted by the bulbul, so that he rarely caught a specimen of this species which had not a piece snipped out of the hind wings. He offered one to a bulbul which he had in a cage, and it was greedily devoured, whilst it was only by repeated persecution that he succeeded in inducing the bird to touch a Danais.[1]

Besides these three families of butterflies, there are certain groups of the great genus Papilio—the true swallow-tailed butterflies—which have all the characteristics of uneatable insects. They have a special coloration, usually red and black (at least in the females), they fly slowly, they are very abundant, and they possess a peculiar odour somewhat like that of the Heliconidæ. One of these groups is common in tropical America, another in tropical Asia, and it is curious that, although not very closely allied, they have each the same red and black colours, and are very distinct from all the other butterflies of their respective countries. There is reason to believe also that many of the brilliantly coloured and weak-flying diurnal moths, like the fine tropical Agaristidæ and burnet-moths, are similarly protected, and that their conspicuous colours serve as a warning of inedibility. The common burnet-moth (Anthrocera filipendula) and the equally conspicuous ragwort-moth (Euchelia jacobeæ) have been proved to be distasteful to insect-eating creatures.

[1] *Nature*, vol. iii. p. 165. Professor Meldola observed that specimens of Danais and Euplæa in collections were less subject to the attacks of mites (*Proc. Ent. Soc.*, 1877, p. xii.); and this was corroborated by Mr. Jenner Weir. *Entomologist*, 1882, vol. xv. p. 160.

The most interesting and most conclusive example of warning coloration is, however, furnished by caterpillars, because in this case the facts have been carefully ascertained experimentally by competent observers. In the year 1866, when Mr. Darwin was collecting evidence as to the supposed effect of sexual selection in bringing about the brilliant coloration of the higher animals, he was struck by the fact that many caterpillars have brilliant and conspicuous colours, in the production of which sexual selection could have no place. We have numbers of such caterpillars in this country, and they are characterised not only by their gay colours but by not concealing themselves. Such are the mullein and the gooseberry caterpillars, the larvæ of the spurge hawk-moth, of the buff-tip, and many others. Some of these caterpillars are wonderfully conspicuous, as in the case of that noticed by Mr. Bates in South America, which was four inches long, banded across with black and yellow, and with bright red head, legs, and tail. Hence it caught the eye of any one who passed by, even at the distance of many yards.

Mr. Darwin asked me to try and suggest some explanation of this coloration; and, having been recently interested in the question of the warning coloration of butterflies, I suggested that this was probably a similar case,—that these conspicuous caterpillars were distasteful to birds and other insect-eating creatures, and that their bright non-protective colours and habit of exposing themselves to view, enabled their enemies to distinguish them at a glance from the edible kinds and thus learn not to touch them; for it must be remembered that the bodies of caterpillars while growing are so delicate, that a wound from a bird's beak would be perhaps as fatal as if they were devoured.[1] At this time not a single experiment or observation had been made on the subject, but after I had brought the matter before the Entomological Society, two gentlemen, who kept birds and other tame animals, undertook to make experiments with a variety of caterpillars.

Mr. Jenner Weir was the first to experiment with ten species of small birds in his aviary, and he found that none of them would eat the following smooth-skinned conspicuous cater-

[1] See Darwin's *Descent of Man*, p. 325.

pillars—Abraxas grossulariata, Diloba cæruleocephala, Anthrocera filipendula, and Cucullia verbasci. He also found that they would not touch any hairy or spiny larvæ, and he was satisfied that it was not the hairs or the spines, but the unpleasant taste that caused them to be rejected, because in one case a young smooth larva of a hairy species, and in another case the pupa of a spiny larva, were equally rejected. On the other hand, all green or brown caterpillars as well as those that resemble twigs were greedily devoured.[1]

Mr. A. G. Butler also made experiments with some green lizards (Lacerta viridis), which greedily ate all kinds of food, including flies of many kinds, spiders, bees, butterflies, and green caterpillars; but they would not touch the caterpillar of the gooseberry-moth (Abraxas grossulariata), or the imago of the burnet-moth (Anthrocera filipendula). The same thing happened with frogs. When the gooseberry caterpillars were first given to them, "they sprang forward and licked them eagerly into their mouths; no sooner, however, had they done so, than they seemed to become aware of the mistake that they had made, and sat with gaping mouths, rolling their tongues about, until they had got quit of the nauseous morsels, which seemed perfectly uninjured, and walked off as briskly as ever." Spiders seemed equally to dislike them. This and another conspicuous caterpillar (Halia wavaria) were rejected by two species—the geometrical garden spider (Epeira diadema) and a hunting spider.[2]

Some further experiments with lizards were made by Professor Weismann, quite confirming the previous observations; and in 1886 Mr. E. B. Poulton of Oxford undertook a considerable series of experiments, with many other species of larvæ and fresh kinds of lizards and frogs. Mr. Poulton then reviewed the whole subject, incorporating all recorded facts, as well as some additional observations made by Mr. Jenner Weir in 1886. More than a hundred species of larvæ or of perfect insects of various orders have now been made the subject of experiment, and the results completely confirm my original suggestion. In almost every case the protectively coloured larvæ have been greedily eaten by all kinds of insectivorous

[1] *Transactions of the Entomological Society of London*, 1869, p. 21.
[2] *Ibid.*, p. 27.

animals, while, in the immense majority of cases, the conspicuous, hairy, or brightly coloured larvæ have been rejected by some or all of them. In some instances the inedibility of the larvæ extends to the perfect insect, but not in others. In the former cases the perfect insect is usually adorned with conspicuous colours, as the burnet and ragwort moths; but in the case of the buff-tip, the moth resembles a broken piece of rotten stick, yet it is partly inedible, being refused by lizards. It is, however, very doubtful whether these are its chief enemies, and its protective form and colour may be needed against insectivorous birds or mammals.

Mr. Samuel H. Scudder, who has largely bred North American butterflies, has found so many of the eggs and larvæ destroyed by hymenopterous and dipterous parasites that he thinks at least nine-tenths, perhaps a greater proportion, never reach maturity. Yet he has never found any evidence that such parasites attack either the egg or the larva of the inedible Danais archippus, so that in this case the insect is distasteful to its most dangerous foes in all the stages of its existence, a fact which serves to explain its great abundance and its extension over almost the whole world.[1]

One case has been found of a protectively coloured larva,—one, moreover, which in all its habits shows that it trusts to concealment to escape its enemies—which was yet always rejected by lizards after they had seized it, evidently under the impression that from its colour it would be eatable. This is the caterpillar of the very common moth Mania typica; and Mr. Poulton thinks that, in this case, the unpleasant taste is an incidental result of some physiological processes in the organism, and is itself a merely useless character. It is evident that the insect would not conceal itself so carefully as it does if it had not some enemies, and these are probably birds or small mammals, as its food-plants are said to be dock and willow-herb, not suggestive of places frequented by lizards; and it has been found by experiment that lizards and birds have not always the same likes and dislikes. The case is interesting, because it shows that nauseous fluids sometimes occur sporadically, and may thus be intensified by natural selection when required for the purpose

[1] *Nature*, vol. iii. p. 147.

of protection. Another exceptional case is that of the very conspicuous caterpillar of the spurge hawk-moth (Deilephila euphorbiæ), which was at once eaten by a lizard, although, as it exposes itself on its food-plant in the daytime and is very abundant in some localities, it must almost certainly be disliked by birds or by some animals who would otherwise devour it. If disturbed while feeding it is said to turn round with fury and eject a quantity of green liquid, of an acid and disagreeable smell similar to that of the spurge milk, only worse.[1]

These facts, and Mr. Poulton's evidence that some larvæ rejected by lizards at first will be eaten if the lizards are very hungry, show that there are differences in the amount of the distastefulness, and render it probable that if other food were wanting many of these conspicuous insects would be eaten. It is the abundance of the eatable kinds that gives value to the inedibility of the smaller number; and this is probably the reason why so many insects rely on protective colouring rather than on the acquisition of any kind of defensive weapons. In the long run the powers of attack and defence must balance each other. Hence we see that even the powerful stings of bees and wasps only protect them against some enemies, since a tribe of birds, the bee-eaters, have been developed which feed upon them, and some frogs and lizards do so occasionally.

The preceding outline will sufficiently explain the characteristics of "warning coloration" and the end it serves in nature. There are many other curious modifications of it, but these will be best appreciated after we have discussed the remarkable phenomenon of "mimicry," which is bound up with and altogether depends upon "warning colour," and is in some cases the chief indication we have of the possession of some offensive weapon to secure the safety of the species imitated.

Mimicry.

This term has been given to a form of protective resemblance, in which one species so closely resembles another in external form and colouring as to be mistaken for it, although the two may not be really allied and often belong to distinct

[1] Stainton's *Manual of Butterflies and Moths*, vol. i. p. 93; E. B. Poulton, *Proceedings of the Zool. Soc. of London*, 1887, pp. 191-274.

families or orders. One creature seems disguised in order to be made like another; hence the terms "mimic" and mimicry, which imply no voluntary action on the part of the imitator. It has long been known that such resemblances do occur, as, for example, the clear-winged moths of the families Sesiidæ and Ægeriidæ, many of which resemble bees, wasps, ichneumons, or saw-flies, and have received names expressive of the resemblance; and the parasitic flies (Volucella) which closely resemble bees, on whose larvæ the larvæ of the flies feed.

The great bulk of such cases remained, however, unnoticed, and the subject was looked upon as one of the inexplicable curiosities of nature, till Mr. Bates studied the phenomenon among the butterflies of the Amazon, and, on his return home, gave the first rational explanation of it.[1] The facts are, briefly, these. Everywhere in that fertile region for the entomologist the brilliantly coloured Heliconidæ abound, with all the characteristics which I have already referred to when describing them as illustrative of "warning coloration." But along with them other butterflies were occasionally captured, which, though often mistaken for them, on account of their close resemblance in form, colour, and mode of flight, were found on examination to belong to a very distinct family, the Pieridæ. Mr. Bates notices fifteen distinct species of Pieridæ, belonging to the genera Leptalis and Euterpe, each of which closely imitates some one species of Heliconidæ, inhabiting the same region and frequenting the same localities. It must be remembered that the two families are altogether distinct in structure. The larvæ of the Heliconidæ are tubercled or spined, the pupæ suspended head downwards, and the imago has imperfect fore-legs in the male; while the larvæ of the Pieridæ are smooth, the pupæ are suspended with a brace to keep the head erect, and the forefeet are fully developed in both sexes. These differences are as large and as important as those between pigs and sheep, or between swallows and sparrows; while English entomologists will best understand the case by supposing that a species of Pieris in this country was coloured and shaped like a small tortoise-shell, while another species on the Continent was equally like a Camberwell beauty—so like in both

[1] See *Transactions of the Linnean Society*, vol. xxiii. pp. 495-566, coloured plates.

cases as to be mistaken when on the wing, and the difference only to be detected by close examination. As an example of the resemblance, woodcuts are given of one pair in which the colours are simple, being olive, yellow, and black, while the

FIG. 23.—Methona psidii (Heliconidæ). Leptalis orise (Pieridæ).

very distinct neuration of the wings and form of the head and body can be easily seen.

Besides these Pieridæ, Mr. Bates found four true Papilios, seven Erycinidæ, three Castnias (a genus of day-flying moths), and fourteen species of diurnal Bombycidæ, all imitating some species of Heliconidæ which inhabited the same district; and it is to be especially noted that none of these insects were so abundant as the Heliconidæ they resembled, generally they were far less common, so that Mr. Bates estimated the proportion in some cases as not one to a thousand. Before giving an account of the numerous remarkable cases of mimicry in other parts of the world, and between various groups of insects and of higher animals, it will be well to explain briefly the use and purport of the phenomenon, and also the mode by which it has been brought about.

How Mimicry has been Produced.

The fact has been now established that the Heliconidæ possess an offensive odour and taste, which lead to their being almost entirely free from attack by insectivorous creatures; they possess a peculiar form and mode of flight, and do not seek concealment; while their colours—although very varied, ranging from deep blue-black, with white, yellow, or vivid red bands and spots, to the most delicate semitransparent wings adorned with pale brown or yellow markings—are yet always very distinctive, and unlike those of all the other families of butterflies in the same country. It is, therefore, clear that if any other butterflies in the same region, which are eatable and suffer great persecution from insectivorous animals, should come to resemble any of these uneatable species so closely as to be mistaken for them by their enemies, they will obtain thereby immunity from persecution. This is the obvious and sufficient reason why the imitation is useful, and therefore why it occurs in nature. We have now to explain how it has probably been brought about, and also why a still larger number of persecuted groups have not availed themselves of this simple means of protection.

From the great abundance of the Heliconidæ[1] all over tropical America, the vast number of their genera and species, and their marked distinctions from all other butterflies, it follows that they constitute a group of high antiquity, which in the course of ages has become more and more specialised, and owing to its peculiar advantages has now become a dominant and aggressive race. But when they first arose from some ancestral species or group which, owing to the food

[1] These butterflies are now divided into two sub-families, one of which is placed with the Danaidæ; but to avoid confusion I shall always speak of the American genera under the old term Heliconidæ.

of the larvæ or some other cause, possessed disagreeable juices that caused them to be disliked by the usual enemies of their kind, they were in all probability not very different either in form or coloration from many other butterflies. They would at that time be subject to repeated attacks by insect-eaters, and, even if finally rejected, would often receive a fatal injury. Hence arose the necessity for some distinguishing mark, by which the devourers of butterflies in general might learn that these particular butterflies were uneatable; and every variation leading to such distinction, whether by form, colour, or mode of flight, was preserved and accumulated by natural selection, till the ancestral Heliconoids became well distinguished from eatable butterflies, and thenceforth comparatively free from persecution. Then they had a good time of it. They acquired lazy habits, and flew about slowly. They increased abundantly and spread all over the country, their larvæ feeding on many plants and acquiring different habits; while the butterflies themselves varied greatly, and colour being useful rather than injurious to them, gradually diverged into the many coloured and beautifully varied forms we now behold.

But, during the early stages of this process, some of the Pieridæ, inhabiting the same district, happened to be sufficiently like some of the Heliconidæ to be occasionally mistaken for them. These, of course, survived while their companions were devoured. Those among their descendants that were still more like Heliconidæ again survived, and at length the imitation would become tolerably perfect. Thereafter, as the protected group diverged into distinct species of many different colours, the imitative group would occasionally be able to follow it with similar variations,—a process that is going on now, for Mr. Bates informs us that in each fresh district he visited he found closely allied representative species or varieties of Heliconidæ, and along with them species of Leptalis (Pieridæ), which had varied in the same way so as still to be exact imitations. But this process of imitation would be subject to check by the increasing acuteness of birds and other animals which, whenever the eatable Leptalis became numerous, would surely find them out, and would then probably attack both these and their friends the Heliconidæ in order to devour the former and reject the latter. The Pieridæ would, however, usually be less numerous, because their larvæ are often protectively coloured and therefore edible, while the larvæ of the Heliconidæ are adorned with warning colours, spines, or tubercles, and are uneatable. It seems probable that the larvæ and pupæ of the Heliconidæ were the first to acquire the protective distastefulness, both because in this stage they are more defenceless and more liable to fatal injury, and also because we now find many instances in which the larvæ are distasteful while the perfect insects are eatable, but I believe none in which the reverse is the case. The larvæ of the Pieridæ are now beginning to acquire offensive juices, but have not yet obtained the corresponding conspicuous colours; while the perfect insects remain eatable, except perhaps in some Eastern groups, the under sides of whose wings are brilliantly coloured although this is the part which is exposed when at rest.

It is clear that if a large majority of the larvæ of Lepidoptera, as well as the perfect insects, acquired these distasteful properties, so as seriously to diminish the food supply of insectivorous and nestling birds, these latter would be forced by necessity to acquire corresponding tastes, and to eat with pleasure what some of them now eat only under pressure of hunger; and variation and natural selection would soon bring about this change.

Many writers have denied the possibility of such wonderful resemblances being produced by the accumulation of fortuitous variations, but if the reader will call to mind the large amount of variability that has been shown to exist in all organisms, the exceptional power of rapid increase possessed by insects, and the tremendous struggle for existence always going on, the difficulty will vanish, especially when we remember that nature has the same fundamental groundwork to act upon in the two groups, general similarity of forms, wings of similar texture and outline, and probably some original similarity of colour and marking. Yet there is evidently considerable difficulty in the process, or with these great resources at her command nature would have produced more of these mimicking forms than she has done. One reason of this deficiency probably is, that the imitators, being always fewer in number, have not been able to keep pace with the variations of the much more numerous imitated form; another reason may be the ever-increasing acuteness of the enemies, which have again and again detected the imposture and exterminated the feeble race before it has had time to become further modified. The result of this growing acuteness of enemies has been, that those mimics that now survive exhibit, as Mr. Bates well remarks, "a palpably intentional likeness that is perfectly staggering," and also "that those features of the portrait are most attended to by nature which produce the most effective deception when the insects are seen in nature." No one, in fact, can understand the perfection of the imitation who has not seen these species in their native wilds. So complete is it in general effect that in almost every box of butterflies, brought from tropical America by amateurs, are to be found some species of the mimicking Pieridæ, Erycinidæ, or moths, and the mimicked Heliconidæ, placed together under the impression that they are the same species. Yet more extraordinary, it sometimes deceives the very insects themselves. Mr. Trimen states that the male Danais chrysippus is sometimes deceived by the female Diadema bolina which mimics that species. Dr. Fritz Müller, writing from Brazil to Professor Meldola, says, "One of the most interesting of our mimicking butterflies is Leptalis melite. The female alone of this species imitates one of our common white Pieridæ, which she copies so well that even her own male is often deceived; for I have repeatedly seen the male pursuing the mimicked species, till, after closely approaching and becoming aware of his error, he suddenly returned."[1] This is evidently not a case of true mimicry, since the species imitated is not protected; but it may be that the less abundant Leptalis is able to mingle with the female Pieridæ and thus obtain partial immunity from attack. Mr. Kirby of the insect department of the British Museum informs me that there are several species of South American Pieridæ which the female Leptalis melite very nearly resembles. The case, however, is interesting as showing that the butterflies are themselves deceived by a resemblance which is not so great as that of some mimicking species.

[1] R. Meldola in *Ann. and Mag. of Nat. Hist.*, Feb. 1878, p. 158.

Other Examples of Mimicry among Lepidoptera.

In tropical Asia, and eastward to the Pacific Islands, the Danaidæ take the place of the Heliconidæ of America, in their abundance, their conspicuousness, their slow flight, and their being the subjects of mimicry. They exist under three principal forms or genera. The genus Euplæa is the most abundant both in species and individuals, and consists of fine broad-winged butterflies of a glossy or metallic blue-black colour, adorned with pure white, or rich blue, or dusky markings situated round the margins of the wings. Danais has generally more lengthened wings, of a semitransparent greenish or a rich brown colour, with radial or marginal pale spots; while the fine Hestias are of enormous size, of a papery or semitransparent white colour, with dusky or black spots and markings. Each of these groups is mimicked by various species of the genus Papilio, usually with such accuracy that it is impossible to distinguish them on the wing.[1] Several species of Diadema, a genus of butterflies allied to our Vanessas, also mimic species of Danais, but in this case the females only are affected, a subject which will be discussed in another chapter.

Another protected group in the Eastern tropics is that of the beautiful day-flying moths forming the family Agaristidæ. These are usually adorned with the most brilliant colours or conspicuous markings, they fly slowly in forests among the butterflies and other diurnal insects, and their great abundance sufficiently indicates their possession of some distastefulness which saves them from attack. Under these conditions we may expect to find other moths which are not so protected imitating them, and this is the case. One of the common and wide-ranging species (Opthalmis lincea), found in the islands from Amboyna to New Ireland, is mimicked in a wonderful manner by one of the Liparidæ (the family to which our common "tussock" and "vapourer" moths belong). This is a new species collected at Amboyna during the voyage of the *Challenger*, and has been named Artaxa simulans. Both

1 See *Trans. Linn. Soc.*, vol. xxv. Wallace, on Variation of Malayan Papilionidæ; and, Wallace's *Contributions to Natural Selection*, chaps. iii. and iv., where full details are given.

insects are black, with the apex of the fore wings ochre coloured, and the outer half of the hind wings bright orange. The accompanying woodcuts (for the use of which I am indebted to Mr. John Murray of the *Challenger* Office) well exhibit their striking resemblance to each other.

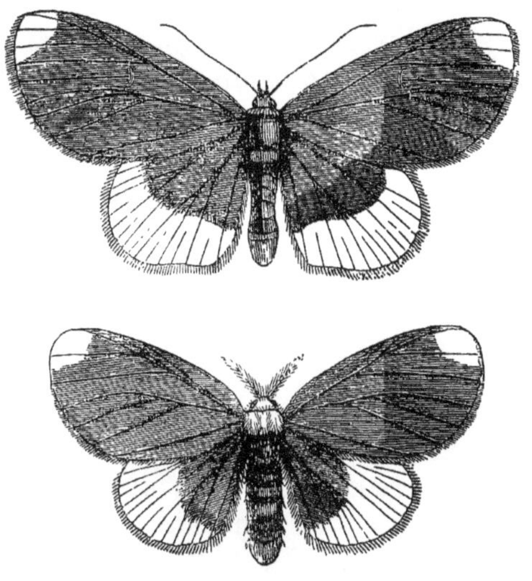

FIG. 24.—Opthalmis lincea (Agaristidæ). Artaxa simulans (Liparidæ).

In Africa exactly similar phenomena recur, species of Papilio and of Diadema mimicking Danaidæ or Acræidæ with the most curious accuracy. Mr. Trimen, who studied this subject in South Africa, has recorded eight species or varieties of Diadema, and eight of Papilio, which each mimic some species of Danais; while eight species or varieties of Panopæa (another genus of Nymphalidæ), three of Melanitis (Eurytelidæ), and two of Papilio, resemble with equal accuracy some species of Acræa.[1] He has also independently observed the main facts on which the explanation of the phenomenon rests,—the unpleasant odour of the Danais and Acræa, extending to their larvæ and pupæ; their great abundance, slow flight, and disregard of concealment; and he states that while lizards, mantidæ, and dragonflies all hunt butterflies, and the rejected wings are to be found abundantly at some of their

[1] See *Trans. Linn. Soc.*, vol. xxvi., with two coloured plates illustrating cases of mimicry.

feeding-places, those of the two genera Danais and Acræa were never among them.

The two groups of the great genus Papilio (the true swallow-tailed butterflies) which have been already referred to as having the special characteristics of uneatable insects, have also their imitators in other groups; and thus, the belief in their inedibility—derived mainly from their style of warning coloration and their peculiar habits—is confirmed. In South America, several species of the "Æneas" group of these butterflies are mimicked by Pieridæ and by day-flying moths of the genera Castnia and Pericopis. In the East, Papilio hector, P. diphilus, and P. liris, all belonging to the inedible group, are mimicked by the females of other species of Papilio belonging to very distinct groups; while in Northern India and China, many fine day-flying moths (Epicopeia) have acquired the strange forms and peculiar colours of some of the large inedible Papilios of the same regions.

In North America, the large and handsome Danais archippus, with rich reddish-brown wings, is very common; and it is closely imitated by Limenitis misippus, a butterfly allied to our "white admiral," but which has acquired a colour quite distinct from that of the great bulk of its allies. In the same country there is a still more interesting case. The beautiful dark bronzy green butterfly, Papilio philenor, is inedible both in larva and perfect insect, and it is mimicked by the equally dark Limenitis ursula. There is also in the Southern and Western States a dark female form of the yellow Papilio turnus, which in all probability obtains protection from its general resemblance to P. philenor. Mr. W. H. Edwards has found, by extensive experiment, that both the dark and yellow females produce their own kinds, with very few exceptions; and he thinks that the dark form has the advantage in the more open regions and in the prairies, where insectivorous birds abound. But in open country the dark form would be quite as conspicuous as the yellow form, if not more so, so that the resemblance to an inedible species would be there more needed.[1]

The only probable case of mimicry in this country is that of the moth, Diaphora mendica, whose female only is white,

[1] Edwards's *Butterflies of North America*, second series, part vi.

while the larva is of protective colours, and therefore almost certainly edible. A much more abundant moth, of about the same size and appearing about the same time, is Spilosoma menthrasti, also white, but in this case both it and its larva have been proved to be inedible. The white colour of the female Diaphora, although it must be very conspicuous at night, may, therefore, have been acquired in order to resemble the uneatable Spilosoma, and thus gain some protection.[1]

Mimicry among Protected (Uneatable) Genera.

Before giving some account of the numerous other cases of warning colours and of mimicry that occur in the animal kingdom, it will be well to notice a curious phenomenon which long puzzled entomologists, but which has at length received a satisfactory explanation.

We have hitherto considered, that mimicry could only occur when a comparatively scarce and much persecuted species obtained protection by its close external resemblance to a much more abundant uneatable species inhabiting its own district; and this rule undoubtedly prevails among the great majority of mimicking species all over the world. But Mr. Bates also found a number of pairs of species of different genera of Heliconidæ, which resembled each other quite as closely as did the other mimicking species he has described; and since all these insects appear to be equally protected by their inedibility, and to be equally free from persecution, it was not easy to see why this curious resemblance existed, or how it had been brought about. That it is not due to close affinity is shown by the fact that the resemblance occurs most frequently between the two distinct sub-families into which (as Mr. Bates first pointed out) the Heliconidæ are naturally divided on account of very important structural differences. One of these sub-families (the true Heliconinæ) consists of two genera only, Heliconius and Eueides, the other (the Danaoid Heliconinæ) of no less than sixteen genera; and, in the instances of mimicry we are now discussing, one of the pairs or

[1] Professor Meldola informs me that he has recorded another case of mimicry among British moths, in which Acidalia subsericata imitates Asthena candidata. See *Ent. Mo. Mag.*, vol. iv. p. 163.

triplets that resemble each other is usually a species of the large and handsome genus Heliconius, the others being species of the genera Mechanitis, Melinæa, or Tithorea, though several species of other Danaoid genera also imitate each other. The following lists will give some idea of the number of these curious imitative forms, and of their presence in every part of the Neotropical area. The bracketed species are those that resemble each other so closely that the difference is not perceptible when they are on the wing.

In the Lower Amazon region are found—

{ Heliconius sylvana.
  Melinæa egina.

{ Heliconius numata.
  Melinæa mneme.
  Tithorea harmonia.

{ Methona psidii.
  Thyridia ino.

{ Ceratina ninonia.
  Melinæa mnasias.

In Central America are found—

Nicaragua { Heliconius zuleika.
  Melinæa hezia.
  Mechanitis sp.

{ Heliconius formosus.
  Tithorea penthias.

Guatemala { Heliconius telchina.
  Melinæa imitata.

In the Upper Amazon region—

{ Heliconius pardalinus.
  Melinæa pardalis.

{ Heliconius aurora.
  Melinæa lucifer.

In New Grenada—

{ Heliconius ismenius.
  Melinæa messatis.

{ Heliconius messene.
  Melinæa mesenina.
  (?) Mechanitis sp.

{ Heliconius hecalesia.
  Tithorea hecalesina.

{ Heliconius hecuba.
  Tithorea bonplandi.

In Eastern Peru and Bolivia—

Heliconius aristona.
Melinæa cydippe.
(?) Mechanitis mothone.

In Pernambuco—

Heliconius ethra.
Mechanitis nesæa.

In Rio Janeiro—

Heliconius eucrate.
Mechanitis lysimnia.

In South Brazil—

Thyridia megisto.
Ituna ilione.

Acræa thalia.
Eueides pavana.

Besides these, a number of species of Ithomia and Napeogenes, and of Napeogenes and Mechanitis, resemble each other with equal accuracy, so that they are liable to be mistaken

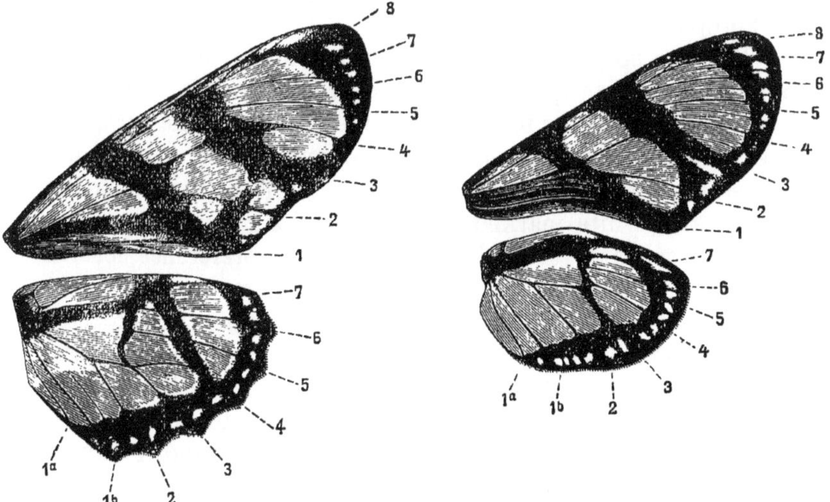

Fig. 25.—Wings of Ituna Ilione, ♂. Wings of Thyridia megisto, ♂.

for each other when on the wing; and no doubt many other equally remarkable cases are yet unnoticed.

The figures above of the fore and hind wings of two of these mimicking species, from Dr. Fritz Müller's original paper in *Kosmos*, will serve to show the considerable amount of difference, in the important character of the neuration of the wings, between these butterflies, which really belong to very distinct and not at all closely allied genera. Other important characters are—(1) The existence of a small basal cell in the hind wings of Ituna which is wanting in Thyridia; (2) the division of the cell between the veins 1*b* and 2 of the hind wings in the former genus, while it is undivided in the latter; and (3) the existence in Thyridia of scent-producing tufts of hair on the upper edge of the hind wing, while in Ituna these are wanting; but in place of them are extensible processes at the end of the abdomen, also emitting a powerful scent. These differences characterise two marked subdivisions of the Danaoid Heliconinæ, each containing several distinct genera; and these subdivisions are further distinguished by very different forms of larvæ, that to which Ituna belongs having from two to four long threadlike tentacles on the back, while in that containing Thyridia these are always absent. The former usually feed on Asclepiadeæ, the latter on Solanaceæ or Scrophulariaceæ.

The two species figured, though belonging to such distinct and even remote genera, have acquired almost identical tints and markings so as to be deceptively alike. The surface of the wings is, in both, transparent yellowish, with black transverse bands and white marginal spots, while both have similar black- and white-marked bodies and long yellow antennæ. Dr. Müller states that they both show a preference for the same flowers growing on the edges of the forest paths.[1]

We will now proceed to give the explanation of these curious similarities, which have remained a complete puzzle for twenty years. Mr. Bates, when first describing them, suggested that they might be due to some form of parallel variation dependent on climatic influences; and I myself adduced other cases of coincident local modifications of colour, which did not appear to be explicable by any form of mimicry.[2] But we neither of us hit upon the simple explanation given by Dr. Fritz Müller in 1879.

His theory is founded on the assumed, but probable,

[1] From Professor Meldola's translation of Dr. F. Müller's paper, in *Proc. Ent. Soc. Lond.*, 1879, p. xx.
[2] *Island Life*, p. 255.

fact, that insect-eating birds only learn by experience to distinguish the edible from the inedible butterflies, and in doing so necessarily sacrifice a certain number of the latter. The quantity of insectivorous birds in tropical America is enormous; and the number of young birds which every year have to learn wisdom by experience, as regards the species of butterflies to be caught or to be avoided, is so great that the sacrifice of life of the inedible species must be considerable, and, to a comparatively weak or scarce species, of vital importance. The number thus sacrificed will be fixed by the quantity of young birds, and by the number of experiences requisite to cause them to avoid the inedible species for the future, and not at all by the numbers of individuals of which each species consists. Hence, if two species are so much alike as to be mistaken for one another, the fixed number annually sacrificed by inexperienced birds will be divided between them, and both will benefit. But if the two species are very unequal in numbers, the benefit will be comparatively slight for the more abundant species, but very great for the rare one. To the latter it may make all the difference between safety and destruction.

To give a rough numerical example. Let us suppose that in a given limited district there are two species of Heliconidæ, one consisting of only 1000, the other of 100,000 individuals, and that the quota required annually in the same district for the instruction of young insectivorous birds is 500. By the larger species this loss will be hardly felt; to the smaller it will mean the most dreadful persecution resulting in a loss of half the total population. But, let the two species become superficially alike, so that the birds see no difference between them. The quota of 500 will now be taken from a combined population of 101,000 butterflies, and if proportionate numbers of each suffer, then the weak species will only lose five individuals instead of 500 as it did before. Now we know that the different species of Heliconidæ are not equally abundant, some being quite rare; so that the benefit to be derived in these latter cases would be very important. A slight inferiority in rapidity of flight or in powers of eluding attack might also be a cause of danger to an inedible species of scanty numbers, and in this case too the being merged in another much more abundant species, by similarity of external appearance, would be an advantage.

The question of fact remains. Do young birds pursue and capture these distasteful butterflies till they have learned by bitter experience what species to avoid? On this point Dr. Müller has fortunately been able to obtain some direct evidence, by capturing several Acræas and Heliconidæ which had evidently been seized by birds but had afterwards escaped, as they had pieces torn out of the wing, sometimes symmetrically out of both wings, showing that the insect had been seized when at rest and with the two pairs of wings in contact. There is, however, a general impression that this knowledge is hereditary, and does not need to be acquired by young birds; in support of which view Mr. Jenner Weir states that his birds always disregarded inedible caterpillars. When, day by day, he threw into his aviary various larvæ, those which were edible were eaten immediately, those which were inedible were no more noticed than if a pebble had been thrown before the birds.

The cases, however, are not strictly comparable. The birds were not young birds of the first year; and, what is more important, edible larvæ have a comparatively simple coloration, being always brown or green and smooth. Uneatable larvæ, on the other hand, comprise all that are of conspicuous colours and are hairy or spiny. But with butterflies there is no such simplicity of contrast. The eatable butterflies comprise not only brown or white species, but hundreds of Nymphalidæ, Papilionidæ, Lycænidæ, etc., which are gaily coloured and of an immense variety of patterns. The colours and patterns of the inedible kinds are also greatly varied, while they are often equally gay; and it is quite impossible to suppose that any amount of instinct or inherited habit (if such a thing exists) could enable young insectivorous birds to distinguish all the species of one kind from all those of the other. There is also some evidence to show that animals do learn by experience what to eat and what to avoid. Mr. Poulton was assured by Rev. G. J. Bursch that very young chickens peck at insects which they afterwards avoid. Lizards, too, often seized larvæ which they were unable to eat and ultimately rejected.

Although the Heliconidæ present, on the whole, many varieties of coloration and pattern, yet, in proportion to the number of distinct species in each district, the types of coloration are few and very well marked, and thus it becomes easier for a bird or other animal to learn that all belonging to such types are uneatable. This must be a decided advantage to the family in question, because, not only do fewer individuals of each species need to be sacrificed in order that their enemies may learn the lesson of their inedibility, but they are more easily recognised at a distance, and thus escape even pursuit. There is thus a kind of mimicry between closely allied species as well as between species of distinct genera, all tending to the same beneficial end. This may be seen in the four or five distinct species of the genus Heliconius which all have the same peculiar type of coloration—a yellow band across the upper wings and radiating red stripes on the lower,—and are all found in the same forests of the Lower Amazon; in the numerous very similar species of Ithomia with transparent wings, found in every locality of the same region; and in the very numerous species of Papilio of the "Æneas" group, all having a similar style of marking, the resemblance being especially close in the females. The very uniform type of colouring of the blue-black Euplæas and of the fulvous Acræas is of the same character.[1] In all these cases the similarity of the allied species is so great, that, when they are on the wing at some distance off, it is difficult to distinguish one species from another. But this close external resemblance is not always a sign of very near affinity; for minute examination detects differences in the form and scalloping of the wings, in the markings on the body, and in those on the under surface of the wings, which do not usually characterise the closest allies. It is to be further noted, that the presence of groups of very similar species of the same genus, in one locality, is not at all a common phenomenon among unprotected groups. Usually the species of a genus found in one locality are each well marked and belong to somewhat distinct types, while the

[1] This extension of the theory of mimicry was pointed out by Professor Meldola in the paper already referred to; and he has answered the objections to Dr. F. Müller's theory with great force in the *Annals and Mag. of Nat. Hist.*, 1882, p. 417.

closely allied forms—those that require minute examination to discriminate them as distinct species—are most generally found in separate areas, and are what are termed representative forms.

The extension we have now given to the theory of mimicry is important, since it enables us to explain a much wider range of colour phenomena than those which were first imputed to mimicry. It is in the richest butterfly region in the world—the Amazon valley—that we find the most abundant evidence of the three distinct sets of facts, all depending on the same general principle. The form of mimicry first elucidated by Mr. Bates is characterised by the presence in each locality of certain butterflies, or other insects, themselves edible and belonging to edible groups, which derived protection from having acquired a deceptive resemblance to some of the inedible butterflies in the same localities, which latter were believed to be wholly free from the attacks of insectivorous birds. Then came the extension of the principle, by Dr. F. Müller, to the case of species of distinct genera of the inedible butterflies resembling each other quite as closely as in the former cases, and like them always found in the same localities. They derive mutual benefit from becoming, in appearance, one species, from which a certain toll is taken annually to teach the young insectivorous birds that they are uneatable. Even when the two or more species are approximately equal in numbers, they each derive a considerable benefit from thus combining their forces; but when one of the species is scarce or verging on extinction, the benefit becomes exceedingly great, being, in fact, exactly apportioned to the need of the species.

The third extension of the same principle explains the grouping of allied species of the same genera of inedible butterflies into sets, each having a distinct type of coloration, and each consisting of a number of species which can hardly be distinguished on the wing. This must be useful exactly in the same way as in the last case, since it divides the inevitable toll to insectivorous birds and other animals among a number of species. It also explains the fact of the great similarity of many species of inedible insects in the same locality—a similarity which does not obtain to anything like the same extent among the edible species. The explanation of the various phenomena of resemblance and mimicry, presented by the distasteful butterflies, may now be considered tolerably complete.

Mimicry in other Orders of Insects.

A very brief sketch of these phenomena will be given, chiefly to show that the same principle prevails throughout nature, and that, wherever a rather extensive group is protected, either by distastefulness or offensive weapons, there are usually some species of edible and inoffensive groups that gain protection by imitating them. It has been already stated that the Telephoridæ, Lampyridæ, and other families of soft-winged beetles, are distasteful; and as they abound in all parts of the world, and especially in the tropics, it is not surprising that insects of many other groups should imitate them. This is especially the case with the longicorn beetles, which are much persecuted by insectivorous birds; and everywhere in tropical regions some of these are to be found so completely disguised as to be mistaken for species of the protected groups. Numbers of these imitations have been already recorded by Mr. Bates and myself, but I will here refer to a few others.

In the recently published volumes on the Longicorn and Malacoderm beetles of Central America [1] there are numbers of beautifully coloured figures of the new species; and on looking over them we are struck by the curious resemblance of some of the Longicorns to species of the Malacoderm group. In some cases we discover perfect mimics, and on turning to the descriptions we always find these pairs to come from the same locality. Thus the Otheostethus melanurus, one of the Prionidæ, imitates the malacoderm, Lucidota discolor, in form, peculiar coloration, and size, and both are found at Chontales in Nicaragua, the species mimicked having, however, as is usual, a wider range. The curious and very rare little longicorn, Tethlimmena aliena, quite unlike its nearest allies in the same country, is an exact copy on a somewhat smaller scale of a malacoderm, Lygistopterus amabilis, both

[1] Godman and Salvin's *Biologia Centrali-Americana, Insecta, Coleoptera,* vol. iii. part ii., and vol. v.

found at Chontales. The pretty longicorn, Callia albicornis, closely resembles two species of malacoderms (Silis chalybeipennis and Colyphus signaticollis), all being small beetles with red head and thorax and bright blue elytra, and all three have been found at Panama. Many other species of Callia also resemble other malacoderms; and the longicorn genus Lycidola has been named from its resemblance to various species of the Lycidæ, one of the species here figured (Lycidola belti) being a good mimic of Calopteron corrugatum and of several other allied species, all being of about the same size and found at Chontales. In these cases, and in most others, the longicorn beetles have lost the general form and aspect of their allies to take on the appearance of a distinct tribe. Some other groups of beetles, as the Elateridæ and Eucnemidæ, also deceptively mimic malacoderms.

Wasps and bees are often closely imitated by insects of other orders. Many longicorn beetles in the tropics exactly mimic wasps, bees, or ants. In Borneo a large black wasp, whose wings have a broad white patch near the apex (Mygnimia aviculus), is closely imitated by a heteromerous beetle (Coloborhombus fasciatipennis), which, contrary to the general habit of beetles, keeps its wings expanded in order to show the white patch on their apex, the wing-coverts being reduced to small oval scales, as shown in the figure. This is a most remarkable instance of mimicry, because the beetle has had to acquire so many characters which are unknown among its allies (except in another species from Java)—the expanded wings, the white band on them, and the oval scale-like elytra.[1] Another remarkable case has been noted by Mr. Neville Goodman, in Egypt, where a common hornet (Vespa orientalis) is exactly imitated in colour, size, shape, attitude when at rest, and mode of flight, by a beetle of the genus Laphria.[2]

The tiger-beetles (Cicindelidæ) are also the subjects of mimicry by more harmless insects. In the Malay Islands I found a heteromerous beetle which exactly resembled a Therates, both being found running on the trunks of trees. A longicorn (Collyrodes Lacordairei) mimics Collyris, another genus of the same family; while in the Philippine Islands

[1] *Trans. Ent. Soc.*, 1885, p. 369.

[2] *Proc. Cambridge Phil. Soc.*, vol. iii. part ii., 1877.

there is a cricket (Condylodeira tricondyloides), which so closely resembles a tiger-beetle of the genus Tricondyla

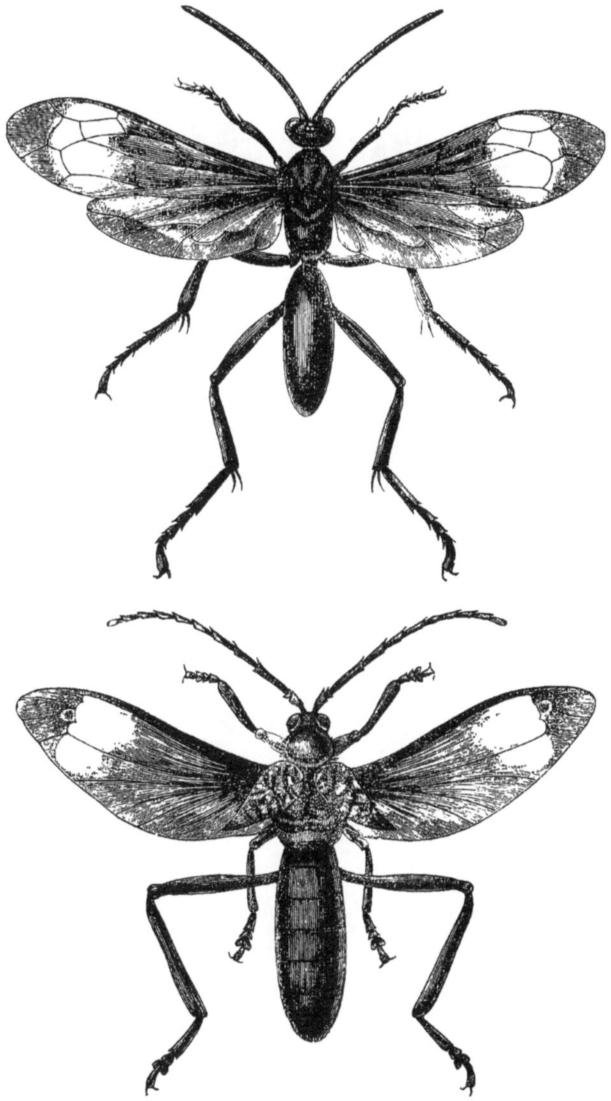

FIG. 26.—Mygnimia aviculus (Wasp). Coloborhombus fasciatipennis (Beetle).

that the experienced entomologist, Professor Westwood, at first placed it in his cabinet among those beetles.

One of the characters by which some beetles are protected is excessive hardness of the elytra and integuments. Several genera of weevils (Curculionidæ) are thus saved from attack, and these are often mimicked by species of softer and more

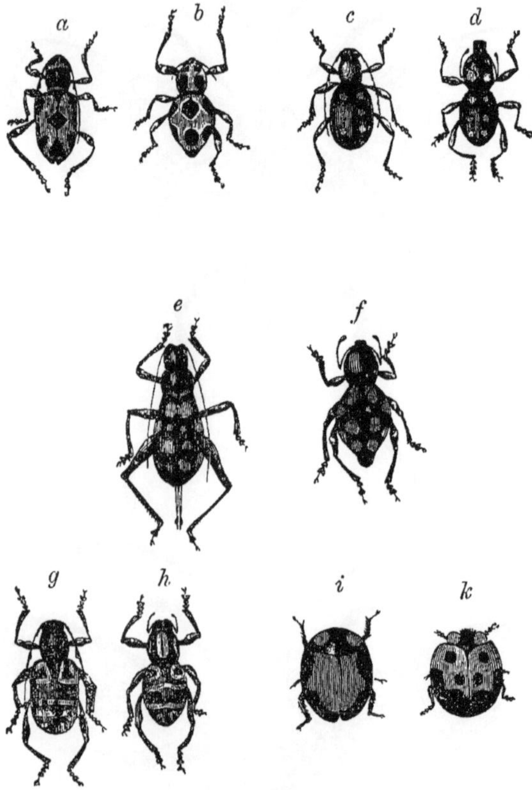

FIG. 27.

a. Doliops sp. (Longicorn) mimics Pachyrhynchus orbifæ, (*b*) (a hard curculio).
c. Doliops curculionoides mimics (*d*) Pachyrhynchus sp.
e. Scepastus pachyrhynchoides (a grasshopper) mimics (*f*) Apocyrtus sp. (a hard curculio).
g. Doliops sp. mimics (*h*) Pachyrhynchus sp.
i. Phoraspis (grasshopper) mimics (*k*) a Coccinella.
All the above are from the Philippines. The exact correspondence of the colours of the insects themselves renders the mimicry much more complete in nature than it appears in the above figures.

eatable groups. In South America, the genus Heilipus is one of these hard groups, and both Mr. Bates and M. Roelofs, a Belgian entomologist, have noticed that species of other genera exactly mimic them. So, in the Philippines, there is a group of Curculionidæ, forming the genus Pachyrhynchus, in which all the species are adorned with the most brilliant metallic colours, banded and spotted in a curious manner, and are very smooth and hard. Other genera of Curculionidæ (Desmidophorus, Alcides), which are usually very differently coloured, have species in the Philippines which mimic the Pachyrhynchi; and there are also several longicorn beetles (Aprophata, Doliops, Acronia, and Agnia), which also mimic them. Besides these, there are some longicorns and cetonias which reproduce the same colours and markings; and there is even a cricket (Scepastus pachyrhynchoides), which has taken on the form and peculiar coloration of these beetles in order to escape from enemies, which then avoid them as uneatable.[1] The figures on the opposite page exhibit several other examples of these mimicking insects.

Innumerable other cases of mimicry occur among tropical insects; but we must now pass on to consider a few of the very remarkable, but much rarer instances, that are found among the higher animals.

Mimicry among the Vertebrata.

Perhaps the most remarkable cases yet known are those of certain harmless snakes which mimic poisonous species. The genus Elaps, in tropical America, consists of poisonous snakes which do not belong to the viper family (in which are included the rattlesnakes and most of those which are poisonous), and which do not possess the broad triangular head which characterises the latter. They have a peculiar style of coloration, consisting of alternate rings of red and black, or red, black, and yellow, of different widths and grouped in various ways in the different species; and it is a style of coloration which does not occur in any other group of snakes in the world. But in the same regions are found three genera of harmless snakes, belonging to other families, some few species of which mimic the poisonous Elaps, often so exactly that it is with difficulty one can be distinguished from the other. Thus Elaps fulvius in Guatemala is imitated by the harmless Pliocerus equalis; Elaps corallinus in Mexico is mimicked by the

[1] *Compte-Rendu de la Société Entomologique de Belgaue*, series ii., No. 59, 1878.

harmless Homalocranium semicinctum; and Elaps lemniscatus in Brazil is copied by Oxyrhopus trigeminus; while in other parts of South America similar cases of mimicry occur, sometimes two harmless species imitating the same poisonous snake.

A few other instances of mimicry in this group have been recorded. There is in South Africa an egg-eating snake (Dasypeltis scaber), which has neither fangs nor teeth, yet it is very like the Berg adder (Clothos atropos), and when alarmed renders itself still more like by flattening out its head and darting forward with a hiss as if to strike a foe.[1] Dr. A. B. Meyer has also discovered that, while some species of the genus Callophis (belonging to the same family as the American Elaps) have large poison fangs, other species of the same genus have none; and that one of the latter (C. gracilis) resembles a poisonous species (C. intestinalis) so closely, that only an exact comparison will discover the difference of colour and marking. A similar kind of resemblance is said to exist between another harmless snake, Megærophis flaviceps, and the poisonous Callophis bivirgatus; and in both these cases the harmless snake is less abundant than the poisonous one, as occurs in all examples of true mimicry.[2]

In the genus Elaps, above referred to, the very peculiar style of colour and marking is evidently a "warning colour" for the purpose of indicating to snake-eating birds and mammals that these species are poisonous; and this throws light on the long-disputed question of the use of the rattle of the rattlesnake. This reptile is really both sluggish and timid, and is very easily captured by those who know its habits. If gently tapped on the head with a stick, it will coil itself up and lie still, only raising its tail and rattling. It may then be easily caught. This shows that the rattle is a warning to its enemies that it is dangerous to proceed to extremities; and the creature has probably acquired this structure and habit because it frequents open or rocky districts where protective colour is needful to save it from being pounced upon by buzzards or other snake-eaters. Quite parallel in function is the expanded hood of the Indian cobra, a

[1] *Nature*, vol. xxxiv. p. 547.

[2] *Proceedings of the Zool. Soc. of London*, 1870, p. 369.

poisonous snake which belongs also to the Elapidæ. This is, no doubt, a warning to its foes, not an attempt to terrify its prey; and the hood has been acquired, as in the case of the rattlesnake, because, protective coloration being on the whole useful, some mark was required to distinguish it from other protectively coloured, but harmless, snakes. Both these species feed on active creatures capable of escaping if their enemy were visible at a moderate distance.

Mimicry among Birds.

The varied forms and habits of birds do not favour the production among them of the phenomena of warning colours or of mimicry; and the extreme development of their instincts and reasoning powers, as well as their activity and their power of flight, usually afford them other means of evading their enemies. Yet there are a few imperfect, and one or two very perfect cases of true mimicry to be found among them. The less perfect examples are those presented by several species of cuckoos, an exceedingly weak and defenceless group of birds. Our own cuckoo is, in colour and markings, very like a sparrow-hawk. In the East, several of the small black cuckoos closely resemble the aggressive drongo-shrikes of the same country, and the small metallic cuckoos are like glossy starlings; while a large ground-cuckoo of Borneo (Carpococcyx radiatus) resembles one of the fine pheasants (Euplocamus) of the same country, both in form and in its rich metallic colours.

More perfect cases of mimicry occur between some of the dull-coloured orioles in the Malay Archipelago and a genus of large honey-suckers—the Tropidorhynchi or "Friar-birds." These latter are powerful and noisy birds which go in small flocks. They have long, curved, and sharp beaks, and powerful grasping claws; and they are quite able to defend themselves, often driving away crows and hawks which venture to approach them too nearly. The orioles, on the other hand, are weak and timid birds, and trust chiefly to concealment and to their retiring habits to escape persecution. In each of the great islands of the Austro-Malayan region there is a distinct species of Tropidorhynchus, and there is always along with it an oriole that exactly mimics it. All the Tropidorhynchi have a patch of bare black skin round the eyes, and a ruff of curious pale recurved feathers on the nape, whence their name of Friar-birds, the ruff being supposed to resemble the cowl of a friar. These peculiarities are imitated in the orioles by patches of feathers of corresponding colours; while the different tints of the two species in each island are exactly the same. Thus in Bouru both are earthy brown; in Ceram they are both washed with yellow ochre; in Timor the under surface is pale and the throat nearly white, and Mr. H. O. Forbes has recently discovered another pair in the island of Timor Laut. The close resemblance of these several pairs of birds, of widely different families, is quite comparable with that of many of the insects already described. It is so close that the preserved specimens have even deceived naturalists; for, in the great French work, *Voyage de l'Astrolabe*, the oriole of Bouru is actually described and figured as a honey-sucker; and Mr. Forbes tells us that, when his birds were submitted to Dr. Sclater for description, the oriole and the honey-sucker were, previous to close examination, considered to be the same species.

Objections to the Theory of Mimicry.

To set forth adequately the varied and surprising facts of mimicry would need a large and copiously illustrated volume; and no more interesting subject could be taken up by a naturalist who has access to our great collections and can devote the necessary time to search out the many examples of mimicry that lie hidden in our museums. The brief sketch of the subject that has been here given will, however, serve to indicate its nature, and to show the weakness of the objections that were at first made to it. It was urged that the action of "like conditions," with "accidental resemblances" and "reversion to ancestral types," would account for the facts. If, however, we consider the actual phenomena as here set forth, and the very constant conditions under which they occur, we shall see how utterly inadequate are these causes, either singly or combined. These constant conditions are—

1. That the imitative species occur in the same area and occupy the very same station as the imitated.
2. That the imitators are always the more defenceless.

3. That the imitators are always less numerous in individuals.
4. That the imitators differ from the bulk of their allies.
5. That the imitation, however minute, is *external* and *visible* only, never extending to internal characters or to such as do not affect the external appearance.

These five characteristic features of mimicry show us that it is really an exceptional form of protective resemblance. Different species in the same group of organisms may obtain protection in different ways: some by a general resemblance to their environment; some by more exactly imitating the objects that surround them—bark, or leaf, or flower; while others again gain an equal protection by resembling some species which, from whatever cause, is almost as free from attack as if it were a leaf or a flower. This immunity may depend on its being uneatable, or dangerous, or merely strong; and it is the resemblance to such creatures for the purpose of sharing in their safety that constitutes mimicry.

Concluding Remarks on Warning Colours and Mimicry.

Colours which have been acquired for the purpose of serving as a warning of inedibility, or of the possession of dangerous offensive weapons, are probably more numerous than have been hitherto supposed; and, if so, we shall be able to explain a considerable amount of colour in nature for which no use has hitherto been conjectured. The brilliant and varied colours of sea-anemones and of many coral animals will probably come under this head, since we know that many of them possess the power of ejecting stinging threads from various parts of their bodies which render them quite uneatable to most animals. Mr. Gosse describes how, on putting an Anthea into a tank containing a half-grown bullhead (Cottus bubalis) which had not been fed for some time, the fish opened his mouth and sucked in the morsel, but instantly shot it out again. He then seized it a second time, and after rolling it about in his mouth for a moment shot it out again, and then darted away to hide himself in a hole. Some tropical fishes, however, of the genera Tetrodon, Pseudoscarus, Astracion, and a few others, seem to have acquired the power of feeding on corals and medusæ; and the beautiful bands and spots and bright colours with which they are frequently adorned, may be either protective when feeding in the submarine coral groves, or may, in some cases, be warning colours to show that they themselves are poisonous and uneatable.

A remarkable illustration of the wide extension of warning colours, and their very definite purpose in nature, is afforded by what may now be termed "Mr. Belt's frog." Frogs in all parts of the world are, usually, protectively coloured with greens or browns; and the little tree-frogs are either green like the leaves they rest upon, or curiously mottled to imitate bark or dead leaves. But there are a certain number of very gaily coloured frogs, and these do not conceal themselves as frogs usually do. Such was the small toad found by Darwin at Bahia Blanca, which was intense black and bright vermilion, and crawled about in the sunshine over dry sand-hills and arid plains. And in Nicaragua, Mr. Belt found a little frog gorgeously dressed in a livery of red and blue, which did not attempt concealment and was very abundant, a combination of characters which convinced him that it was uneatable. He, therefore, took a few specimens home with him and gave them to his fowls and ducks, but none would touch them. At last, by throwing down pieces of meat, for which there was a great competition among the poultry, he managed to entice a young duck into snatching up one of the little frogs. Instead of swallowing it, however, the duck instantly threw it out of its mouth, and went about jerking its head as if trying to get rid of some unpleasant taste.[1]

The power of predicting what will happen in a given case is always considered to be a crucial test of a true theory, and if so, the theory of warning colours, and with it that of mimicry, must be held to be well established. Among the creatures which probably have warning colours as a sign of inedibility are, the brilliantly coloured nudibranchiate molluscs, those curious annelids the Nereis and the Aphrodite or sea-mouse, and many other marine animals. The brilliant colours of the scallops (Pecten) and some other bivalve shells are perhaps

[1] *The Naturalist in Nicaragua*, p. 321.

an indication of their hardness and consequent inedibility, as in the case of the hard beetles; and it is not improbable that some of the phosphorescent fishes and other marine organisms may, like the glow-worm, hold out their lamp as a warning to enemies.[1] In Queensland there is an exceedingly poisonous spider, whose bite will kill a dog, and cause severe illness with excruciating pain in man. It is black, with a bright vermilion patch on the middle of the body; and it is so well recognised by this conspicuous coloration that even the spider-hunting wasps avoid it.[2]

Locusts and grasshoppers are generally of green protective tints, but there are many tropical species most gaudily decorated with red, blue, and black colours. On the same general grounds as those by which Mr. Belt predicted the inedibility of his conspicuous frog, we might safely predict the same for these insects; but we have fortunately a proof that they are so protected, since Mr. Charles Horne states that one of the bright coloured Indian locusts was invariably rejected when offered to birds and lizards.[3]

The examples now given lead us to the conclusion that colours acquired for the purpose of serving as a danger-signal to enemies are very widespread in nature, and, with the corresponding colours of the species which mimic them, furnish us with a rational explanation of a considerable portion of the coloration of animals which is outside the limits of those colours that have been acquired for either protection or recognition. There remains, however, another set of colours, chiefly among the higher animals, which, being connected with some of the most interesting and most disputed questions in natural history, must be discussed in a separate chapter.

[1] Mr. Belt first suggested this use of the light of the Lampyridæ (fireflies and glow-worms)—*Naturalist in Nicaragua*, p. 320. Mr. Verrill and Professor Meldola made the same suggestion in the case of medusæ and other phosphorescent marine organisms (*Nature*, vol. xxx. pp. 281, 289).

[2] W. E. Armit, in *Nature*, vol. xviii. p. 642.

[3] *Proc. Ent. Soc.*, 1869, p. xiii.

CHAPTER X

COLOURS AND ORNAMENTS CHARACTERISTIC OF SEX

Sex colours in the mollusca and crustacea—In insects—In butterflies and moths—Probable causes of these colours—Sexual selection as a supposed cause—Sexual coloration of birds—Cause of dull colours of female birds—Relation of sex colour to nesting habits—Sexual colours of other vertebrates—Sexual selection by the struggles of males—Sexual characters due to natural selection—Decorative plumage of males and its effect on the females—Display of decorative plumage by the males—A theory of animal coloration—The origin of accessory plumes—Development of accessory plumes and their display—The effect of female preference will be neutralised by natural selection—General laws of animal coloration—Concluding remarks.

In the preceding chapters we have dealt chiefly with the coloration of animals as distinctive of the several species; and we have seen that, in an enormous number of cases, the colours can be shown to have a definite purpose, and to be useful either as a means of protection or concealment, of warning to enemies, or of recognition by their own kind. We have now to consider a subordinate but very widespread phenomenon—the differences of colour or of ornamental appendages in the two sexes. These differences are found to have special relations with the three classes of coloration above referred to, in many cases confirming the explanation already given of their purport and use, and furnishing us with important aid in formulating a general theory of animal coloration.

In comparing the colours of the two sexes we find a perfect gradation, from absolute identity of colour up to such extreme difference that it is difficult to believe that the two forms can belong to the same species; and this diversity in the colours of the sexes does not bear any constant relation to affinity or systematic position. In both insects and birds we find examples of complete identity and extreme diversity of the sexes; and these differences occur sometimes in the same tribe or family, and sometimes even in the same genus.

It is only among the higher and more active animals that sexual differences of colour acquire any prominence. In the mollusca the two sexes, when separated, are always alike in colour, and only very rarely present slight differences in the form of the shell. In the extensive group of crustacea the two sexes as a rule are identical in colour, though there are often differences in the form of the prehensile organs; but in a very few cases there are differences of colour also. Thus, in a Brazilian species of shore-crab (Gelasimus) the female is grayish-brown, while in the male the posterior part of the cephalo-thorax is pure white, with the anterior part of a rich green. This colour is only acquired by the males when they become mature, and is liable to rapid change in a few minutes to dusky tints.[1] In some of the fresh-water fleas (Daphnoidæ) the males are ornamented with red and blue spots, while in others similar colours occur in both sexes. In spiders also, though as a rule the two sexes are alike in colour, there are a few exceptions, the males being ornamented with brilliant colours on the abdomen, while the female is dull coloured.

Sexual Coloration in Insects.

It is only when we come to the winged insects that we find any large amount of peculiarity in sexual coloration, and even here it is only developed in certain orders. Flies (Diptera), field-bugs (Hemiptera), cicadas (Homoptera), and the grasshoppers, locusts, and crickets (Orthoptera) present very few and unimportant sexual differences of colour; but the last two groups have special musical organs very fully developed in the males of some of the species, and these no doubt enable the sexes to discover and recognise each other. In some cases, however, when the female is protectively coloured, as in the well-known leaf-insects already referred to (p. 207), the male

[1] Darwin's *Descent of Man*, p. 271.

is smaller and much less protectively formed and coloured. In the bees and wasps (Hymenoptera) it is also the rule that the sexes are alike in colour, though there are several cases among solitary bees where they differ; the female being black, and the male brown in Anthophora retusa, while in Andræna fulva the female is more brightly coloured than the male. Of the great order of beetles (Coleoptera) the same thing may be said. Though often so rich and varied in their colours the sexes are usually alike, and Mr. Darwin was only able to find about a dozen cases in which there was any conspicuous difference between them.[1] They exhibit, however, numerous sexual characters, in the length of the antennæ, and in horns, legs, or jaws remarkably enlarged or curiously modified in the male sex.

It is in the family of dragonflies (order Neuroptera) that we first meet with numerous cases of distinctive sexual coloration. In some of the Agrionidæ the males have the bodies rich blue and the wings black, while the females have the bodies green and the wings transparent. In the North American genus Hetærina the males alone have a carmine spot at the base of each wing; but in some other genera the sexes hardly differ at all.

The great order of Lepidoptera, including the butterflies and moths, affords us the most numerous and striking examples of diversity of sexual colouring. Among the moths the difference is usually but slight, being manifested in a greater intensity of the colour of the smaller winged male; but in a few cases there is a decided difference, as in the ghost-moth (Hepialus humuli), in which the male is pure white, while the female is yellow with darker markings. This may be a recognition colour, enabling the female more readily to discover her mate; and this view receives some support from the fact that in the Shetland Islands the male is almost as yellow as the female, since it has been suggested that at midsummer, when this moth appears, there is in that high latitude sufficient twilight all night to render any special coloration unnecessary.[2]

Butterflies present us with a wonderful amount of sexual

[1] Darwin's *Descent of Man*, p. 294, and footnote.

[2] *Nature*, 1871, p. 489.

difference of colour, in many cases so remarkable that the two sexes of the same species remained for many years under different names and were thought to be quite distinct species. We find, however, every gradation from perfect identity to complete diversity, and in some cases we are able to see a reason for this difference. Beginning with the most extraordinary cases of diversity—as in Diadema misippus, where the male is black, ornamented with a large white spot on each wing margined with rich changeable blue, while the female is orange-brown with black spots and stripes—we find the explanation in the fact that the female mimics an uneatable Danais, and thus gains protection while laying its eggs on low plants in company with that insect. In the allied species, Diadema bolina, the females are also very different from the males, but are of dusky brown tints, evidently protective and very variable, some specimens having a general resemblance to the uneatable Euplæas; so that we see here some of the earlier stages of both forms of protection. The remarkable differences in some South American Pieridæ are similarly explained. The males of Pieris pyrrha, P. lorena, and several others, are white with a few black bands and marginal spots like so many of their allies, while the females are gaily coloured with yellow and brown, and exactly resemble some species of the uneatable Heliconidæ of the same district. Similarly, in the Malay Archipelago, the female of Diadema anomala is glossy metallic blue, while the male is brown; the reason for this reversal of the usual rule being, that the female exactly mimics the brilliant colouring of the common and uneatable Euplæa midamus, and thus secures protection. In the fine Adolias dirtea, the male is black with a few specks of ochre-yellow and a broad marginal band of rich metallic greenish-blue, while the female is brownish-black entirely covered with rows of ochre-yellow spots. This latter coloration does not appear to be protective when the insect is seen in the cabinet, but it really is so. I have observed the female of this butterfly in Sumatra, where it settles on the ground in the forest, and its yellow spots so harmonise with the flickering gleams of sunlight on the dead leaves that it can only be detected with the greatest difficulty.

A hundred other cases might be quoted in which the female is either more obscurely coloured than the male, or gains protection by imitating some inedible species; and any one who has watched these female insects flying slowly along in search of the plants on which to deposit their eggs, will understand how important it must be to them not to attract the attention of insect-eating birds by too conspicuous colours. The number of birds which capture insects on the wing is much greater in tropical regions than in Europe; and this is perhaps the reason why many of our showy species are alike, or almost alike, in both sexes, while they are protectively coloured on the under side which is exposed to view when they are at rest. Such are our peacock, tortoise-shell, and red admiral butterflies; while in the tropics we more commonly find that the females are less conspicuous on the upper surface even when protectively coloured beneath.

We may here remark, that the cases already quoted prove clearly that either male or female may be modified in colour apart from the opposite sex. In Pieris pyrrha and its allies the male retains the usual type of coloration of the whole genus, while the female has acquired a distinct and peculiar style of colouring. In Adolias dirtea, on the other hand, the female appears to retain something like the primitive colour and markings of the two sexes, modified perhaps for more perfect protection; while the male has acquired more and more intense and brilliant colours, only showing his original markings by the few small yellow spots that remain near the base of the wings. In the more gaily coloured Pieridæ, of which our orange-tip butterfly may be taken as a type, we see in the female the plain ancestral colours of the group, while the male has acquired the brilliant orange tip to its wings, probably as a recognition mark.

In those species in which the under surface is protectively coloured, we often find the upper surface alike in both sexes, the tint of colour being usually more intense in the male. But in some cases this leads to the female being more conspicuous, as in some of the Lycænidæ, where the female is bright blue and the male of a blue so much deeper and soberer in tint as to appear the less brilliantly coloured of the two.

Probable Causes of these Colours.

In the production of these varied results there have probably been several causes at work. There seems to be a constant tendency in the male of most animals—but especially of birds and insects—to develop more and more intensity of colour, often culminating in brilliant metallic blues or greens or the most splendid iridescent hues; while, at the same time, natural selection is constantly at work, preventing the female from acquiring these same tints, or modifying her colours in various directions to secure protection by assimilating her to her surroundings, or by producing mimicry of some protected form. At the same time, the need for recognition must be satisfied; and this seems to have led to diversities of colour in allied species, sometimes the female, sometimes the male undergoing the greatest change according as one or other could be modified with the greatest ease, and so as to interfere least with the welfare of the race. Hence it is that sometimes the males of allied species vary most, as in the different species of Epicalia; sometimes the females, as in the magnificent green species of Ornithoptera and the "Æneas" group of Papilio.

The importance of the two principles—the need of protection and recognition—in modifying the comparative coloration of the sexes among butterflies, is beautifully illustrated in the case of the groups which are protected by their distastefulness, and whose females do not, therefore, need the protection afforded by sober colours.

In the great families, Heliconidæ and Acræidæ, we find that the two sexes are almost always alike; and, in the very few exceptions, that the female, though differently, is not less gaily or less conspicuously coloured. In the Danaidæ the same general rule prevails, but the cases in which the male exhibits greater intensity of colour than the female are perhaps more numerous than in the other two families. There is, however, a curious difference in this respect between the Oriental and the American groups of distasteful Papilios with warning colours, both of which are the subjects of mimicry. In the Eastern groups—of which P. hector and P. coon may be taken as types—the two sexes are nearly alike, the male being sometimes more intensely coloured and with fewer pale markings; but in the American groups—represented by P. æneas, P. sesostris, and allies—there is a wonderful diversity, the males having a rich green or bluish patch on the fore wings, while the females have a band or spots of pure white, not always corresponding in position to the green spot of the males. There are, however, transitional forms, by which a complete series can be traced, from close similarity to great diversity of colouring between the sexes; and this may perhaps be only an extreme example of the intenser colour and more concentrated markings which are a very prevalent characteristic of male butterflies.

There are, in fact, many indications of a regular succession of tints in which colour development has occurred in the various groups of butterflies, from an original grayish or brownish neutral tint. Thus in the "Æneas" group of Papilios we have the patch on the upper wings yellowish in P. triopas, olivaceous in P. bolivar, bronzy-gray with a white spot in P. erlaces, more greenish and buff in P. iphidamas, gradually changing to the fine blue of P. brissonius, and the magnificent green of P. sesostris. In like manner, the intense crimson spots of the lower wings can be traced step by step from a yellow or buff tint, which is one of the most widespread colours in the whole order. The greater purity and intensity of colour seem to be usually associated with more pointed wings, indicating greater vigour and more rapid flight.

Sexual Selection as a supposed Cause of Colour Development.

Mr. Darwin, as is well known, imputed most of the brilliant colours and varied patterns of butterflies' wings to sexual selection—that is, to a constant preference, by female butterflies, for the more brilliant males; the colours thus produced being sometimes transmitted to the males alone, sometimes to both sexes. This view has always seemed to me to be unsupported by evidence, while it is also quite inadequate to account for the facts. The only direct evidence, as set forth with his usual fairness by Mr. Darwin himself, is opposed to his views. Several entomologists assured him that, in moths, the females evince not the least choice of their partners; and Dr. Wallace of Colchester, who has largely bred the fine Bombyx cynthia, confirmed this statement. Among butterflies, several males often pursue one female, and Mr. Darwin says, that, unless the female exerts a choice the pairing must be left to chance. But, surely, it may be the most vigorous or most persevering male that is chosen, not necessarily one more brightly or differently coloured, and this will be true "natural selection." Butterflies have been noticed to prefer some coloured flowers to others; but that does not prove, or even render probable, any preference for the colour itself, but only for flowers of certain colours, on account of the more agreeable or more abundant nectar obtained from them. Dr. Schulte called Mr. Darwin's attention to the fact, that in the Diadema bolina the brilliant blue colour surrounding the white spots is only visible when we look towards the insect's head, and this is true of many of the iridescent colours of butterflies, and probably depends upon the direction of the striæ on the scales. It is suggested, however, that this display of colour will be seen by the female as the male is approaching her, and that it has been developed by sexual selection.[1] But in the majority of cases the males *follow* the female, hovering over her in a position which would render it almost impossible for her to see the particular colours or patterns on his upper surface; to do so the female should mount higher than the male, and fly towards him—being the seeker instead of the sought, and this is quite opposed to the actual facts. I cannot, therefore, think that this suggestion adds anything whatever to the evidence for sexual selection of colour by female butterflies. This question will, however, be again touched upon after we have considered the phenomena of sexual colour among the vertebrata.

Sexual Coloration of Birds.

The general rule among vertebrates, as regards colour, is, for the two sexes to be alike. This prevails, with only a few exceptions, in fishes, reptiles, and mammalia; but in birds diversity of sexual colouring is exceedingly frequent, and is, not improbably, present in a greater or less degree in more

[1] Darwin in *Nature*, 1880, p. 237.

than half of the known species. It is this class, therefore, that will afford us the best materials for a discussion of the problem, and that may perhaps lead us to a satisfactory explanation of the causes to which sexual colour is due.

The most fundamental characteristic of birds, from our present point of view, is a greater intensity of colour in the male. This is the case in hawks and falcons; in many thrushes, warblers, and finches; in pigeons, partridges, rails, plovers, and many others. When the plumage is highly protective or of dull uniform tints, as in many of the thrushes and warblers, the sexes are almost or quite identical in colour; but when any rich markings or bright tints are acquired, they are almost always wanting or much fainter in the female, as we see in the black-cap among warblers, and the chaffinch among finches.

It is in tropical regions, where from a variety of causes colour has been developed to its fullest extent, that we find the most remarkable examples of sexual divergence of colour. The most gorgeously coloured birds known are the birds of paradise, the chatterers, the tanagers, the humming-birds, and the pheasant-tribe, including the peacocks. In all these the females are much less brilliant, and, in the great majority of cases, exceptionally plain and dull coloured birds. Not only are the remarkable plumes, crests, and gorgets of the birds of paradise entirely wanting in the females, but these latter are usually without any bright colour at all, and rank no higher than our thrushes in ornamental plumage. Of the humming-birds the same may be said, except that the females are often green, and sometimes slightly metallic, but from their small size and uniform tints are never conspicuous. The glorious blues and purples, the pure whites and intense crimsons of the male chatterers are represented in the females by olive-greens or dull browns, as are the infinitely varied tints of the male tanagers. And in pheasants, the splendour of plumage which characterises the males is entirely absent in the females, which, though often ornamental, have always comparatively sober and protective tints. The same thing occurs with many other groups. In the Eastern tropics are many brilliant birds belonging to the families of the warblers, flycatchers, shrikes, etc., but the female is always much less brilliant than the male and often quite dull coloured.

Cause of Dull Colours of Female Birds.

The reason of this phenomenon is not difficult to find, if we consider the essential conditions of a bird's existence, and the most important function it has to fulfil. In order that the species may be continued, young birds must be produced, and the female birds have to sit assiduously on their eggs. While doing this they are exposed to observation and attack by the numerous devourers of eggs and birds, and it is of vital importance that they should be protectively coloured in all those parts of the body which are exposed during incubation. To secure this end all the bright colours and showy ornaments which decorate the male have not been acquired by the female, who often remains clothed in the sober hues which were probably once common to the whole order to which she belongs. The different amounts of colour acquired by the females have no doubt depended on peculiarities of habits and of environment, and on the powers of defence or of concealment possessed by the species. Mr. Darwin has taught us that natural selection cannot produce absolute, but only relative perfection; and as a protective colour is only one out of many means by which the female birds are able to provide for the safety of their young, those which are best endowed in other respects will have been allowed to acquire more colour than those with whom the struggle for existence is more severe.

Relation of Sex Colour to Nesting Habits.

This principle is strikingly illustrated by the existence of considerable numbers of birds in which both sexes are similarly and brilliantly coloured,—in some cases as brilliantly as the males of many of the groups above referred to. Such are the extensive families of the kingfishers, the woodpeckers, the toucans, the parrots, the turacos, the hangnests, the starlings, and many other smaller groups, all the species of which are conspicuously or brilliantly coloured, while in all of them the females are either coloured exactly like the males, or, when differently coloured, are equally conspicuous. When searching for some cause for this singular apparent exception to the rule of female protective colouring, I came upon a fact which beautifully explains it; for in all these cases, without exception, the species either nests in holes in the ground or in trees, or builds a domed or covered nest, so as completely to conceal the sitting-bird. We have here a case exactly parallel to that of the butterflies protected by distastefulness, whose females are either exactly like the males, or, if different, are equally conspicuous. We can hardly believe that so exact a parallel should exist between such remote classes of animals, except under the influence of a general law; and, in the need of protection by all defenceless animals, and especially by most female insects and birds, we have such a law, which has been proved to have influenced the colours of a considerable proportion of the animal kingdom.[1]

The general relation which exists between the mode of nesting and the coloration of the sexes in those groups of birds which need protection from enemies, may be thus expressed: When both sexes are brilliant or conspicuous, the nest is such as to conceal the sitting-bird; but when the male is brightly coloured and the female sits exposed on the nest, she is always less brilliant and generally of quite sober and protective hues.

It must be understood that the mode of nesting has influenced the colour, not that the colour has determined the mode of nesting; and this, I believe, has been generally, though not perhaps universally, the case. For we know that colour varies more rapidly, and can be more easily modified and fixed by selection, than any other character; whereas habits, especially when connected with structure, and when they pervade a whole group, are much more persistent and more difficult to change, as shown by the habit of the dog turning round two or three times before lying down, believed to be that of the wild ancestral form which thus smoothed down the herbage so as to form a comfortable bed. We see, too, that the general mode of nesting is characteristic of whole families differing widely in size, form, and colours. Thus, all the kingfishers and their allies in every part of the world nest

[1] See the author's *Contributions to Natural Selection*, chap. vii. in which these facts were first brought forward.

in holes, usually in banks, but sometimes in trees. The motmots and the puff-birds (Bucconidæ) build in similar places; while the toucans, barbets, trogons, woodpeckers, and parrots all make their nests in hollow trees. This habit, pervading all the members of extensive families, must therefore be extremely ancient, more especially as it evidently depends in some degree on the structure of the birds, the bills, and especially the feet, of all these groups being unfitted for the construction of woven arboreal nests.[1] But in all these families the colour varies greatly from species to species, being constant only in the one character of the similarity of the sexes, or, at all events, in their being equally conspicuous even though differently coloured.

When I first put forward this view of the connection between the mode of nesting and the coloration of female birds, I expressed the law in somewhat different terms, which gave rise to some misunderstanding, and led to numerous criticisms and objections. Several cases were brought forward in which the females were far less brilliant than the males, although the nest was covered. This is the case with the Maluridæ, or superb warblers of Australia, in which the males are very brilliant during the pairing season and the females quite plain, yet they build domed nests. Here, there can be little doubt, the covered nest is a protection from rain or from some special enemies to the eggs; while the birds themselves are protectively coloured in both sexes, except for a short time during the breeding season when the male acquires brilliant colours; and this is probably connected with the fact of their inhabiting the open plains and thin scrub of Australia, where protective colours are as generally advantageous as they are in our north-temperate zones.

As I have now stated the law, I do not think there are any exceptions to it, while there are an overwhelming number of cases which give it a strong support. It has been objected that the domed nests of many birds are as conspicuous as the birds themselves would be, and would, therefore, be of no use as a protection to the birds and young. But, as a matter of fact, they do protect from attack, for hawks or crows do not pluck such nests to pieces, as in doing so they would be

[1] On this point see the author's *Contributions to Natural Selection*, chap. v. i.

exposed to the attack of the whole colony; whereas a hawk or falcon could carry off a sitting-bird or the young at a swoop, and entirely avoid attack. Moreover, each kind of covered nest is doubtless directed against the attacks of the most dangerous enemies of the species, the purse-like nests, often a yard long, suspended from the extremity of thin twigs, being useful against the attacks of snakes, which, if they attempted to enter them, would be easily made to lose their hold and fall to the ground. Such birds as jays, crows, magpies, hawks, and other birds of prey, have also been urged as an exception; but these are all aggressive birds, able to protect themselves, and thus do not need any special protection for their females during nidification. Some birds which build in covered nests are comparatively dull coloured, like many of the weaver birds, but in others the colours are more showy, and in all the sexes are alike; so that none of these are in any way opposed to the rule. The golden orioles have, however, been adduced as a decided exception, since the females are showy and build in an open nest. But even here the females are less brilliant than the males, and are sometimes greenish or olivaceous on the upper surface; while they very carefully conceal their nests among dense foliage, and the male is sufficiently watchful and pugnacious to drive off most intruders.

On the other hand, how remarkable it is that the only small and brightly coloured birds of our own country in which the male and female are alike—the tits and starlings—either build in holes or construct covered nests; while the beautiful hangnests (Icteridæ) of South America, which always build covered or purse-shaped nests, are equally showy in both sexes, in striking contrast with the chatterers and tanagers of the same country, whose females are invariably less conspicuous than the males. On a rough estimate, there are about 1200 species of birds in the class of showy males and females, with concealed nidification; while there are probably, from an equally rough estimate, about the same number in the contrasted class of showy males and dull females, with open nests. This will leave the great bulk of known birds in the classes of those which are more or less protectively coloured in both sexes; or which, from their organisation and habits, do not require special protective coloration, such as many of the birds of prey, the larger waders, and the oceanic birds.

There are a few very curious cases in which the female bird is actually more brilliant than the male, and which yet have open nests. Such are the dotterel (Eudromias morinellus), several species of phalarope, an Australian creeper (Climacteris erythropus), and a few others; but in every one of these cases the relation of the sexes in regard to nidification is reversed, the male performing the duties of incubation, while the female is the stronger and more pugnacious. This curious case, therefore, quite accords with the general law of coloration.[1]

Sexual Colours of other Vertebrates.

We may consider a few of the cases of sexual colouring of other classes of vertebrates, as given by Mr. Darwin. In fishes, though the sexes are usually alike, there are several species in which the males are more brightly coloured, and have more elongated fins, spines, or other appendages, and in some few cases the colours are decidedly different. The males often fight together, and are altogether more vivacious and excitable than the females during the breeding season; and with this we may connect a greater intensity of coloration.

In frogs and toads the colours are usually alike, or a little more intense in the males, and the same may be said of most snakes. It is in lizards that we first meet with considerable sexual differences, many of the species having gular pouches, frills, dorsal crests, or horns, either confined to the males, or more developed in them than in the females, and these ornaments are often brightly coloured. In most cases, however, the tints of lizards are protective, the male being usually a little more intense in coloration; and the difference in extreme cases may be partly due to the need of protection for the female, which, when laden with eggs, must be less active and less able to escape from enemies than the male, and may, therefore, have retained more protective colours, as so many insects and birds have certainly done.[2]

In mammalia there is often a somewhat greater intensity

[1] Seebohm's *History of British Birds*, vol. ii., introduction, p. xiii.

[2] For details see Darwin's *Descent of Man*, chap. xii.

of colour in the male, but rarely a decided difference. The female of the great red kangaroo, however, is a delicate gray; while in the Lemur macaco of Madagascar the male is jet-black and the female brown. In many monkeys also there are some differences of colour, especially on the face. The sexual weapons and ornaments of male mammalia, as horns, crests, manes, and dewlaps, are well known, and are very numerous and remarkable. Having thus briefly reviewed the facts, we will now consider the theories to which they have given rise.

Sexual Selection by the Struggles of Males.

Among the higher animals it is a very general fact that the males fight together for the possession of the females. This leads, in polygamous animals especially, to the stronger or better armed males becoming the parents of the next generation, which inherits the peculiarities of the parents; and thus vigour and offensive weapons are continually increased in the males, resulting in the strength and horns of the bull, the tusks of the boar, the antlers of the stag, and the spurs and fighting instinct of the gamecock. But almost all male animals fight together, though not specially armed; even hares, moles, squirrels, and beavers fight to the death, and are often found to be scarred and wounded. The same rule applies to almost all male birds; and these battles have been observed in such different groups as humming-birds, finches, goatsuckers, woodpeckers, ducks, and waders. Among reptiles, battles of the males are known to occur in the cases of crocodiles, lizards, and tortoises; among fishes, in those of salmon and sticklebats. Even among insects the same law prevails; and male spiders, beetles of many groups, crickets, and butterflies often fight together.

From this very general phenomenon there necessarily results a form of natural selection which increases the vigour and fighting power of the male animal, since, in every case, the weaker are either killed, wounded, or driven away. This selection would be more powerful if males were always in excess of females, but after much research Mr. Darwin could not obtain any satisfactory evidence that this was the case. The same effect, however, is produced in some cases by constitution or habits; thus male insects usually emerge first from the pupa, and among migrating birds the males arrive first both in this country and in North America. The struggle is thus intensified, and the most vigorous males are the first to have offspring. This in all probability is a great advantage, as the early breeders have the start in securing food, and the young are strong enough to protect themselves while the later broods are being produced.

It is to this form of male rivalry that Mr. Darwin first applied the term "sexual selection." It is evidently a real power in nature; and to it we must impute the development of the exceptional strength, size, and activity of the male, together with the possession of special offensive and defensive weapons, and of all other characters which arise from the development of these or are correlated with them. But he has extended the principle into a totally different field of action, which has none of that character of constancy and of inevitable result that attaches to natural selection, including male rivalry; for by far the larger portion of the phenomena, which he endeavours to explain by the direct action of sexual selection, can only be so explained on the hypothesis that the immediate agency is female choice or preference. It is to this that he imputes the origin of all secondary sexual characters other than weapons of offence and defence, of all the ornamental crests and accessory plumes of birds, the stridulating sounds of insects, the crests and beards of monkeys and other mammals, and the brilliant colours and patterns of male birds and butterflies. He even goes further, and imputes to it a large portion of the brilliant colour that occurs in both sexes, on the principle that variations occurring in one sex are sometimes transmitted to the same sex only, sometimes to both, owing to peculiarities in the laws of inheritance. In this extension of sexual selection to include the action of female choice or preference, and in the attempt to give to that choice such wide-reaching effects, I am unable to follow him more than a very little way; and I will now state some of the reasons why I think his views are unsound.

Sexual Characters due to Natural Selection.

Besides the acquisition of weapons by the male for the purpose of fighting with other males, there are some other sexual characters which may have been produced by natural selection. Such are the various sounds and odours which are peculiar to the male, and which serve as a call to the female or as an indication of his presence. These are evidently a valuable addition to the means of recognition of the two sexes, and are a further indication that the pairing season has arrived; and the production, intensification, and differentiation of these sounds and odours are clearly within the power of natural selection. The same remark will apply to the peculiar calls of birds, and even to the singing of the males. These may well have originated merely as a means of recognition between the two sexes of a species, and as an invitation from the male to the female bird. When the individuals of a species are widely scattered, such a call must be of great importance in enabling pairing to take place as early as possible, and thus the clearness, loudness, and individuality of the song becomes a useful character, and therefore the subject of natural selection. Such is especially the case with the cuckoo, and with all solitary birds, and it may have been equally important at some period of the development of all birds. The act of singing is evidently a pleasurable one; and it probably serves as an outlet for superabundant nervous energy and excitement, just as dancing, singing, and field sports do with us. It is suggestive of this view that the exercise of the vocal power seems to be complementary to the development of accessory plumes and ornaments, all our finest singing birds being plainly coloured, and with no crests, neck or tail plumes to display; while the gorgeously ornamented birds of the tropics have no song, and those which expend much energy in display of plumage, as the turkey, peacocks, birds of paradise, and humming-birds, have comparatively an insignificant development of voice. Some birds have, in the wings or tail, peculiarly developed feathers which produce special sounds. In some of the little manakins of Brazil, two or three of the wing-feathers are curiously shaped and stiffened in the male, so that the bird is able to produce with them a peculiar snapping or cracking sound; and the tail-feathers of several species of snipe are so narrowed as to produce distinct drumming, whistling, or switching sounds when the birds descend rapidly from a great height. All these are probably recognition and call notes, useful to each species in relation to the most important function of their lives, and thus capable of being developed by the agency of natural selection.

Decorative Plumage of Birds and its Display.

Mr. Darwin has devoted four chapters of his *Descent of Man* to the colours of birds, their decorative plumage, and its display at the pairing season; and it is on this latter circumstance that he founds his theory, that both the plumage and the colours have been developed by the preference of the females, the more ornamented males becoming the parents of each successive generation. Any one who reads these most interesting chapters will admit, that the fact of the display is demonstrated; and it may also be admitted, as highly probable, that the female is pleased or excited by the display. But it by no means follows that slight differences in the shape, pattern, or colours of the ornamental plumes are what lead a female to give the preference to one male over another; still less that all the females of a species, or the great majority of them, over a wide area of country, and for many successive generations, prefer exactly the same modification of the colour or ornament.

The evidence on this matter is very scanty, and in most cases not at all to the point. Some peahens preferred an old pied peacock; albino birds in a state of nature have never been seen paired with other birds; a Canada goose paired with a Bernicle gander; a male widgeon preferred a pintail duck to its own species; a hen canary preferred a male greenfinch to either linnet, goldfinch, siskin, or chaffinch. These cases are evidently exceptional, and are not such as generally occur in nature; and they only prove that the female does exert some choice between very different males, and some observations on birds in a state of nature prove the same thing; but there is no evidence that slight variations in the colour or plumes, in the way of increased intensity or complexity, are what determines the choice. On the other hand, Mr. Darwin gives much evidence that it is *not* so determined. He tells us that Messrs. Hewitt, Tegetmeier, and Brent, three of the highest authorities and best observers, "do not believe that the females prefer certain males on account of the beauty of their plumage." Mr. Hewitt was convinced "that the female almost invariably prefers the most vigorous, defiant, and mettlesome male;" and Mr. Tegetmeier, "that a gamecock, though disfigured by being dubbed, and with his hackles trimmed, would be accepted as readily as a male retaining all his natural ornaments."[1] Evidence is adduced that a female pigeon will sometimes take an antipathy to a particular male without any assignable cause; or, in other cases, will take a strong fancy to some one bird, and will desert her own mate for him; but it is not stated that superiority or inferiority of plumage has anything to do with these fancies. Two instances are indeed given, of male birds being rejected, which had lost their ornamental plumage; but in both cases (a widow-finch and a silver pheasant) the long tail-plumes are the indication of sexual maturity. Such cases do not support the idea that males with the tail-feathers a trifle longer, or the colours a trifle brighter, are generally preferred, and that those which are only a little inferior are as generally rejected,—and this is what is absolutely needed to establish the theory of the development of these plumes by means of the choice of the female.

It will be seen, that female birds have unaccountable likes and dislikes in the matter of their partners, just as we have ourselves, and this may afford us an illustration. A young man, when courting, brushes or curls his hair, and has his moustache, beard, or whiskers in perfect order, and no doubt his sweetheart admires them; but this does not prove that she marries him on account of these ornaments, still less that hair, beard, whiskers, and moustache were developed by the continued preferences of the female sex. So, a girl likes to see her lover well and fashionably dressed, and he always dresses as well as he can when he visits her; but we cannot conclude from this that the whole series of male costumes, from the brilliantly coloured, puffed, and slashed doublet and hose of the Elizabethan period, through the gorgeous coats, long waistcoats, and pigtails of the early Georgian era, down to the funereal dress-suit of the present day, are the direct result of female preference. In like manner, female birds may be

[1] *Descent of Man*, pp. 417, 418, 420.

charmed or excited by the fine display of plumage by the males; but there is no proof whatever that slight differences in that display have any effect in determining their choice of a partner.

Display of Decorative Plumage.

The extraordinary manner in which most birds display their plumage at the time of courtship, apparently with the full knowledge that it is beautiful, constitutes one of Mr. Darwin's strongest arguments. It is, no doubt, a very curious and interesting phenomenon, and indicates a connection between the exertion of particular muscles and the development of colour and ornament; but, for the reasons just given, it does not prove that the ornament has been developed by female choice. During excitement, and when the organism develops superabundant energy, many animals find it pleasurable to exercise their various muscles, often in fantastic ways, as seen in the gambols of kittens, lambs, and other young animals. But at the time of pairing, male birds are in a state of the most perfect development, and possess an enormous store of vitality; and under the excitement of the sexual passion they perform strange antics or rapid flights, as much probably from an internal impulse to motion and exertion as with any desire to please their mates. Such are the rapid descent of the snipe, the soaring and singing of the lark, and the dances of the cock-of-the-rock and of many other birds.

It is very suggestive that similar strange movements are performed by many birds which have no ornamental plumage to display. Goatsuckers, geese, carrion vultures, and many other birds of plain plumage have been observed to dance, spread their wings or tails, and perform strange love-antics. The courtship of the great albatross, a most unwieldy and dull coloured bird, has been thus described by Professor Moseley: "The male, standing by the female on the nest, raises his wings, spreads his tail and elevates it, throws up his head with the bill in the air, or stretches it straight out, or forwards, as far as he can, and then utters a curious cry."[1] Mr. Jenner Weir informs me that "the male blackbird is full of action, spreads out his glossy wing and tail, turns his rich golden

[1] *Notes of a Naturalist on the Challenger.*

beak towards the female, and chuckles with delight," while he has never seen the more plain coloured thrush demonstrative to the female. The linnet distends his rosy breast, and slightly expands his brown wings and tail; while the various gay coloured Australian finches adopt such attitudes and postures as, in every case, to show off their variously coloured plumage to the best advantage.[1]

A Theory of Animal Coloration.

Having rejected Mr. Darwin's theory of female choice as incompetent to account for the brilliant colours and markings of the higher animals, the preponderance of these colours and markings in the male sex, and their display during periods of activity or excitement, I may be asked what explanation I have to offer as a preferable substitute. In my *Tropical Nature* I have already indicated such a theory, which I will now briefly explain, supporting it by some additional facts and arguments, which appear to me to have great weight, and for which I am mainly indebted to a most interesting and suggestive posthumous work by Mr. Alfred Tylor.[2]

The fundamental or ground colours of animals are, as has been shown in preceding chapters, very largely protective, and it is not improbable that the primitive colours of all animals were so. During the long course of animal development other modes of protection than concealment by harmony of colour arose, and thenceforth the normal development of colour due to the complex chemical and structural changes ever going on in the organism, had full play; and the colours thus produced were again and again modified by natural selection for purposes of warning, recognition, mimicry, or special protection, as has been already fully explained in the preceding chapters.

Mr. Tylor has, however, called attention to an important principle which underlies the various patterns or ornamental markings of animals—namely, that diversified coloration follows the chief lines of structure, and changes at points, such as the joints, where function changes. He says, "If we take highly decorated species—that is, animals marked by

[1] *Descent of Man*, pp. 401, 402.

[2] *Coloration in Animals and Plants*, London, 1886.

alternate dark or light bands or spots, such as the zebra, some deer, or the carnivora, we find, first, that the region of the spinal column is marked by a dark stripe; secondly, that the regions of the appendages, or limbs, are differently marked; thirdly, that the flanks are striped or spotted, along or between the regions of the lines of the ribs; fourthly, that the shoulder and hip regions are marked by curved lines; fifthly, that the pattern changes, and the direction of the lines, or spots, at the head, neck, and every joint of the limbs; and lastly, that the tips of the ears, nose, tail, and feet, and the eye are emphasised in colour. In spotted animals the greatest length of the spot is generally in the direction of the largest development of the skeleton."

This structural decoration is well seen in many insects. In caterpillars, similar spots and markings are repeated in each segment, except where modified for some form of protection. In butterflies, the spots and bands usually have reference to the form of the wing and the arrangement of the nervures; and there is much evidence to show that the primitive markings are always spots in the cells, or between the nervures, or at the junctions of nervures, the extension and coalescence of these spots forming borders, bands, or blotches, which have become modified in infinitely varied ways for protection, warning, or recognition. Even in birds, the distribution of colours and markings follows generally the same law. The crown of the head, the throat, the ear-coverts, and the eyes have usually distinct tints in all highly coloured birds; the region of the furcula has often a distinct patch of colour, as have the pectoral muscles, the uropygium or root of the tail, and the under tail-coverts.[1]

Mr. Tylor was of opinion that the primitive form of ornamentation consisted of spots, the confluence of these in certain directions forming lines or bands; and, these again, sometimes coalescing into blotches, or into more or less uniform tints covering a large portion of the surface of the body. The young lion and tiger are both spotted; and in the Java hog (Sus vittatus) very young animals are banded, but have spots over the shoulders and thighs. These spots run into stripes

[1] *Coloration of Animals*, Pl. X, p. 90; and Pls. II, III, and IV, pp. 30, 40, 42.

as the animal grows older; then the stripes expand, and at last, meeting together, the adult animal becomes of a uniform dark brown colour. So many of the species of deer are spotted when young, that Darwin concludes the ancestral form, from which all deer are derived, must have been spotted. Pigs and tapirs are banded or spotted when young; an imported young specimen of Tapirus Bairdi was covered with white spots in longitudinal rows, here and there forming short stripes.[1] Even the horse, which Darwin supposes to be descended from a striped animal, is often spotted, as in dappled horses; and great numbers show a tendency to spottiness, especially on the haunches.

Ocelli may also be developed from spots, or from bars, as pointed out by Mr. Darwin. Spots are an ordinary form of marking in disease, and these spots sometimes run together, forming blotches. There is evidence that colour markings are in some way dependent on nerve distribution. In the disease known as frontal herpes, an eruption occurs which corresponds exactly to the distribution of the ophthalmic division of the fifth cranial nerve, mapping out all its little branches even to the one which goes to the tip of the nose. In a Hindoo suffering from herpes the pigment was destroyed in the arm along the course of the ulnar nerve, with its branches along both sides of one finger and the half of another. In the leg the sciatic and scaphenous nerves were partly mapped out, giving to the patient the appearance of an anatomical diagram.[2]

These facts are very interesting, because they help to explain the general dependence of marking on structure which has been already pointed out. For, as the nerves everywhere follow the muscles, and these are attached to the various bones, we see how it happens, that the tracts in which distinct developments of colour appear, should so often be marked out by the chief divisions of the bony structure in vertebrates, and by the segments in the annulosa. There is, however, another correspondence of even greater interest and importance. Brilliant colours usually appear just in proportion to the

[1] See coloured Fig. in *Proc. Zool. Soc.*, 1871, p. 626.

[2] A. Tylor's *Coloration*, p. 40; and Photograph in Hutchinson's *Illustrations of Clinical Surgery*, quoted by Tylor.

development of tegumentary appendages. Among birds the most brilliant colours are possessed by those which have developed frills, crests, and elongated tails like the humming-birds; immense tail-coverts like the peacock; enormously expanded wing-feathers, as in the argus-pheasant; or magnificent plumes from the region of the coracoids in many of the birds of paradise. It is to be noted, also, that all these accessory plumes spring from parts of the body which, in other species, are distinguished by patches of colour; so that we may probably impute the development of colour and of accessory plumage to the same fundamental cause.

Among insects, the most brilliant and varied coloration occurs in the butterflies and moths, groups in which the wing-membranes have received their greatest expansion, and whose specialisation has been carried furthest in the marvellous scaly covering which is the seat of the colour. It is suggestive, that the only other group in which functional wings are much coloured is that of the dragonflies, where the membrane is exceedingly expanded. In like manner, the colours of beetles, though greatly inferior to those of the lepidoptera, occur in a group in which the anterior pair of wings has been thickened and modified in order to protect the vital parts, and in which these wing-covers (elytra), in the course of development in the different groups, must have undergone great changes, and have been the seat of very active growth.

The Origin of Accessory Plumes.

Mr. Darwin supposes, that these have in almost every case been developed by the preference of female birds for such males as possessed them in a higher degree than others; but this theory does not account for the fact that these plumes usually appear in a few definite parts of the body. We require some cause to initiate the development in one part rather than in another. Now, the view that colour has arisen over surfaces where muscular and nervous development is considerable, and the fact that it appears especially upon the accessory or highly developed plumes, leads us to inquire whether the same cause has not primarily determined the development of these plumes. The immense tuft of golden plumage in the best known birds of paradise (Paradisea apoda and P. minor) springs from a very small area on the side of the breast. Mr. Frank E. Beddard, who has kindly examined a specimen for me, says that "this area lies upon the pectoral muscles, and near to the point where the fibres of the muscle converge towards their attachment to the humerus. The plumes arise, therefore, close to the most powerful muscle of the body, and near to where the activities of that muscle would be at a maximum. Furthermore, the area of attachment of the plumes is just above the point where the arteries and nerves for the supply of the pectoral muscles, and neighbouring regions, leave the interior of the body. The area of attachment of the plume is, also, as you say in your letter, just above the junction of the coracoid and sternum." Ornamental plumes of considerable size rise from the same part in many other species of paradise birds, sometimes extending laterally in front, so as to form breast shields. They also occur in many humming-birds, and in some sun-birds and honey-suckers; and in all these cases there is a wonderful amount of activity and rapid movement, indicating a surplus of vitality, which is able to manifest itself in the development of these accessory plumes.[1]

In a quite distinct set of birds, the gallinaceæ, we find the ornamental plumage usually arising from very different parts, in the form of elongated tail-feathers or tail-coverts, and of ruffs or hackles from the neck. Here the wings are comparatively little used, the most constant activities depending on the legs, since the gallinaceæ are pre-eminently walking, running, and scratching birds. Now the magnificent train of the peacock—the grandest development of accessory plumes in this order—springs from an oval or circular area, about three inches in diameter, just above the base of the tail, and, therefore, situated over the lower part of the spinal column near the insertion of the powerful muscles which move the hind limbs and elevate the tail. The very frequent presence of neck-ruffs or breast-shields in the males of birds with accessory plumes may be partly due to selection, because they must serve as a protection in their mutual combats, just as does the lion's or the horse's mane. The enormously lengthened plumes of the bird of paradise and of the peacock can, however, have no such use,

[1] For activity and pugnacity of humming-birds, see *Tropical Nature*, pp. 130, 213.

but must be rather injurious than beneficial in the bird's ordinary life. The fact that they have been developed to so great an extent in a few species is an indication of such perfect adàptation to the conditions of existence, such complete success in the battle for life, that there is, in the adult male at all events, a surplus of strength, vitality, and growth-power which is able to expend itself in this way without injury. That such is the case is shown by the great abundance of most of the species which possess these wonderful superfluities of plumage. Birds of paradise are among the commonest birds in New Guinea, and their loud voices can be often heard when the birds themselves are invisible in the depths of the forest; while Indian sportsmen have described the peafowl as being so abundant, that from twelve to fifteen hundred have been seen within an hour at one spot; and they range over the whole country from the Himalayas to Ceylon. Why, in allied species, the development of accessory plumes has taken different forms, we are unable to say, except that it may be due to that individual variability which has served as the starting-point for so much of what seems to us strange in form, or fantastic in colour, both in the animal and vegetable world.

Development of Accessory Plumes and their Display.

If we have found a *vera causa* for the origin of ornamental appendages of birds and other animals in a surplus of vital energy, leading to abnormal growths in those parts of the integument where muscular and nervous action are greatest, the continuous development of these appendages will result from the ordinary action of natural selection in preserving the most healthy and vigorous individuals, and the still further selective agency of sexual struggle in giving to the very strongest and most energetic the parentage of the next generation. And, as all the evidence goes to show that, so far as female birds exercise any choice, it is of "the most vigorous, defiant, and mettlesome male," this form of sexual selection will act in the same direction, and help to carry on the process of plume development to its culmination. That culmination will be reached when the excessive length or abundance of the plumes begins to be injurious to the bearer of them; and it may be this check to the further lengthening of the peacock's train that has led to the broadening of the feathers at the ends, and the consequent production of the magnificent eye-spots which now form its crowning ornament.

The display of these plumes will result from the same causes which led to their production. Just in proportion as the feathers themselves increased in length and abundance, the skin-muscles which serve to elevate them would increase also; and the nervous development as well as the supply of blood to these parts being at a maximum, the erection of the plumes would become a habit at all periods of nervous or sexual excitement. The display of the plumes, like the existence of the plumes themselves, would be the chief external indication of the maturity and vigour of the male, and would, therefore, be necessarily attractive to the female. We have, thus, no reason for imputing to her any of those æsthetic emotions which are excited in us, by the beauty of form, colour, and pattern of these plumes; or the still more improbable æsthetic tastes, which would cause her to choose her mate on account of minute differences in their forms, colours, or patterns.

As co-operating causes in the production of accessory ornamental plumes, I have elsewhere suggested[1] that crests and other erectile feathers may have been useful in making the bird more formidable in appearance, and thus serving to frighten away enemies; while long tail or wing feathers might serve to distract the aim of a bird of prey. But though this might be of some use in the earlier stages of their development, it is probably of little importance compared with the vigour and pugnacity of which the plumes are the indication, and which enable most of their possessors to defend themselves against the enemies which are dangerous to weaker and more timid birds. Even the tiny humming-birds are said to attack birds of prey that approach too near to their nests.

The Effect of Female Preference will be Neutralised by Natural Selection.

The various facts and arguments now briefly set forth, afford an explanation of the phenomena of male ornament,

[1] *Tropical Nature*, p. 209. In Chapter V of this work the views here advocated were first set forth, and the reader is referred there for further details.

as being due to the general laws of growth and development, and make it unnecessary to call to our aid so hypothetical a cause as the cumulative action of female preference. There remains, however, a general argument, arising from the action of natural selection itself, which renders it almost inconceivable that female preference could have been effective in the way suggested; while the same argument strongly supports the view here set forth. Natural selection, as we have seen in our earlier chapters, acts perpetually and on an enormous scale in weeding out the "unfit" at every stage of existence, and preserving only those which are in all respects the very best. Each year, only a small percentage of young birds survive to take the place of the old birds which die; and the survivors will be those which are best able to maintain existence from the egg onwards, an important factor being that their parents should be well able to feed and protect them, while they themselves must in turn be equally able to feed and protect their own offspring. Now this extremely rigid action of natural selection must render any attempt to select mere ornament utterly nugatory, unless the most ornamented always coincide with "the fittest" in every other respect; while, if they do so coincide, then any selection of ornament is altogether superfluous. If the most brightly coloured and fullest plumaged males are *not* the most healthy and vigorous, have *not* the best instincts for the proper construction and concealment of the nest, and for the care and protection of the young, they are certainly not the fittest, and will not survive, or be the parents of survivors. If, on the other hand, there *is* generally this correlation—if, as has been here argued, ornament is the natural product and direct outcome of superabundant health and vigour, then no other mode of selection is needed to account for the presence of such ornament. The action of natural selection does not indeed disprove the existence of female selection of ornament as ornament, but it renders it entirely ineffective; and as the direct evidence for any such female selection is almost *nil*, while the objections to it are certainly weighty, there can be no longer any reason for upholding a theory which was provisionally useful in calling attention to a most curious and suggestive body of facts, but which is now no longer tenable.

The term "sexual selection" must, therefore, be restricted to the direct results of male struggle and combat. This is really a form of natural selection, and is a matter of direct observation; while its results are as clearly deducible as those of any of the other modes in which selection acts. And if this restriction of the term is needful in the case of the higher animals it is much more so with the lower. In butterflies the weeding out by natural selection takes place to an enormous extent in the egg, larva, and pupa states; and perhaps not more than one in a hundred of the eggs laid produces a perfect insect which lives to breed. Here, then, the impotence of female selection, if it exist, must be complete; for, unless the most brilliantly coloured males are those which produce the best protected eggs, larvæ, and pupæ, and unless the particular eggs, larvæ, and pupæ, which are able to survive, are those which produce the most brilliantly coloured butterflies, any choice the female might make must be completely swamped. If, on the other hand, there *is* this correlation between colour development and perfect adaptation to conditions in all stages, then this development will necessarily proceed by the agency of natural selection and the general laws which determine the production of colour and of ornamental appendages.[1]

General Laws of Animal Coloration.

The condensed account which has now been given of the phenomena of colour in the animal world will sufficiently show the wonderful complexity and extreme interest of the subject; while it affords an admirable illustration of the importance of the great principle of utility, and of the effect of the theories of natural selection and development in giving a new interest

[1] The Rev. O. Pickard-Cambridge, who has devoted himself to the study of spiders, has kindly sent me the following extract from a letter, written in 1869, in which he states his views on this question:—

"I myself doubt that particular application of the Darwinian theory which attributes male peculiarities of form, structure, colour, and ornament to female appetency or predilection. There is, it seems to me, undoubtedly something in the male organisation of a special, and sexual nature, which, of its own vital force, develops the remarkable male peculiarities so commonly seen, and of no imaginable use to that sex. In as far as these peculiarities show a great vital power, they point out to us the finest and strongest individuals of the sex, and show us which of them would most certainly appropriate to themselves the best and greatest number of females, and leave behind them the strongest and greatest number of

to the most familiar facts of nature. Much yet remains to be done, both in the observation of new facts as to the relations between the colours of animals and their habits or economy, and, more especially, in the elucidation of the laws of growth which determine changes of colour in the various groups; but so much is already known that we are able, with some confidence, to formulate the general principles which have brought about all the beauty and variety of colour which everywhere delight us in our contemplation of animated nature. A brief statement of these principles will fitly conclude our exposition of the subject.

1. Colour may be looked upon as a necessary result of the highly complex chemical constitution of animal tissues and fluids. The blood, the bile, the bones, the fat, and other tissues have characteristic, and often brilliant colours, which we cannot suppose to have been determined for any special purpose, as colours, since they are usually concealed. The external organs, with their various appendages and integuments, would, by the same general laws, naturally give rise to a greater variety of colour.

2. We find it to be the fact that colour increases in variety and intensity as external structures and dermal appendages become more differentiated and developed. It is on scales, hair, and especially on the more highly specialised feathers, that colour is most varied and beautiful; while among insects colour is most fully developed in those whose wing membranes are most expanded, and, as in the lepidoptera, are clothed with highly specialised scales. Here, too, we find an additional mode of colour production in transparent lamellæ or in fine surface striæ which, by the laws of interference, produce the wonderful metallic hues of so many birds and insects.

progeny. And here would come in, as it appears to me, the proper application of Darwin's theory of Natural Selection; for the possessors of greatest vital power being those most frequently produced and reproduced, the external signs of it would go on developing in an ever-increasing exaggeration, only to be checked where it became really detrimental in some respect or other to the individual."

This passage, giving the independent views of a close observer—one, moreover, who has studied the species of an extensive group of animals both in the field and in the laboratory—very nearly accords with my own conclusions above given; and, so far as the matured opinions of a competent naturalist have any weight, afford them an important support.

3. There are indications of a progressive change of colour, perhaps in some definite order, accompanying the development of tissues or appendages. Thus spots spread and fuse into bands, and when a lateral or centrifugal expansion has occurred—as in the termination of the peacocks' train feathers, the outer web of the secondary quills of the Argus pheasant, or the broad and rounded wings of many butterflies—into variously shaded or coloured ocelli. The fact that we find gradations of colour in many of the more extensive groups, from comparatively dull or simple to brilliant and varied hues, is an indication of some such law of development, due probably to progressive local segregation in the tissues of identical chemical or organic molecules, and dependent on laws of growth yet to be investigated.

4. The colours thus produced, and subject to much individual variation, have been modified in innumerable ways for the benefit of each species. The most general modification has been in such directions as to favour concealment when at rest in the usual surroundings of the species, sometimes carried on by successive steps till it has resulted in the most minute imitation of some inanimate object or exact mimicry of some other animal. In other cases bright colours or striking contrasts have been preserved, to serve as a warning of inedibility or of dangerous powers of attack. Most frequent of all has been the specialisation of each distinct form by some tint or marking for purposes of easy recognition, especially in the case of gregarious animals whose safety largely depends upon association and mutual defence.

5. As a general rule the colours of the two sexes are alike; but in the higher animals there appears a tendency to deeper or more intense colouring in the male, due probably to his greater vigour and excitability. In many groups in which this superabundant vitality is at a maximum, the development of dermal appendages and brilliant colours has gone on increasing till it has resulted in a great diversity between the sexes; and in most of these cases there is evidence to show that natural selection has caused the female to retain the primitive and more sober colours of the group for purposes of protection.

Concluding Remarks.

The general principles of colour development now sketched out enable us to give some rational explanation of the wonderful amount of brilliant colour which occurs among tropical animals. Looking on colour as a normal product of organisation, which has either been allowed free play, or has been checked and modified for the benefit of the species, we can see at once that the luxuriant and perennial vegetation of the tropics, by affording much more constant means of concealment, has rendered brilliant colour less hurtful there than in the temperate and colder regions. Again, this perennial vegetation supplies abundance of both vegetable and insect food throughout the year, and thus a greater abundance and greater variety of the forms of life are rendered possible, than where recurrent seasons of cold and scarcity reduce the possibilities of life to a minimum. Geology furnishes us with another reason, in the fact, that throughout the tertiary period tropical conditions prevailed far into the temperate regions, so that the possibilities of colour development were still greater than they are at the present time. The tropics, therefore, present to us the results of animal development in a much larger area and under more favourable conditions than prevail to-day. We see in them samples of the productions of an earlier and a better world, from an animal point of view; and this probably gives a greater variety and a finer display of colour than would have been produced, had conditions always been what they are now. The temperate zones, on the other hand, have recently suffered the effects of a glacial period of extreme severity, with the result that almost the only gay coloured birds they now possess are summer visitors from tropical or sub-tropical lands. It is to the unbroken and almost unchecked course of development from remote geological times that has prevailed in the tropics, favoured by abundant food and perennial shelter, that we owe such superb developments as the frills and crests and jewelled shields of the humming-birds, the golden plumes of the birds of paradise, and the resplendent train of the peacock. This last exhibits to us the culmination of that marvel and mystery of animal colour which is so well expressed by a poet-artist in the following lines. The marvel will ever remain to the sympathetic student of nature, but I venture to hope that in the preceding chapters I have succeeded in lifting—if only by one of its corners—the veil of mystery which has for long shrouded this department of nature.

On a Peacock's Feather.

In Nature's workshop but a shaving,
  Of her poem but a word,
But a tint brushed from her palette,
  This feather of a bird!
Yet set it in the sun glance,
  Display it in the shine,
Take graver's lens, explore it,
  Note filament and line,
Mark amethyst to sapphire,
  And sapphire to gold,
And gold to emerald changing
  The archetype unfold!
Tone, tint, thread, tissue, texture,
  Through every atom scan,
Conforming still, developing,
  Obedient to plan.
This but to form a pattern
  On the garment of a bird!
What then must be the poem,
  This but its lightest word!
Sit before it; ponder o'er it,
  'Twill thy mind advantage more,
Than a treatise, than a sermon,
  Than a library of lore.

CHAPTER XI

THE SPECIAL COLOURS OF PLANTS: THEIR ORIGIN AND PURPOSE

The general colour relations of plants—Colours of fruits—The meaning of nuts—Edible or attractive fruits—The colours of flowers—Modes of securing cross-fertilisation—The interpretation of the facts—Summary of additional facts bearing on insect fertilisation—Fertilisation of flowers by birds—Self-fertilisation of flowers—Difficulties and contradictions—Intercrossing not necessarily advantageous—Supposed evil results of close interbreeding—How the struggle for existence acts among flowers—Flowers the product of insect agency—Concluding remarks on colour in nature.

THE colours of plants are both less definite and less complex than are those of animals, and their interpretation on the principle of utility is, on the whole, more direct and more easy. Yet here, too, we find that in our investigation of the uses of the various colours of fruits and flowers, we are introduced to some of the most obscure recesses of nature's workshop, and are confronted with problems of the deepest interest and of the utmost complexity.

So much has been written on this interesting subject since Mr. Darwin first called attention to it, and its main facts have become so generally known by means of lectures, articles, and popular books, that I shall give here a mere outline sketch, for the purpose of leading up to a discussion of some of the more fundamental problems which arise out of the facts, and which have hitherto received less attention than they deserve.

The General Colour Relations of Plants.

The green colour of the foliage of leafy plants is due to the existence of a substance called chlorophyll, which is almost universally developed in the leaves under the action of light. It is subject to definite chemical changes during the processes of growth and of decay, and it is owing to these changes that we have the delicate tints of spring foliage, and the more varied, intense, and gorgeous hues of autumn. But these all belong to the class of intrinsic or normal colours, due to the chemical constitution of the organism; as colours they are unadaptive, and appear to have no more relation to the wellbeing of the plants themselves than do the colours of gems and minerals. We may also include in the same category those algæ and fungi which have bright colours—the "red snow" of the arctic regions, the red, green, or purple seaweeds, the brilliant scarlet, yellow, white, or black agarics, and other fungi. All these colours are probably the direct results of chemical composition or molecular structure, and, being thus normal products of the vegetable organism, need no special explanation from our present point of view; and the same remark will apply to the varied tints of the bark of trunks, branches, and twigs, which are often of various shades of brown and green, or even vivid reds or yellows.

There are, however, a few cases in which the need of protection, which we have found to be so important an agency in modifying the colours of animals, has also determined those of some of the smaller members of the vegetable kingdom. Dr. Burchell found a mesembryanthemum in South Africa like a curiously shaped pebble, closely resembling the stones among which it grew;[1] and Mr. J. P. Mansel Weale states that in the same country one of the Asclepiadeæ has tubers growing above ground among stones which they exactly resemble, and that, when not in leaf, they are for this reason quite invisible.[2] It is clear that such resemblances must be highly useful to these plants, inhabiting an arid country abounding in herbivorous mammalia, which,

[1] Burchell's *Travels*, vol. i. p. 10.
[2] *Nature*, vol. iii. p. 507.

in times of drought or scarcity, will devour everything in the shape of a fleshy stem or tuber.

True mimicry is very rare in plants, though adaptation to like conditions often produces in foliage and habit a similarity that is deceiving. Euphorbias growing in deserts often closely resemble cacti. Seaside plants and high alpine plants of different orders are often much alike; and innumerable resemblances of this kind are recorded in the names of plants, as Veronica epacridea (the veronica like an epacris), Limnanthemum nymphæoides (the limnanthemum like a nymphæa), the resembling species in each case belonging to totally distinct families. But in these cases, and in most others that have been observed, the essential features of true mimicry are absent, inasmuch as the one plant cannot be supposed to derive any benefit from its close resemblance to the other, and this is still more certain from the fact that the two species usually inhabit different localities. A few cases exist, however, in which there does seem to be the necessary accordance and utility. Mr. Mansel Weale mentions a labiate plant (Ajuga ophrydis), the only species of the genus Ajuga in South Africa, which is strikingly like an orchid of the same country; while a balsam (Impatiens capensis), also a solitary species of the genus in that country, is equally like an orchid, growing in the same locality and visited by the same insects. As both these genera of plants are specialised for insect fertilisation, and both of the plants in question are isolated species of their respective genera, we may suppose that, when they first reached South Africa they were neglected by the insects of the country; but, being both remotely like orchids in form of flower, those varieties that approached nearest to the familiar species of the country were visited by insects and cross-fertilised, and thus a closer resemblance would at length be brought about. Another case of close general resemblance, is that of our common white dead-nettle (Lamium album) to the stinging-nettle (Urtica dioica); and Sir John Lubbock thinks that this is a case of true mimicry, the dead-nettle being benefited by being mistaken by grazing animals for the stinging-nettle.[1]

[1] *Flowers, Fruits, and Leaves*, p. 128 (Fig. 79).

Colours of Fruits.

It is when we come to the essential parts of plants on which their perpetuation and distribution depends, that we find colour largely utilised for a distinct purpose in flowers and fruits. In the former we find attractive colours and guiding marks to secure cross-fertilisation by insects; in the latter attractive or protective coloration, the first to attract birds or other animals when the fruits are intended to be eaten, the second to enable them to escape being eaten when it would be injurious to the species. The colour phenomena of fruits being much the most simple will be considered first.

The perpetuation and therefore the very existence of each species of flowering plant depend upon its seeds being preserved from destruction and more or less effectually dispersed over a considerable area. The dispersal is effected either mechanically or by the agency of animals. Mechanical dispersal is chiefly by means of air-currents, and large numbers of seeds are specially adapted to be so carried, either by being clothed with down or pappus, as in the well-known thistle and dandelion seeds; by having wings or other appendages, as in the sycamore, birch, and many other trees; by being thrown to a considerable distance by the splitting of the seed-vessel, and by many other curious devices.[1] Very large numbers of seeds, however, are so small and light that they can be carried enormous distances by gales of wind, more especially as most of this kind are flattened or curved, so as to expose a large surface in proportion to their weight. Those which are carried by animals have their surfaces, or that of the seed-vessel, armed with minute hooks, or some prickly covering which attaches itself to the hair of mammalia or the feathers of birds, as in the burdock, cleavers, and many other species. Others again are sticky, as in Plumbago europæa, mistletoe, and many foreign plants.

All the seeds or seed-vessels which are adapted to be dispersed in any of these ways are of dull protective tints, so that when they fall on the ground they are almost indistinguishable; besides which, they are usually small, hard, and

[1] For a popular sketch of these, see Sir J. Lubbock's *Flowers, Fruits, and Leaves*, or any general botanical work.

altogether unattractive, never having any soft, juicy pulp; while the edible seeds often bear such a small proportion to the hard, dry envelopes or appendages, that few animals would care to eat them.

The Meaning of Nuts.

There is, however, another class of fruits or seeds, usually termed nuts, in which there is a large amount of edible matter, often very agreeable to the taste, and especially attractive and nourishing to a large number of animals. But when eaten, the seed is destroyed and the existence of the species endangered. It is evident, therefore, that it is by a kind of accident that these nuts are eatable; and that they are not intended to be eaten is shown by the special care nature seems to have taken to conceal or to protect them. We see that all our common nuts are green when on the tree, so as not easily to be distinguished from the leaves; but when ripe they turn brown, so that when they fall on to the ground they are equally indistinguishable among the dead leaves and twigs, or on the brown earth. Then they are almost always protected by hard coverings, as in hazel-nuts, which are concealed by the enlarged leafy involucre, and in the large tropical brazil-nuts and cocoa-nuts by such a hard and tough case as to be safe from almost every animal. Others have an external bitter rind, as in the walnut; while in the chestnuts and beechnuts two or three fruits are enclosed in a prickly involucre.

Notwithstanding all these precautions, nuts are largely devoured by mammalia and birds; but as they are chiefly the product of trees or shrubs of considerable longevity, and are generally produced in great profusion, the perpetuation of the species is not endangered. In some cases the devourers of nuts may aid in their dispersal, as they probably now and then swallow the seed whole, or not sufficiently crushed to prevent germination; while squirrels have been observed to bury nuts, many of which are forgotten and afterwards grow in places they could not have otherwise reached.[1] Nuts, especially the larger kinds which are so well protected by their hard, nearly globular cases, have their dispersal facilitated by rolling down hill, and more especially

[1] *Nature*, vol. xv. p. 117.

by floating in rivers and lakes, and thus reaching other localities. During the elevation of land areas this method would be very effective, as the new land would always be at a lower level than that already covered with vegetation, and therefore in the best position for being stocked with plants from it.

The other modes of dispersal of seeds are so clearly adapted to their special wants, that we feel sure they must have been acquired by the process of variation and natural selection. The hooked and sticky seeds are always those of such herbaceous plants as are likely, from their size, to come in contact with the wool of sheep or the hair of cattle; while seeds of this kind never occur on forest trees, on aquatic plants, or even on very dwarf creepers or trailers. The winged seed-vessels or seeds, on the other hand, mostly belong to trees and to tall shrubs or climbers. We have, therefore, a very exact adaptation to conditions in these different modes of dispersal; while, when we come to consider individual cases, we find innumerable other adaptations, some of which the reader will find described in the little work by Sir John Lubbock already referred to.

Edible or Attractive Fruits.

It is, however, when we come to true fruits (in a popular sense) that we find varied colours evidently intended to attract animals, in order that the fruits may be eaten, while the seeds pass through the body undigested and are then in the fittest state for germination. This end has been gained in a great variety of ways, and with so many corresponding adaptations as to leave no doubt as to the value of the result. Fruits are pulpy or juicy, and usually sweet, and form the favourite food of innumerable birds and some mammals. They are always coloured so as to contrast with the foliage or surroundings, red being the most common as it is certainly the most conspicuous colour, but yellow, purple, black, or white being not uncommon. The edible portion of fruits is developed from different parts of the floral envelopes, or of the ovary, in the various orders and genera. Sometimes the calyx becomes enlarged and fleshy, as in the apple and pear tribe; more often the integuments of the ovary itself are enlarged, as in the plum, peach, grape, etc.; the receptacle is enlarged and forms the fruit of the strawberry; while the mulberry, pine-apple, and fig are examples of compound fruits formed in various ways from a dense mass of flowers.

In all cases the seeds themselves are protected from injury by various devices. They are small and hard in the straw-berry, raspberry, currant, etc., and are readily swallowed among the copious pulp. In the grape they are hard and bitter; in the rose (hip) disagreeably hairy; in the orange tribe very bitter; and all these have a smooth, glutinous exterior which facilitates their being swallowed. When the seeds are larger and are eatable, they are enclosed in an excessively hard and thick covering, as in the various kinds of "stone" fruit (plums, peaches, etc.), or in a very tough core, as in the apple. In the nutmeg of the Eastern Archipelago we have a curious adaptation to a single group of birds. The fruit is yellow, somewhat like an oval peach, but firm and hardly eatable. This splits open and shows the glossy black covering of the seed or nutmeg, over which spreads the bright scarlet arillus or "mace," an adventitious growth of no use to the plant except to attract attention. Large fruit pigeons pluck out this seed and swallow it entire for the sake of the mace, while the large nutmeg passes through their bodies and germinates; and this has led to the wide distribution of wild nutmegs over New Guinea and the surrounding islands.

In the restriction of bright colour to those edible fruits the eating of which is beneficial to the plant, we see the undoubted result of natural selection; and this is the more evident when we find that the colour never appears till the fruit is ripe—that is, till the seeds within it are fully matured and in the best state for germination. Some brilliantly coloured fruits are poisonous, as in our bitter-sweet (Solanum dulcamara), cuckoo-pint (Arum) and the West Indian manchineel. Many of these are, no doubt, eaten by animals to whom they are harmless; and it has been suggested that even if some animals are poisoned by them the plant is benefited, since it not only gets dispersed, but finds, in the decaying body of its victim, a rich manure heap.[1] The particular colours of fruits are not, so far as we know, of any use to them other

[1] Grant Allen's *Colour Sense*, p. 113.

than as regards conspicuousness, hence a tendency to *any* decided colour has been preserved and accumulated as serving to render the fruit easily visible among its surroundings of leaves or herbage. Out of 134 fruit-bearing plants in Mongredien's *Trees and Shrubs*, and Hooker's *British Flora*, the fruits of no less than sixty-eight, or rather more than half, are red, forty-five are black, fourteen yellow, and seven white. The great prevalence of red fruits is almost certainly due to their greater conspicuousness having favoured their dispersal, though it may also have arisen in part from the chemical changes of chlorophyll during ripening and decay producing red tints as in many fading leaves. Yet the comparative scarcity of yellow in fruits, while it is the most common tint of fading leaves, is against this supposition.

There are, however, a few instances of coloured fruits which do not seem to be intended to be eaten; such are the colocynth plant (Cucumis colocynthus), which has a beautiful fruit the size and colour of an orange, but nauseous beyond description to the taste. It has a hard rind, and may perhaps be dispersed by being blown along the ground, the colour being an adventitious product; but it is quite possible, notwithstanding its repulsiveness to us, that it may be eaten by some animals. With regard to the fruit of another plant, Calotropis procera, there is less doubt, as it is dry and full of thin, flat-winged seeds, with fine silky filaments, eminently adapted for wind-dispersal; yet it is of a bright yellow colour, as large as an apple, and therefore very conspicuous. Here, therefore, we seem to have colour which is a mere by-product of the organism and of no use to it; but such cases are exceedingly rare, and this rarity, when compared with the great abundance of cases in which there is an obvious purpose in the colour, adds weight to the evidence in favour of the theory of the attractive coloration of edible fruits in order that birds and other animals may assist in their dispersal. Both the above-named plants are natives of Palestine and the adjacent arid countries.[1]

The Colours of Flowers.

Flowers are much more varied in their colours than fruits,

[1] Canon Tristram's *Natural History of the Bible*, pp. 483, 484.

as they are more complex and more varied in form and structure; yet there is some parallelism between them in both respects. Flowers are frequently adapted to attract insects as fruits are to attract birds, the object being in the former to secure cross-fertilisation, in the latter dispersal; while just as colour is an index of the edibility of fruits which supply pulp or juice to birds, so are the colours of flowers an indication of the presence of nectar or of pollen which are devoured by insects.

The main facts and many of the details, as to the relation of insects to flowers, were discovered by Sprengel in 1793. He noticed the curious adaptation of the structure of many flowers to the particular insects which visit them; he proved that insects do cross-fertilise flowers, and he believed that this was the object of the adaptations, while the presence of nectar and pollen ensured the continuance of their visits; yet he missed discovering the *use* of this cross-fertilisation. Several writers at a later period obtained evidence that cross-fertilisation of plants was a benefit to them; but the wide generality of this fact and its intimate connection with the numerous and curious adaptations discovered by Sprengel, was first shown by Mr. Darwin, and has since been demonstrated by a vast mass of observations, foremost among which are his own researches on orchids, primulas, and other plants.[1]

By an elaborate series of experiments carried on for many years Mr. Darwin demonstrated the great value of cross-fertilisation in increasing the rapidity of growth, the strength and vigour of the plant, and in adding to its fertility. This effect is produced immediately, not as he expected would be the case, after several generations of crosses. He planted seeds from cross-fertilised and self-fertilised plants on two sides of the same pot exposed to exactly similar conditions, and in most cases the difference in size and vigour was amazing, while the plants from cross-fertilised parents also produced more and finer seeds. These experiments entirely confirmed the experience of breeders of animals already referred to (p. 160), and led him to enunciate his famous aphorism,

[1] For a complete historical account of this subject with full references to all the works upon it, see the Introduction to Hermann Müller's *Fertilisation of Flowers*, translated by D'Arcy W. Thompson.

"Nature abhors perpetual self-fertilisation.[1] In this principle we appear to have a sufficient reason for the various contrivances by which so many flowers secure cross-fertilisation, either constantly or occasionally. These contrivances are so numerous, so varied, and often so highly complex and extraordinary, that they have formed the subject of many elaborate treatises, and have also been amply popularised in lectures and handbooks. It will be unnecessary, therefore, to give details here, but the main facts will be summarised in order to call attention to some difficulties of the theory which seem to require further elucidation.

Modes of securing Cross-Fertilisation.

When we examine the various modes in which the cross-fertilisation of flowers is brought about, we find that some are comparatively simple in their operation and needful adjustments, others highly complex. The simple methods belong to four principal classes:—(1) By dichogamy—that is, by the anthers and the stigma becoming mature or in a fit state for fertilisation at slightly different times on the same plant. The result of this is that, as plants in different stations, on different soils, or exposed to different aspects flower earlier or later, the mature pollen of one plant can only fertilise some plant exposed to somewhat different conditions or of different constitution, whose stigma will be mature at the same time; and this difference has been shown by Darwin to be that which is adapted to secure the fullest benefit of cross-fertilisation. This occurs in Geranium pratense, Thymus serpyllum, Arum maculatum, and many others. (2) By the flower being self-sterile with its own pollen, as in the crimson flax. This absolutely prevents self-fertilisation. (3) By the stamens and anthers being so placed that the pollen cannot fall upon the stigma, while it does fall upon a visiting insect which carries it to the stigma of another flower. This effect is produced in a variety of very simple ways, and is often aided by the motion of the stamens which bend down out of the way of the stigmas before the pollen is ripe, as in Malva sylvestris (see Fig. 28). (4) By the male and female flowers being on

[1] For the full detail of his experiments, see *Cross- and Self-Fertilisation of Plants*, 1876.

different plants, forming the class Diœcia of Linnaeus. In these cases the pollen may be carried to the stigmas either by the wind or by the agency of insects.

Now these four methods are all apparently very simple, and easily produced by variation and selection. They are applicable to flowers of any shape, requiring only such size and colour as to attract insects, and some secretion of nectar to ensure their repeated visits, characters common to the great majority of flowers. All these methods are common, except perhaps the second; but there are many flowers in which the pollen from another plant is prepotent over the pollen from the same flower, and this has nearly the same effect as self-sterility if the flowers are frequently crossed by insects. We cannot help asking, therefore, why have other and much more elaborate methods been needed? And how have the more complex arrangements of so many flowers been brought about? Before attempting to answer these questions, and in order that the reader may appreciate the difficulty of the problem and the nature of the facts to be explained, it will be necessary to give a summary of the more elaborate modes of securing cross-fertilisation.

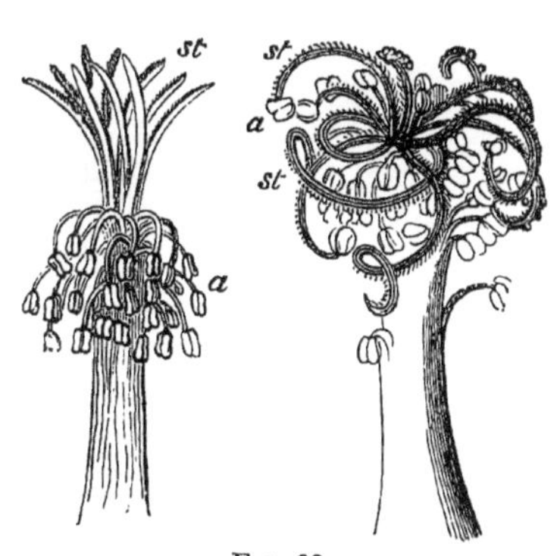

FIG. 28.
Malva sylvestris, adapted for insect-fertilisation. Malva rotundifolia, adapted for self-fertilisation.

(1) We first have dimorphism and heteromorphism, the phenomena of which have been already sketched in our seventh chapter.

Here we have both a mechanical and a physiological modification, the stamens and pistil being variously modified in length and position, while the different stamens in the same flower have widely different degrees of fertility when applied to the same stigma,—a phenomenon which, if it were not so well established, would have appeared in the highest degree improbable. The most remarkable case is that of the three different forms of the loosestrife (Lythrum salicaria) here figured (Fig. 29 on next page).

(2) Some flowers have irritable stamens which, when their

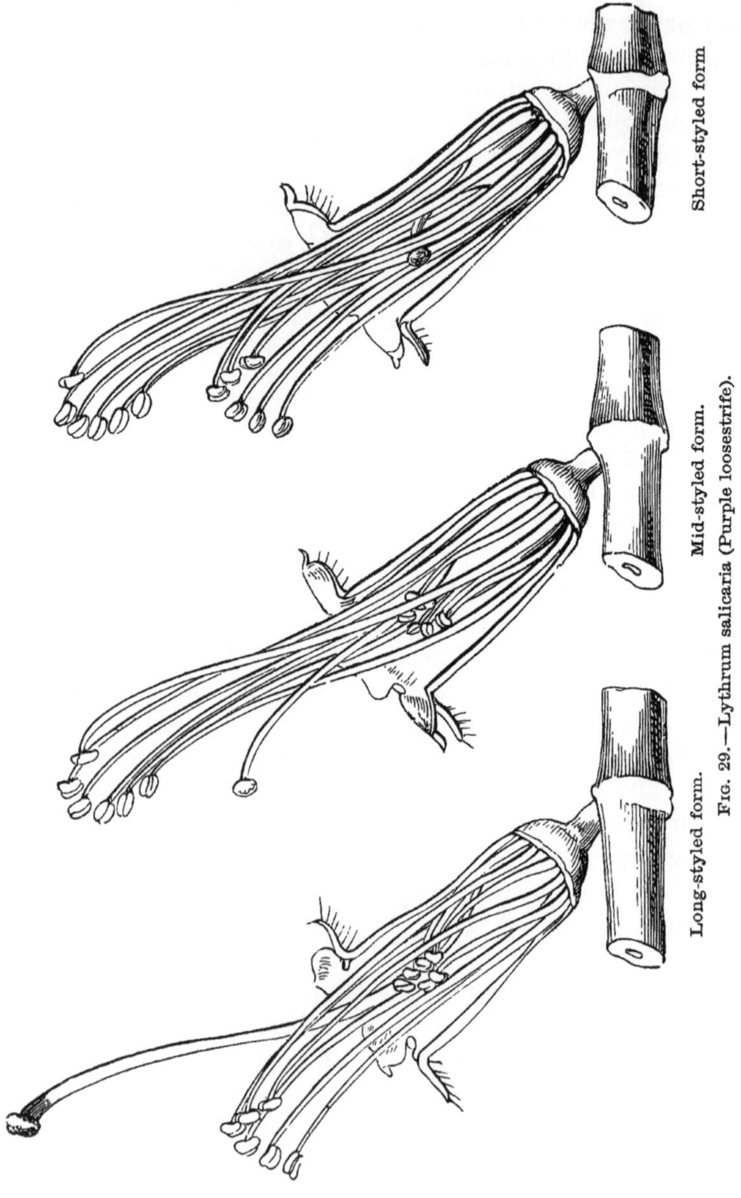

Fig. 29.—Lythrum salicaria (Purple loosestrife).

bases are touched by an insect, spring up and dust it with pollen. This occurs in our common berberry.

(3) In others there are levers or processes by which the anthers are mechanically brought down on to the head or back of an insect entering the flower, in such a position as to be carried to the stigma of the next flower it visits. This may be well seen in many species of Salvia and Erica.

(4) In some there is a sticky secretion which, getting on to the proboscis of an insect, carries away the pollen, and applies it to the stigma of another flower. This occurs in our common milkwort (Polygala vulgaris).

(5) In papilionaceous plants there are many complex adjustments, such as the squeezing out of pollen from a receptacle on to an insect, as in Lotus corniculatus, or the sudden springing out and exploding of the anthers so as thoroughly to dust the insect, as in Medicago falcata, this occurring after the stigma has touched the insect and taken off some pollen from the last flower.

(6) Some flowers or spathes form closed boxes in which insects find themselves entrapped, and when they have fertilised the flower, the fringe of hairs opens and allows them to escape. This occurs in many species of Arum and Aristolochia.

(7) Still more remarkable are the traps in the flower of Asclepias which catch flies, butterflies, and wasps by the legs, and the wonderfully complex arrangements of the orchids. One of these, our common Orchis pyramidalis, may be briefly described to show how varied and beautiful are the arrangements to secure cross-fertilisation. The broad trifid lip of the flower offers a support to the moth which is attracted by its sweet odour, and two ridges at the base guide the proboscis with certainty to the narrow entrance of the nectary. When the proboscis has reached the end of the spur, its basal portion depresses the little hinged rostellum that covers the saddle-shaped sticky glands to which the pollen masses (pollinia) are attached. On the proboscis being withdrawn, the two pollinia stand erect and parallel, firmly attached to the proboscis. In this position, however, they would be useless, as they would miss the stigmatic surface of the next flower visited by the moth. But as soon as the proboscis is withdrawn, the two pollen masses begin to diverge till they are exactly as far apart as are the stigmas of the flower; and then commences a second move-

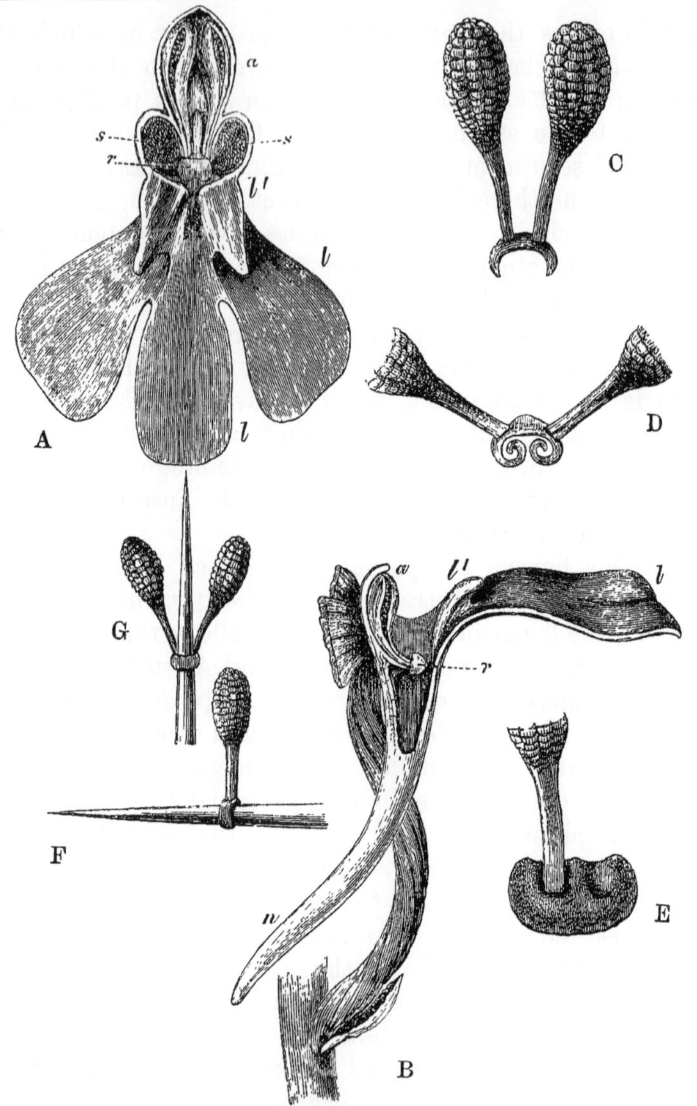

FIG. 30.—Orchis pyramidalis.

DESCRIPTION OF FIGURE.

a . . . anther.	*r* . . rostellum	*l'* . guiding ridges on labellum.
s,s . . . stigma.	*l* . . labellum or lip.	*n* . nectary.

A. Front view, with all the sepals and petals removed, except the labellum.
B. Side view, with all the sepals and petals removed and the upper part of the flower [bisected.
C. The two pollinia attached to the saddle-shaped viscid disc.
D. The disc after the first act of contraction.
E. The disc seen from above with one pollinium removed.
F. The pollinia removed by the insertion of a needle into the nectary.
G. The same pollinia after depression has taken place.

ment which brings them down till they project straight forward nearly at right angles to their first position, so as exactly to hit against the stigmatic surfaces of the next flower visited on which they leave a portion of their pollen. The whole of these motions take about half a minute, and in that time the moth will usually have flown to another plant, and thus effect the most beneficial kind of cross-fertilisation.[1] This description will be better understood by referring to the illustration opposite, from Darwin's *Fertilisation of Orchids* (Fig. 30).

The Interpretation of these Facts.

Having thus briefly indicated the general character of the more complex adaptations for cross-fertilisation, the details of which are to be found in any of the numerous works on the subject,[2] we find ourselves confronted with the very puzzling question—Why were these innumerable highly complex adaptations produced, when the very same result may be effected—and often is effected—by extremely simple means? Supposing, as we must do, that all flowers were once of simple and regular forms, like a buttercup or a rose, how did such irregular and often complicated flowers as the papilionaceous or pea family, the labiates or sage family, and the infinitely varied and fantastic orchids ever come into existence? No cause has yet been suggested but the need of attracting insects to cross-fertilise them; yet the attractiveness of regular flowers with bright colours and an ample supply of nectar is equally great, and cross-fertilisation can be quite as effectively secured in these by any of the four simple methods already described. Before attempting to suggest a possible solution of this difficult problem, we have yet to pass in review a large body of curious adaptations connected with insect fertilisation, and will first call attention to that portion of the phenomena which throw some light upon the special colours of flowers in their relation to the various kinds of insects which visit them. For these facts we are largely in-

[1] See Darwin's *Fertilisation of Orchids* for the many extraordinary and complex arrangements in these plants.

[2] The English reader may consult Sir John Lubbock's *British Wild Flowers in Relation to Insects*, and H. Müller's great and original work, *The Fertilisation of Flowers*.

debted to the exact and long-continued researches of Professor Hermann Müller.

Summary of Additional Facts bearing on Insect Fertilisation.

1. That the size and colour of a flower are important factors in determining the visits of insects, is shown by the general fact of more insects visiting conspicuous than inconspicuous flowers. As a single instance, the handsome Geranium palustre was observed by Professor Müller to be visited by sixteen different species of insects, the equally showy G. pratense by thirteen species, while the smaller and much less conspicuous G. molle was visited by eight species, and G. pusillum by only one. In many cases, however, a flower may be very attractive to only a few species of insects; and Professor Müller states, as the result of many years' assiduous observation, that "a species of flower is the more visited by insects the more conspicuous it is."

2. Sweet odour is usually supplementary to the attraction of colour. Thus it is rarely present in the largest and most gaudily coloured flowers which inhabit open places, such as poppies, pæonies, sunflowers, and many others; while it is often the accompaniment of inconspicuous flowers, as the mignonette; of such as grow in shady places, as the violet and primrose; and especially of white or yellowish flowers, as the white jasmine, clematis, stephanotis, etc.

3. White flowers are often fertilised by moths, and very frequently give out their scent only by night, as in our butterfly-orchis (Habenaria chlorantha); and they sometimes open only at night, as do many of the evening primroses and other flowers. These flowers are often long tubed in accordance with the length of the moths' probosces, as in the genus Pancratium, our butterfly orchis, white jasmine, and a host of others.

4. Bright red flowers are very attractive to butterflies, and are sometimes specially adapted to be fertilised by them, as in many pinks (Dianthus deltoides, D. superbus, D. atrorubens), the corn-cockle (Lychnis Githago), and many others. Blue flowers are especially attractive to bees and other hymenoptera (though they frequent flowers of all colours), no less than sixty-seven species of this order having been observed to visit the common "sheep's-bit" (Jasione montana). Dull yellow or

brownish flowers, some of which smell like carrion, are attractive to flies, as the Arum and Aristolochia; while the dull purplish flowers of the Scrophularia are specially attractive to wasps.

5. Some flowers have neither scent nor nectar, and yet attract insects by sham nectaries! In the herb-paris (Paris quadrifolia) the ovary glistens as if moist, and flies alight on it and carry away pollen to another flower; while in grass of parnassus (Parnassia palustris) there are a number of small stalked yellow balls near the base of the flower, which look like drops of honey but are really dry. In this case there is a little nectar lower down, but the special attraction is a sham; and as there are fresh broods of insects every year, it takes time for them to learn by experience, and thus enough are always deceived to effect cross-fertilisation.[1] This is analogous to the case of the young birds, which have to learn by experience the insects that are inedible, as explained at page 253.

6. Many flowers change their colour as soon as fertilised; and this is beneficial, as it enables bees to avoid wasting time in visiting those blossoms which have been already fertilised and their nectar exhausted. The common lungwort (Pulmonaria officinalis), is at first red, but later turns blue; and H. Müller observed bees visiting many red flowers in succession, but neglecting the blue. In South Brazil there is a species of Lantana, whose flowers are yellow the first day, orange the second, and purple the third; and Dr. Fritz Müller observed that many butterflies visited the yellow flowers only, some both the yellow and the orange flowers, but none the purple.

7. Many flowers have markings which serve as guides to insects; in some cases a bright central eye, as in the borage and forget-me-not; or lines or spots converging to the centre, as in geraniums, pinks, and many others. This enables insects to go quickly and directly to the opening of the flower, and is equally important in aiding them to obtain a better supply of food, and to fertilise a larger number of flowers.

8. Flowers have been specially adapted to the kinds of

[1] Müller's *Fertilisation of Flowers*, p. 248.

insects that most abound where they grow. Thus the gentians of the lowlands are adapted to bees, those of the high alps to butterflies only; and while most species of Rhinanthus (a genus to which our common "yellow rattle" belongs) are bee-flowers, one high alpine species (R. alpinus) has been also adapted for fertilisation by butterflies only. The reason of this is, that in the high alps butterflies are immensely more plentiful than bees, and flowers adapted to be fertilised by bees can often have their nectar extracted by butterflies without effecting cross-fertilisation. It is, therefore, important to have a modification of structure which shall make butterflies the fertilisers, and this in many cases has been done.[1]

9. Economy of time is very important both to the insects and the flowers, because the fine working days are comparatively few, and if no time is wasted the bees will get more honey, and in doing so will fertilise more flowers. Now, it has been ascertained by several observers that many insects, bees especially, keep to one kind of flower at a time, visiting hundreds of blossoms in succession, and passing over other species that may be mixed with them. They thus acquire quickness in going at once to the nectar, and the change of colour in the flower, or incipient withering when fertilised, enables them to avoid those flowers that have already had their honey exhausted. It is probably to assist the insects in keeping to one flower at a time, which is of vital importance to the perpetuation of the species, that the flowers which bloom intermingled at the same season are usually very distinct both in form and colour. In the sandy districts of Surrey, in the early spring, the copses are gay with three flowers—the primrose, the wood-anemone, and the lesser celandine, forming a beautiful contrast, while at the same time the purple and the white dead-nettles abound on hedge banks. A little later, in the same copses, we have the blue wild hyacinth (Scilla nutans), the red campion (Lychnis dioica), the pure white great starwort (Stellaria Holosteum), and the yellow dead-nettle (Lamium Galeobdolon), all distinct and well-contrasted flowers. In damp meadows in summer we have the ragged robin (Lychnis Floscuculi), the spotted orchis (O. maculata), and the yellow rattle (Rhinanthus.

[1] "Alpenblumen," by D. H. Müller. See *Nature*, vol. xxiii. p. 333.

Crista-galli); while in drier meadows we have cowslips, ox-eye daisies, and buttercups, all very distinct both in form and colour. So in cornfields we have the scarlet poppies, the purple corn-cockle, the yellow corn-marygold, and the blue cornflower; while on our moors the purple heath and the dwarf gorse make a gorgeous contrast. Thus the difference of colour which enables the insect to visit with rapidity and unerring aim a number of flowers of the same kind in succession, serves to adorn our meadows, banks, woods, and heaths with a charming variety of floral colour and form at each season of the year.[1]

Fertilisation of Flowers by Birds.

In the temperate regions of the Northern Hemisphere, insects are the chief agents in cross-fertilisation when this is not effected by the wind; but in warmer regions, and in the Southern hemisphere, birds are found to take a considerable part in the operation, and have in many cases led to modifications in the form and colour of flowers. Each part of the globe has special groups of birds which are flower-haunters. America has the humming-birds (Trochilidæ), and the smaller group of the sugar-birds (Cærebidæ). In the Eastern tropics the sun-birds (Nectarineidæ) take the place of the humming-birds, and another small group, the flower-peckers (Dicæidæ), assist them. In the Australian region there are also two flower-feeding groups, the Meliphagidæ, or honey-suckers, and the brush-tongued lories (Trichoglossidæ). Recent researches by American naturalists have shown that many flowers are fertilised by humming-birds, such as passion-flowers, trumpet-flowers, fuchsias, and lobelias; while some, as the Salvia splendens of Mexico, are specially adapted to their visits. We may thus perhaps explain the number of very large tubular flowers in the tropics, such as the huge brugmansias and bignonias; while in the Andes and in

[1] This peculiarity of local distribution of colour in flowers may be compared, as regards its purpose, with the recognition colours of animals. Just as these latter colours enable the sexes to recognise each other, and thus avoid sterile unions of distinct species, so the distinctive form and colour of each species of flower, as compared with those that usually grow around it, enables the fertilising insects to avoid carrying the pollen of one flower to the stigma of a distinct species.

Chile, where humming-birds are especially plentiful, we find great numbers of red tubular flowers, often of large size and apparently adapted to these little creatures. Such are the beautiful Lapageria and Philesia, the grand Pitcairneas, and the genera Fuchsia, Mitraria, Embothrium, Escallonia, Desfontainea, Eccremocarpus, and many Gesneraceæ. Among the most extraordinary modifications of flower structure adapted

FIG. 31.—Humming-bird fertilising Marcgravia nepenthoides.

to bird fertilisation are the species of Marcgravia, in which the pedicels and bracts of the terminal portion of a pendent bunch of flowers have been modified into pitchers which secrete nectar and attract insects, while birds feeding on the nectar, or insects, have the pollen of the overhanging flowers dusted on their backs, and, carrying it to other flowers, thus cross-fertilise them (see Illustration).

In Australia and New Zealand the fine "glory peas" (Clianthus), the Sophora, Loranthus, many Epacrideæ and Myrtaceæ, and the large flowers of the New Zealand flax

(Phormium tenax), are cross-fertilised by birds; while in Natal the fine trumpet-creeper (Tecoma capensis) is fertilised by Nectarineas.

The great extent to which insect and bird agency is necessary to flowers is well shown by the case of New Zealand. The entire country is comparatively poor in species of insects, especially in bees and butterflies which are the chief flower fertilisers; yet according to the researches of local botanists no less than one-fourth of all the flowering plants are incapable of self-fertilisation, and, therefore, wholly dependent on insect or bird agency for the continuance of the species.

The facts as to the cross-fertilisation of flowers which have now been very briefly summarised, taken in connection with Darwin's experiments proving the increased vigour and fertility given by cross-fertilisation, seem amply to justify his aphorism that "Nature abhors self-fertilisation," and his more precise statement, that, "No plant is perpetually self-fertilised;" and this view has been upheld by Hildebrand, Delpino, and other botanists.[1]

Self-Fertilisation of Flowers.

But all this time we have been only looking at one side of the question, for there exists an abundance of facts which seem to imply, just as surely, the utter uselessness of cross-fertilisation. Let us, then, see what these facts are before proceeding further.

1. An immense variety of plants are habitually self-fertilised, and their numbers probably far exceed those which are habitually cross-fertilised by insects. Almost all the very small or obscure flowered plants with hermaphrodite flowers are of this kind. Most of these, however, may be insect fertilised occasionally, and may, therefore, come under the rule that no species are perpetually self-fertilised.

2. There are many plants, however, in which special arrangements exist to secure self-fertilisation. Sometimes the corolla closes and brings the anthers and stigma into contact; in others the anthers cluster round the stigmas, both maturing together, as in many buttercups, stitchwort (Stellaria media),

[1] See H. Müller's *Fertilisation of Flowers*, p. 18.

sandwort (Spergula), and some willow-herbs (Epilobium); or they arch over the pistil, as in Galium aparine and Alisma Plantago. The style is also modified to bring it into contact with the anthers, as in the dandelion, groundsel, and many other plants.[1] All these, however, may be occasionally cross-fertilised.

3. In other cases precautions are taken to prevent cross-fertilisation, as in the numerous cleistogamous or closed flowers. These occur in no less than fifty-five different genera, belonging to twenty-four natural orders, and in thirty-two of these genera the normal flowers are irregular, and have therefore been specially modified for insect fertilisation.[2] These flowers appear to be degradations of the normal flowers, and are closed up by various modifications of the petals or other parts, so that it is impossible for insects to reach the interior, yet they produce seed in abundance, and are often the chief means by which the species is continued. Thus, in our common dog-violet the perfect flowers rarely produce seed, while the rudimentary cleistogamic flowers do so in abundance. The sweet violet also produces abundance of seed from its cleistogamic flowers, and few from its perfect flowers; but in Liguria it produces only perfect flowers which seed abundantly. No case appears to be known of a plant which has cleistogamic flowers only, but a small rush (Juncus bufonius) is in this condition in some parts of Russia, while in other parts perfect flowers are also produced.[3] Our common henbit dead-nettle (Lamium amplexicaule) produces cleistogamic flowers, as do also some orchids. The advantage gained by the plant is great economy of specialised material, since with very small flowers and very little expenditure of pollen an abundance of seed is produced.

4. A considerable number of plants which have evidently been specially modified for insect fertilisation have, by further

[1] The above examples are taken from Rev. G. Henslow's paper on "Self-Fertilisation of Plants," in *Trans. Linn. Soc.* Second series, *Botany*, vol. i. pp. 317-398, with plate. Mr. H. O. Forbes has shown that the same thing occurs among tropical orchids, in his paper "On the Contrivances for insuring Self-Fertilisation in some Tropical Orchids," *Journ. Linn. Soc.*, xxi. p. 538.

[2] These are the numbers given by Darwin, but I am informed by Mr. Hemsley that many additions have been since made to the list, and that cleistogamic flowers probably occur in nearly all the natural orders.

[3] For a full account of cleistogamic flowers, see Darwin's *Forms of Flowers*, chap. viii.

modification, become quite self-fertile. This is the case with the garden-pea, and also with our beautiful bee-orchis, in which the pollen-masses constantly fall on to the stigmas, and the flower, being thus self-fertilised, produces abundance of capsules and of seed. Yet in many of its close allies insect agency is absolutely required; but in one of these, the fly-orchis, comparatively very little seed is produced, and self-fertilisation would therefore be advantageous to it. When garden-peas were artificially cross-fertilised by Mr. Darwin, it seemed to do them no good, as the seeds from these crosses produced less vigorous plants than seed from those which were self-fertilised; a fact directly opposed to what usually occurs in cross-fertilised plants.

5. As opposed to the theory that there is any absolute need for cross-fertilisation, it has been urged by Mr. Henslow and others that many self-fertilised plants are exceptionally vigorous, such as groundsel, chickweed, sow-thistle, buttercups, and other common weeds; while most plants of world-wide distribution are self-fertilised, and these have proved themselves to be best fitted to survive in the battle of life. More than fifty species of common British plants are very widely distributed, and all are habitually self-fertilised.[1] That self-fertilisation has some great advantage is shown by the fact that it is usually the species which have the smallest and least conspicuous flowers which have spread widely, while the large and showy flowered species of the same genera or families, which require insects to cross-fertilise them, have a much more limited distribution.

6. It is now believed by some botanists that many inconspicuous and imperfect flowers, including those that are wind-fertilised, such as plantains, nettles, sedges, and grasses, do not represent primitive or undeveloped forms, but are degradations from more perfect flowers which were once adapted to insect fertilisation. In almost every order we find some plants which have become thus reduced or degraded for wind or self-fertilisation, as Poterium and Sanguisorba among the Rosaceæ; while this has certainly been the case in the cleistogamic flowers. In most of the above-mentioned plants there are distinct rudiments of petals or other floral organs,

[1] Henslow's "Self-Fertilisation," *Trans. Linn. Soc.* Second series, *Botany*, vol. i. p. 391.

and as the chief use of these is to attract insects, they could hardly have existed in primitive flowers.[1] We know, moreover, that when the petals cease to be required for the attraction of

[1] The Rev. George Henslow, in his *Origin of Floral Structures*, says: "There is little doubt but that all wind-fertilised angiosperms are degradations from insect-fertilised flowers. . . . *Poterium sanguisorba* is anemophilous; and *Sanguisorba officinalis* presumably was so formerly, but has reacquired an entomophilous habit; the whole tribe Poterieæ being, in fact, a degraded group which has descended from Potentilleæ. Plantains retain their corolla but in a degraded form. Junceæ are degraded Lilies; while Cyperaceæ and Gramineæ among monocotyledons may be ranked with Amentiferæ among dicotyledons, as representing orders which have retrograded very far from the entomophilous forms from which they were possibly and probably descended" (p. 266).

"The genus Plantago, like *Thalictrum minus*, Poterium, and others, well illustrate the change from an entomophilous to the anemophilous state. *P. lanceolata* has polymorphic flowers, and is visited by pollen-seeking insects, so that it can be fertilised either by insects or the wind. *P. media* illustrates transitions in point of structure, as the filaments are pink, the anthers motionless, and the pollen grains aggregated, and it is regularly visited by *Bombus terrestris*. On the other hand, the slender filaments, versatile anthers, powdery pollen, and elongated protogynous style are features of other species indicating anemophily; while the presence of a degraded corolla shows its ancestors to have been entomophilous. *P. media*, therefore, illustrates, not a primitive entomophilous condition, but a return to it; just as is the case with *Sanguisorba officinalis* and *Salix Caprea;* but these show no capacity of restoring the corolla, the attractive features having to be borne by the calyx, which is purplish in Sanguisorba, by the pink filaments of Plantago, and by the yellow anthers in the Sallow willow" (p. 271).

"The interpretation, then, I would offer of inconspicuousness and all kinds of degradations is the exact opposite to that of conspicuousness and great differentiations; namely, that species with minute flowers, rarely or never visited by insects, and habitually self-fertilised, have primarily arisen through the neglect of insects, and have in consequence assumed their present floral structures" (p. 282).

In a letter just received from Mr. Henslow, he gives a few additional illustrations of his views, of which the following are the most important: "Passing to Incompletæ, the orders known collectively as 'Cyclospermeæ' are related to Caryophylleæ; and to my mind are degradations from it, of which Orache is anemophilous. Cupuliferæ have an inferior ovary and rudimentary calyx-limb on the top. These, as far as I know, cannot be interpreted except as degradations. The whole of Monocotyledons appear to me (from anatomical reasons especially) to be degradations from Dicotyledons, and primarily through the agency of growth in water. Many subsequently became terrestrial, but retained the effects of their primitive habitat through heredity. The 3-merous perianth of grasses, the parts of the flower being in whorls, point to a degradation from a sub-liliaceous condition."

Mr. Henslow informs me that he has long held these views, but, as far as he knows, alone. Mr. Grant Allen, however, set forth a similar theory in his *Vignettes from Nature* (p. 15) and more fully in *The Colours of Flowers* (chap. v.), where he develops it fully and uses similar arguments to those of Mr. Henslow.

insects, they rapidly diminish in size, lose their bright colour or almost wholly disappear [1]

Difficulties and Contradictions.

The very bare summary that has now been given of the main facts relating to the fertilisation of flowers, will have served to show the vast extent and complexity of the inquiry, and the extraordinary contradictions and difficulties which it presents. We have direct proof of the beneficial results of intercrossing in a great number of cases; we have an overwhelming mass of facts as to the varied and complex structure of flowers evidently adapted to secure this intercrossing by insect agency; yet we see many of the most vigorous plants which spread widely over the globe, with none of these adaptations, and evidently depending on self-fertilisation for their continued existence and success in the battle of life. Yet more extraordinary is it to find numerous cases in which the special arrangements for cross-fertilisation appear to have been a failure, since they have either been supplemented by special means for self-fertilisation, or have reverted back in various degrees to simpler forms in which self-fertilisation becomes the rule. There is also a further difficulty in the highly complex modes by which cross-fertilisation is often brought about; for we have seen that there are several very effective yet very simple modes of securing intercrossing, involving a minimum of change in the form and structure of the flower; and when we consider that the result attained with so much cost of structural modification is by no means an unmixed good, and is far less certain in securing the perpetuation of the species than is self-fertilisation, it is most puzzling to find such complex methods resorted to, sometimes to the extent of special precautions against the possibility of self-fertilisation ever taking place. Let us now see whether any light can be thrown on these various anomalies and contradictions.

Intercrossing not necessarily Advantageous.

No one was more fully impressed than Mr. Darwin with the beneficial effects of intercrossing on the vigour and fertility

[1] H. Müller gives ample proof of this in his *Fertilisation of Flowers.*

of the species or race, yet he clearly saw that it was not always and necessarily advantageous. He says: "The most important conclusion at which I have arrived is, that the mere act of intercrossing by itself does no good. The good depends on the individuals which are crossed differing slightly in constitution, owing to their progenitors having been subjected during several generations to slightly different conditions. This conclusion, as we shall hereafter see, is closely connected with various important physiological problems, such as the benefit derived from slight changes in the conditions of life."[1] Mr. Darwin has also adduced much direct evidence proving that slight changes in the conditions of life are beneficial to both animals and plants, maintaining or restoring their vigour and fertility in the same way as a favourable cross seems to restore it.[2] It is, I believe, by a careful consideration of these two classes of facts that we shall find the clue to the labyrinth in which this subject has appeared to involve us.

Supposed Evil Results of Close Interbreeding.

Just as we have seen that intercrossing is not necessarily good, we shall be forced to admit that close interbreeding is not necessarily bad. Our finest breeds of domestic animals have been thus produced, and by a careful statistical inquiry Mr. George Darwin has shown that the most constant and long-continued intermarriages among the British aristocracy have produced no prejudicial results. The rabbits on Porto Santo are all the produce of a single female; they have lived on the same small island for 470 years, and they still abound there and appear to be vigorous and healthy (see p. 161).

We have, however, on the other hand, overwhelming evidence that in many cases, among our domestic animals and cultivated plants, close interbreeding does produce bad results, and the apparent contradiction may perhaps be explained on the same general principles, and under similar limitations, as were found to be necessary in defining the value of intercrossing. It appears probable, then, that it is not interbreeding in itself that is hurtful, but interbreeding without

1 *Cross- and Self-Fertilisation*, p. 27.
2 *Animals and Plants*, vol. ii. p. 145.

rigid selection or some change of conditions. Under nature, as in the case of the Porto Santo rabbits, the rapid increase of these animals would in a very few years stock the island with a full population, and thereafter natural selection would act powerfully in the preservation only of the healthiest and the most fertile, and under these conditions no deterioration would occur. Among the aristocracy there has been a constant selection of beauty, which is generally synonymous with health, while any constitutional infertility has led to the extinction of the family. With domestic animals the selection practised is usually neither severe enough nor of the right kind. There is no natural struggle for existence, but certain points of form and colour characteristic of the breed are considered essential, and thus the most vigorous or the most fertile are not always those which are selected to continue the stock. In nature, too, the species always extends over a larger area and consists of much greater numbers, and thus a difference of constitution soon arises in different parts of the area, which is wanting in the limited numbers of pure bred domestic animals. From a consideration of these varied facts we conclude that an occasional disturbance of the organic equilibrium is what is essential to keep up the vigour and fertility of any organism, and that this disturbance may be equally well produced either by a cross between individuals of somewhat different constitutions, or by occasional slight changes in the conditions of life. Now plants which have great powers of dispersal enjoy a constant change of conditions, and can, therefore, exist permanently, or at all events, for very long periods, without intercrossing; while those which have limited powers of dispersal, and are restricted to a comparatively small and uniform area, need an occasional cross to keep up their fertility and general vigour. We should, therefore, expect that those groups of plants which are adapted both for cross- and self-fertilisation, which have showy flowers and possess great powers of seed-dispersal, would be the most abundant and most widely distributed; and this we find to be the case, the Compositæ possessing all these characteristics in the highest degree, and being the most generally abundant group of plants with conspicuous flowers in all parts of the world.

How the Struggle for Existence Acts among Flowers.

Let us now consider what will be the action of the struggle for existence under the conditions we have seen to exist.

Everywhere and at all times some species of plants will be dominant and aggressive; while others will be diminishing in numbers, reduced to occupy a smaller area, and generally having a hard struggle to maintain themselves. Whenever a self-fertilising plant is thus reduced in numbers it will be in danger of extinction, because, being limited to a small area, it will suffer from the effects of too uniform conditions which will produce weakness and infertility. But while this change is in progress, any crosses between individuals of slightly different constitution will be beneficial, and all variations favouring either insect agency on the one hand, or wind-dispersal of pollen on the other, will lead to the production of a somewhat stronger and more fertile stock. Increased size or greater brilliancy of the flower, more abundant nectar, sweeter odour, or adaptations for more effectual cross-fertilisation would all be preserved, and thus would be initiated some form of specialisation for insect agency in cross-fertilisation; and in every different species so circumstanced the result would be different, depending as it would on many and complex combinations of variation of parts of the flower, and of the insect species which most abounded in the district.

Species thus favourably modified might begin a new era of development, and, while spreading over a somewhat wider area, give rise to new varieties or species, all adapted in various degrees and modes to secure cross-fertilisation by insect agency. But in course of ages some change of conditions might prove adverse. Either the insects required might diminish in numbers or be attracted by other competing flowers, or a change of climate might give the advantage to other more vigorous plants. Then self-fertilisation with greater means of dispersal might be more advantageous; the flowers might become smaller and more numerous; the seeds smaller and lighter so as to be more easily dispersed by the wind, while some of the special adaptations for insect fertilisation being useless would, by the absence of selection and by the law of economy of growth, be reduced to a rudimentary form. With these modifications the species might extend its range into new districts, thereby obtaining increased vigour by the change of conditions, as appears to have been the case with so many of the small flowered self-fertilised plants. Thus it might continue to exist for a long series of ages, till under other changes—geographical or biological—it might again suffer from competition or from other adverse circumstances, and be at length again confined to a limited area, or reduced to very scanty numbers.

But when this cycle of change had taken place, the species would be very different from the original form. The flower would have been at one time modified to favour the visits of insects and to secure cross-fertilisation by their aid, and when the need for this passed away, some portions of these structures would remain, though in a reduced or rudimentary condition. But when insect agency became of importance a second time, the new modifications would start from a different or more advanced basis, and thus a more complex result might be produced. Owing to the unequal rates at which the reduction of the various parts might occur, some amount of irregularity in the flower might arise, and on a second development towards insect cross-fertilisation this irregularity, if useful, might be increased by variation and selection.

The rapidity and comparative certainty with which such changes as are here supposed do really take place, are well shown by the great differences in floral structure, as regards the mode of fertilisation, in allied genera and species, and even in some cases in varieties of the same species. Thus in the Ranunculaceæ we find the conspicuous part of the flower to be the petals in Ranunculus, the sepals in Helleborus, Anemone, etc., and the stamens in most species of Thalictrum. In all these we have a simple regular flower, but in Aquilegia it is made complex by the spurred petals, and in Delphinium and Aconitum it becomes quite irregular. In the more simple class self-fertilisation occurs freely, but it is prevented in the more complex flowers by the stamens maturing before the pistil. In the Caprifoliaceæ we have small and regular greenish flowers, as in the moschatel (Adoxa); more conspicuous regular open flowers without honey, as in the elder (Sambucus); and tubular flowers increasing in length and irregularity, till in some, like our common honeysuckle, they are adapted for fertilisation by moths only, with abundant honey and delicious perfume to attract them. In the Scrophulariaceæ we find open, almost regular flowers, as Veronica and Verbascum, fertilised by flies and bees, but also self-fertilised; Scrophularia adapted in form and colour to be fertilised by wasps; and the more complex and irregular flowers of Linaria, Rhinanthus, Melampyrum, Pedicularis, etc., mostly adapted to be fertilised by bees.

In the genera Geranium, Polygonum, Veronica, and several others there is a gradation of forms from large and bright to small and obscure coloured flowers, and in every case the former are adapted for insect fertilisation, often exclusively, while in the latter self-fertilisation constantly occurs. In the yellow rattle (Rhinanthus Crista-galli) there are two forms (which have been named *major* and *minor*), the larger and more conspicuous adapted to insect fertilisation only, the smaller capable of self-fertilisation; and two similar forms exist in the eyebright (Euphrasia officinalis). In both these cases there are special modifications in the length and curvature of the style as well as in the size and shape of the corolla; and the two forms are evidently becoming each adapted to special conditions, since in some districts the one, in other districts the other is most abundant.[1]

These examples show us that the kind of change suggested above is actually going on, and has presumably always been going on in nature throughout the long geological epochs during which the development of flowers has been progressing. The two great modes of gaining increased vigour and fertility —intercrossing and dispersal over wider areas—have been resorted to again and again, under the pressure of a constant struggle for existence and the need for adaptation to ever-changing conditions. During all the modifications that ensued, useless parts were reduced or suppressed, owing to the absence of selection and the principle of economy of growth; and thus at each fresh adaptation some rudiments of old structures were

[1] Müller's *Fertilisation of Flowers*, pp. 448, 455. Other cases of recent degradation and readaptation to insect-fertilisation are given by Professor Henslow (see footnote, p. 324).

re-developed, but not unfrequently in a different form and for a distinct purpose.

The chief types of flowering plants have existed during the millions of ages of the whole tertiary period, and during this enormous lapse of time many of them may have been modified in the direction of insect fertilisation, and again into that of self-fertilisation, not once or twice only, but perhaps scores or even hundreds of times; and at each such modification a difference in the environment may have led to a distinct line of development. At one epoch the highest specialisation of structure in adaptation to a single species or group of insects may have saved a plant from extinction; while, at other times, the simplest mode of self-fertilisation, combined with greater powers of dispersal and a constitution capable of supporting diverse physical conditions, may have led to a similar result. With some groups the tendency seems to have been almost continuously to greater and greater specialisation, while with others a tendency to simplification and degradation has resulted in such plants as the grasses and sedges.

We are now enabled dimly to perceive how the curious anomaly of very simple and very complex methods of securing cross-fertilisation—both equally effective—may have been brought about. The simple modes may be the result of a comparatively direct modification from the more primitive types of flowers, which were occasionally, and, as it were, accidentally visited and fertilised by insects; while the more complex modes, existing for the most part in the highly irregular flowers, may result from those cases in which adaptation to insect-fertilisation, and partial or complete degradation to self-fertilisation or to wind-fertilisation, have again and again recurred, each time producing some additional complexity, arising from the working up of old rudiments for new purposes, till there have been reached the marvellous flower structures of the papilionaceous tribes, of the asclepiads, or of the orchids.

We thus see that the existing diversity of colour and of structure in flowers is probably the ultimate result of the ever-recurring struggle for existence, combined with the ever-changing relations between the vegetable and animal kingdoms during countless ages. The constant variability of every part and organ, with the enormous powers of increase possessed by plants, have enabled them to become again and again readjusted to each change of condition as it occurred, resulting in that endless variety, that marvellous complexity, and that exquisite colouring which excite our admiration in the realm of flowers, and constitute them the perennial charm and crowning glory of nature.

Flowers the Product of Insect Agency.

In his *Origin of Species*, Mr. Darwin first stated that flowers had been rendered conspicuous and beautiful in order to attract insects, adding: "Hence we may conclude that, if insects had not been developed on the earth, our plants would not have been decked with beautiful flowers, but would have produced only such poor flowers as we see on our fir, oak, nut, and ash trees, on grasses, docks, and nettles, which are all fertilised through the agency of the wind." The argument in favour of this view is now much stronger than when he wrote; for not only have we reason to believe that most of these wind-fertilised flowers are degraded forms of flowers which have once been insect fertilised, but we have abundant evidence that whenever insect agency becomes comparatively ineffective, the colours of the flowers become less bright, their size and beauty diminish, till they are reduced to such small, greenish, inconspicuous flowers as those of the rupture-wort (Herniaria glabra), the knotgrass (Polygonum aviculare), or the cleistogamic flowers of the violet. There is good reason to believe, therefore, not only that flowers have been developed in order to attract insects to aid in their fertilisation, but that, having been once produced, in however great profusion, if the insect races were all to become extinct, flowers (in the temperate zones at all events) would soon dwindle away, and that ultimately all floral beauty would vanish from the earth.

We cannot, therefore, deny the vast change which insects have produced upon the earth's surface, and which has been thus forcibly and beautifully delineated by Mr. Grant Allen: "While man has only tilled a few level plains, a few great river valleys, a few peninsular mountain slopes, leaving the vast mass of earth untouched by his hand, the insect has spread himself over every land in a thousand shapes, and has made the whole flowering creation subservient to his daily wants. His buttercup, his dandelion, and his meadow-sweet grow thick in every English field. His thyme clothes the hillside; his heather purples the bleak gray moorland. High up among the alpine heights his gentian spreads its lakes of blue; amid the snows of the Himalayas his rhododendrons gleam with crimson light. Even the wayside pond yields him the white crowfoot and the arrowhead, while the broad expanses of Brazilian streams are beautified by his gorgeous water-lilies. The insect has thus turned the whole surface of the earth into a boundless flower-garden, which supplies him from year to year with pollen or honey, and itself in turn gains perpetuation by the baits that it offers for his allurement."[1]

Concluding Remarks on Colour in Nature.

In the last four chapters I have endeavoured to give a general and systematic, though necessarily condensed view of the part which is played by colour in the organic world. We have seen in what infinitely varied ways the need of concealment has led to the modification of animal colours, whether among polar snows or sandy deserts, in tropical forests or in the abysses of the ocean. We next find these general adaptations giving way to more specialised types of coloration, by which each species has become more and more harmonised with its immediate surroundings, till we reach the most curiously minute resemblances to natural objects in the leaf and stick insects, and those which are so like flowers or moss or birds' droppings that they deceive the acutest eye. We have learnt, further, that these varied forms of protective colouring are far more numerous than has been usually suspected, because, what appear to be very conspicuous colours or markings when the species is observed in a museum or in a menagerie, are often highly protective when the creature is seen under the natural conditions of its existence. From these varied classes of facts it seems not improbable that fully one-half of the species in the animal kingdom possess colours which have been more or less adapted to secure for them concealment or protection.

Passing onward we find the explanation of a distinct type

[1] *The Colour Sense*, by Grant Allen, p. 95.

of colour or marking, often superimposed upon protective tints, in the importance of easy recognition by many animals of their fellows, their parents, or their mates. By this need we have been able to account for markings that seem calculated to make the animal conspicuous, when the general tints and well-known habits of the whole group demonstrate the need of concealment. Thus also we are able to explain the constant symmetry in the markings of wild animals, as well as the numerous cases in which the conspicuous colours are concealed when at rest and only become visible during rapid motion.

In striking contrast to ordinary protective coloration we have "warning colours," usually very conspicuous and often brilliant or gaudy, which serve to indicate that their possessors are either dangerous or uneatable to the usual enemies of their tribe. This kind of coloration is probably more prevalent than has been hitherto supposed, because in the case of many tropical animals we are quite unacquainted with their special and most dangerous enemies, and are also unable to determine whether they are or are not distasteful to those enemies. As a kind of corollary to the "warning colours," we find the extraordinary phenomena of "mimicry," in which defenceless species obtain protection by being mistaken for those which, from any cause, possess immunity from attack. Although a large number of instances of warning colour and of mimicry are now recorded, it is probably still an almost unworked field of research, more especially in tropical regions and among the inhabitants of the ocean.

The phenomena of sexual diversities of coloration next engaged our attention, and the reasons why Mr. Darwin's theory of "sexual selection," as regards colour and ornament, could not be accepted were stated at some length, together with the theory of animal coloration and ornament we propose to substitute for it. This theory is held to be in harmony with the general facts of animal coloration, while it entirely dispenses with the very hypothetical and inadequate agency of female choice in producing the detailed colours, patterns, and ornaments, which in so many cases distinguish the male sex.

If my arguments on this point are sound, they will dispose also of Mr. Grant Allen's view of the direct action of the colour sense on the animal integuments.[1] He argues that the colours of insects and birds reproduce generally the colours of the flowers they frequent or the fruits they eat, and he adduces numerous cases in which flower-haunting insects and fruit-eating birds are gaily coloured. This he supposes to be due to the colour-taste, developed by the constant presence of bright flowers and fruits, being applied to the selection of each variation towards brilliancy in their mates; thus in time producing the gorgeous and varied hues they now possess. Mr. Allen maintains that "insects are bright where bright flowers exist in numbers, and dull where flowers are rare or inconspicuous;" and he urges that "we can hardly explain this wide coincidence otherwise than by supposing that a taste for colour is produced through the constant search for food among entomophilous blossoms, and that this taste has reacted upon its possessors through the action of unconscious sexual selection."

The examples Mr. Allen quotes of bright insects being associated with bright flowers seem very forcible, but are really deceptive or erroneous; and quite as many cases could be quoted which prove the very opposite. For example, in the dense equatorial forests flowers are exceedingly scarce, and there is no comparison with the amount of floral colour to be met with in our temperate meadows, woods, and hill-sides. The forests about Para in the lower Amazon are typical in this respect, yet they abound with the most gorgeously coloured butterflies, almost all of which frequent the forest depths, keeping near the ground, where there is the greatest deficiency of brilliant flowers. In contrast with this let us take the Cape of Good Hope—the most flowery region probably that exists upon the globe,—where the country is a complete flower-garden of heaths, pelargoniums, mesembryanthemus, exquisite iridaceous and other bulbs, and numerous flowering shrubs and trees; yet the Cape butterflies are hardly equal, either in number or variety, to those of any country in South Europe, and are utterly insignificant when compared with those of the comparatively flowerless forest-depths of the Amazon or of New Guinea. Neither is there any relation between the colours of other insects and their haunts. Few

[1] *The Colour Sense*, chap. ix.

are more gorgeous than some of the tiger-beetles and the carabi, yet these are all carnivorous; while many of the most brilliant metallic buprestidæ and longicorns are always found on the bark of fallen trees. So with the humming-birds; their brilliant metallic tints can only be compared with metals or gems, and are totally unlike the delicate pinks and purples, yellows and reds of the majority of flowers. Again, the Australian honey-suckers (Meliphagidæ) are genuine flower-haunters, and the Australian flora is more brilliant in colour display than that of most tropical regions, yet these birds are, as a rule, of dull colours, not superior on the average to our grain-eating finches. Then, again, we have the grand pheasant family, including the gold and the silver pheasants, the gorgeous fire-backed and ocellated pheasants, and the resplendent peacock, all feeding on the ground on grain or seeds or insects, yet adorned with the most gorgeous colours.

There is, therefore, no adequate basis of facts for this theory to rest upon, even if there were the slightest reason to believe that not only birds, but butterflies and beetles, take any delight in colour for its own sake, apart from the food-supply of which it indicates the presence. All that has been proved or that appears to be probable is, that they are able to perceive differences of colour, and to associate each colour with the particular flowers or fruits which best satisfy their wants. Colour being in its nature diverse, it has been beneficial for them to be able to distinguish all its chief varieties, as manifested more particularly in the vegetable kingdom, and among the different species of their own group; and the fact that certain species of insects show some preference for a particular colour may be explained by their having found flowers of that colour to yield them a more abundant supply of nectar or of pollen. In those cases in which butterflies frequent flowers of their own colour, the habit may well have been acquired from the protection it affords them.

It appears to me that, in imputing to insects and birds the same love of colour for its own sake and the same æsthetic tastes as we ourselves possess, we may be as far from the truth as were those writers who held that the bee was a good mathematician, and that the honeycomb was constructed throughout to satisfy its refined mathematical instincts; whereas it is now generally admitted to be the result of the simple principle of economy of material applied to a primitive cylindrical cell.[1]

In studying the phenomena of colour in the organic world we have been led to realise the wonderful complexity of the adaptations which bring each species into harmonious relation with all those which surround it, and which thus link together the whole of nature in a network of relations of marvellous intricacy. Yet all this is but, as it were, the outward show and garment of nature, behind which lies the inner structure —the framework, the vessels, the cells, the circulating fluids, and the digestive and reproductive processes,—and behind these again those mysterious chemical, electrical, and vital forces which constitute what we term Life. These forces appear to be fundamentally the same for all organisms, as is the material of which all are constructed; and we thus find behind the outer diversities an inner relationship which binds together the myriad forms of life.

Each species of animal or plant thus forms part of one harmonious whole, carrying in all the details of its complex structure the record of the long story of organic development; and it was with a truly inspired insight that our great philosophical poet apostrophised the humble weed—

Flower in the crannied wall,
I pluck you out of the crannies,
I hold you here, root and all, in my hand,
Little flower—but *if* I could understand
What you are, root and all, and all in all,
I should know what God and man is.

[1] See *Origin of Species*, sixth edition, p. 220.

CHAPTER XII

THE GEOGRAPHICAL DISTRIBUTION OF ORGANISMS

The facts to be explained—The conditions which have determined distribution—The permanence of oceans—Oceanic and continental areas—Madagascar and New Zealand—The teachings of the thousand-fathom line—The distribution of marsupials—The distribution of tapirs—Powers of dispersal as illustrated by insular organisms—Birds and insects at sea—Insects at great altitudes—The dispersal of plants—Dispersal of seeds by the wind—Mineral matter carried by the wind—Objections to the theory of wind-dispersal answered—Explanation of north temperate plants in the southern hemisphere—No proof of glaciation in the tropics—Lower temperature not needed to explain the facts—Concluding remarks.

THE theory which we may now take as established—that all the existing forms of life have been derived from other forms by a natural process of descent with modification, and that this same process has been in action during past geological time—should enable us to give a rational account not only of the peculiarities of form and structure presented by animals and plants, but also of their grouping together in certain areas, and their general distribution over the earth's surface.

In the absence of any exact knowledge of the facts of distribution, a student of the theory of evolution might naturally anticipate that all groups of allied organisms would be found in the same region, and that, as he travelled farther and farther from any given centre, the forms of life would differ more and more from those which prevailed at the starting-point, till, in the remotest regions to which he could penetrate, he would find an entirely new assemblage of animals and plants, altogether unlike those with which he was familiar. He would also anticipate that diversities of climate would always be associated with a corresponding diversity in the forms of life.

Now these anticipations are to a considerable extent justified. Remoteness on the earth's surface is usually an indication of diversity in the fauna and flora, while strongly contrasted climates are always accompanied by a considerable contrast in the forms of life. But this correspondence is by no means exact or proportionate, and the converse propositions are often quite untrue. Countries which are near to each other often differ radically in their animal and vegetable productions; while similarity of climate, together with moderate geographical proximity, are often accompanied by marked diversities in the prevailing forms of life. Again, while many groups of animals—genera, families, and sometimes even orders—are confined to limited regions, most of the families, many genera, and even some species are found in every part of the earth. An enumeration of a few of these anomalies will better illustrate the nature of the problem we have to solve.

As examples of extreme diversity, notwithstanding geographical proximity, we may adduce Madagascar and Africa, whose animal and vegetable productions are far less alike than are those of Great Britain and Japan at the remotest extremities of the great northern continent; while an equal, or perhaps even a still greater, diversity exists between Australia and New Zealand. On the other hand, Northern Africa and South Europe, though separated by the Mediterranean Sea, have faunas and floras which do not differ from each other more than do the various countries of Europe. As a proof that similarity of climate and general adaptability have had but a small part in determining the forms of life in each country, we have the fact of the enormous increase of rabbits and pigs in Australia and New Zealand, of horses and cattle in South America, and of the common sparrow in North America, though in none of these cases are the animals natives of the countries in which they thrive so well. And lastly, in illustration of the fact that allied forms are not always found in adjacent regions, we have the tapirs, which are found only on opposite sides of the globe, in tropical America and the Malayan Islands; the camels of the Asiatic deserts, whose nearest allies are the llamas and alpacas of the Andes; and the marsupials, only found in Australia and on the opposite side of the globe, in America. Yet, again, although mammalia may be said to be universally distributed over the globe, being found abundantly on all the continents and on a great many of the larger islands, yet they are entirely wanting in New Zealand, and in a considerable number of other islands which are, nevertheless, perfectly able to support them when introduced.

Now most of these difficulties can be solved by means of well-known geographical and geological facts. When the productions of remote countries resemble each other, there is almost always continuity of land with similarity of climate between them. When adjacent countries differ greatly in their productions, we find them separated by a sea or strait whose great depth is an indication of its antiquity or permanence. When a group of animals inhabits two countries or regions separated by wide oceans, it is found that in past geological times the same group was much more widely distributed, and may have reached the countries it inhabits from an intermediate region in which it is now extinct. We know, also, that countries now united by land were divided by arms of the sea at a not very remote epoch; while there is good reason to believe that others now entirely isolated by a broad expanse of sea were formerly united and formed a single land area. There is also another important factor to be taken account of in considering how animals and plants have acquired their present peculiarities of distribution,—changes of climate. We know that quite recently a glacial epoch extended over much of what are now the temperate regions of the northern hemisphere, and that consequently the organisms which inhabit those parts must be, comparatively speaking, recent immigrants from more southern lands. But it is a yet more important fact that, down to middle Tertiary times at all events, an equable temperate climate, with a luxuriant vegetation, extended to far within the arctic circle, over what are now barren wastes, covered for ten months of the year with snow and ice. The arctic zone has, therefore, been in past times capable of supporting almost all the forms of life of our temperate regions; and we must take account of this condition of things whenever we have to speculate on the possible migrations of organisms between the old and new continents.

The Conditions which have determined Distribution.

When we endeavour to explain in detail the facts of the existing distribution of organic beings, we are confronted by several preliminary questions, upon the solution of which will depend our treatment of the phenomena presented to us. Upon the theory of descent which we have adopted, all the different species of a genus, as well as all the genera which compose a family or higher group, have descended from some common ancestor, and must therefore, at some remote epoch, have occupied the same area, from which their descendants have spread to the regions they now inhabit. In the numerous cases in which the same group now occupies countries separated by oceans or seas, by lofty mountain-chains, by wide deserts, or by inhospitable climates, we have to consider how the migration which must certainly have taken place has been effected. It is possible that during some portion of the time which has elapsed since the origin of the group the interposing barriers have not been in existence; or, on the other hand, the particular organisms we are dealing with may have the power of overpassing the barriers, and thus reaching their present remote dwelling-places. As this is really the fundamental question of distribution on which the solution of all its more difficult problems depends, we have to inquire, in the first place, what is the nature of, and what are the limits to, the changes of the earth's surface, especially during the Tertiary and latter part of the Secondary periods, as it was during those periods that most of the existing types of the higher animals and plants came into existence; and, in the next place, what are the extreme limits of the powers of dispersal possessed by the chief groups of animals and plants. We will first consider the question of barriers, more especially those formed by seas and oceans.

The Permanence of Oceans.

It was formerly a very general belief, even amongst geologists, that the great features of the earth's surface, no less than the smaller ones, were subject to continual mutations, and that during the course of known geological time the continents and great oceans had again and again changed places with each other. Sir Charles Lyell, in the last edition of his *Principles of Geology* (1872), said: "Continents, therefore, although permanent for whole geological epochs, shift their positions entirely in the course of ages;" and this may be said to have been the orthodox opinion down to the very recent period when, by means of deep-sea soundings, the nature of the ocean bottom was made known. The first person to throw doubt on this view appears to have been the veteran American geologist, Professor Dana. In 1849, in the Report of Wilke's Exploring Expedition, he adduced the argument against a former continent in the Pacific during the Tertiary period, from the absence of all native quadrupeds. In 1856, in articles in the *American Journal*, he discussed the development of the American continent, and argued for its general permanence; and in his *Manual of Geology* in 1863 and later editions, the same views were more fully enforced and were latterly applied to all continents. Darwin, in his *Journal of Researches*, published in 1845, called attention to the fact that all the small islands far from land in the Pacific, Indian, and Atlantic Oceans are either of coralline or volcanic formation. He excepted, however, the Seychelles and St. Paul's rocks; but the former have since been shown to be no exception, as they consist entirely of coral rock; and although Darwin himself spent a few hours on St. Paul's rocks on his outward voyage in the *Beagle*, and believed he had found some portions of them to be of a "cherty," and others of a "felspathic" nature, this also has been shown to be erroneous, and the careful examination of the rocks by the Abbé Renard clearly proves them to be wholly of volcanic origin.[1] We have, therefore, at the present time, absolutely no exception whatever to the remarkable fact that all the oceanic islands of the globe are either of volcanic or coral formation; and there is, further, good reason to believe that those of the latter class in every case rest upon a volcanic foundation.

In his *Origin of Species*, Darwin further showed that no true oceanic island had any native mammals or batrachia

[1] See A. Agassiz, *Three Cruises of the Blake* (Cambridge, Mass., 1888), vol. i. p. 127, footnote.

when first discovered, this fact constituting the test of the class to which an island belongs; whence he argued that none of them had ever been connected with continents, but all had originated in mid-ocean. These considerations alone render it almost certain that the areas now occupied by the great oceans have never, during known geological time, been occupied by continents, since it is in the highest degree improbable that every fragment of those continents should have completely disappeared, and have been replaced by volcanic islands rising out of profound oceanic abysses; but recent research into the depth of the oceans and the nature of the deposits now forming on their floors, adds greatly to the evidence in this direction, and renders it almost a certainty that they represent very ancient if not primæval features of the earth's surface. A very brief outline of the nature of this evidence will be now given.

The researches of the *Challenger* expedition into the nature of the sea-bottom show, that the whole of the land debris brought down by rivers to the ocean (with the exception of pumice and other floating matter), is deposited comparatively near to the shores, and that the fineness of the material is an indication of the distance to which it has been carried. Everything in the nature of gravel and sand is laid down within a very few miles of land, only the finer muddy sediments being carried out for 20 or 50 miles, and the very finest of all, under the most favourable conditions, rarely extending beyond 150, or at the utmost, 300 miles from land into the deep ocean.[1] Beyond these distances, and covering the entire ocean floor, are various oozes formed wholly from the debris of marine organisms; while intermingled with these are found various volcanic products which have been either carried through the air or floated on the surface, and a small but perfectly recognisable quantity of meteoric matter. Ice-borne rocks are also found abundantly scattered over the ocean bottom within a definite distance of the arctic and antarctic circles, clearly marking out the limit of floating icebergs in recent geological times.

[1] Even the extremely fine Mississippi mud is nowhere found beyond a hundred miles from the mouths of the river in the Gulf of Mexico (A. Agassiz, *Three Cruises of the Blake*, vol. i. p. 128).

Now the whole series of marine stratified rocks, from the earliest Palæozoic to the most recent Tertiary beds, consist of materials closely corresponding to the land debris now being deposited within a narrow belt round the shores of all continents; while no rocks have been found which can be identified with the various oozes now forming in the deep abysses of the ocean. It follows, therefore, that all the geological formations have been formed in comparatively shallow water, and always adjacent to the continental land of the period. The great thickness of some of the formations is no indication of a deep sea, but only of slow subsidence during the time that the deposition was in progress. This view is now adopted by many of the most experienced geologists, especially by Dr. Archibald Geikie, Director of the Geological Survey of Great Britain, who, in his lecture on "Geographical Evolution," says: "From all this evidence we may legitimately conclude that the present land of the globe, though consisting in great measure of marine formations, has never lain under the deep sea; but that its site must always have been near land. Even its thick marine limestones are the deposits of comparatively shallow water."[1]

But besides these geological and physical considerations, there is a mechanical difficulty in the way of repeated change of position of oceans and continents which has not yet received the attention it deserves. According to the recent careful estimate by Mr. John Murray, the land area of the globe is to the water area as ·28 to ·72. The mean height of the land above sea-level is 2250 feet, while the mean depth of the ocean is 14,640 feet. Hence the bulk of dry land is 23,450,000 cubic miles, and that of the waters of the ocean 323,800,000 cubic miles; and it follows that if the whole of the solid matter of the earth's surface were reduced to one level, it would be everywhere covered by an ocean about two miles deep. The accompanying diagram will serve to render these figures more intelligible. The length of the sections of land and ocean are in the proportion of their respective areas, while the mean height of the land and the mean depth of the ocean are exhibited on a greatly increased

[1] I have given a full summary of the evidence for the permanence of oceanic and continental areas in my *Island Life*, chap. vi.

vertical scale. If we considered the continents and their adjacent oceans separately they would differ a little, but not very materially, from this diagram; in some cases the proportion of land to ocean would be a little greater, in others a little less.

Now, if we try to imagine a process of elevation and depression by which the sea and land shall completely change places, we shall be met by insuperable difficulties. We must, in the first place, assume a general equality between elevation and subsidence during any given period, because if the elevation over any extensive continental area were not balanced by some subsidence of approximately equal amount,

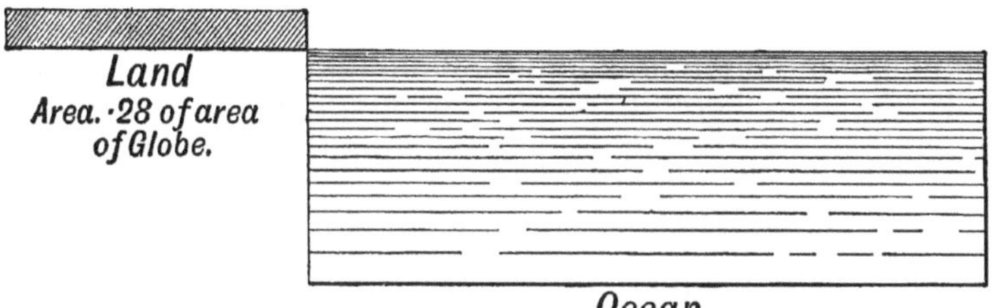

FIG. 32.

an unsupported hollow would be left under the earth's crust. Let us now suppose a continental area to sink, and an adjacent oceanic area to rise, it will be seen that the greater part of the land will disappear long before the new land has approached the surface of the ocean. This difficulty will not be removed by supposing a portion of a continent to subside, and the immediately adjacent portion of the ocean on the other side of the continent to rise, because in almost every case we find that within a comparatively short distance from the shores of all existing continents, the ocean floor sinks rapidly to a depth of from 2000 to 3000 fathoms, and maintains a similar depth, generally speaking, over a large portion of the oceanic areas. In order, therefore, that any area of continental extent be upraised from the great oceans, there must be a subsidence of a land area five or six times as great, unless it can be shown that an extensive elevation of the ocean floor up to and far above the surface could occur without an equivalent depression elsewhere. The fact that the waters of the ocean are sufficient to cover the whole globe to a depth of two miles, is alone sufficient to indicate that the great ocean basins are permanent features of the earth's surface, since any process of alternation of these with the land areas would have been almost certain to result again and again in the total disappearance of large portions, if not of all, of the dry land of the globe. But the continuity of terrestrial life since the Devonian and Carboniferous periods, and the existence of very similar forms in the corresponding deposits of every continent—as well as the occurrence of sedimentary rocks, indicating the proximity of land at the time of their deposit, over a large portion of the surface of all the continents, and in every geological period—assure us that no such disappearance has ever occurred.

Oceanic and Continental Areas.

When we speak of the permanence of oceanic and continental areas as one of the established facts of modern research, we do not mean that existing continents and oceans have always maintained the exact areas and outlines that they now present, but merely, that while all of them have been undergoing changes in outline and extent from age to age, they have yet maintained substantially the same positions, and have never actually changed places with each other. There are, moreover, certain physical and biological facts which enable us to mark out these areas with some confidence.

We have seen that there are a large number of islands which may be classed as oceanic, because they have never formed parts of continents, but have originated in mid-ocean, and have derived their forms of life by migration across the sea. Their peculiarities are seen to be very marked in comparison with those islands which there is good reason to believe are really fragments of more extensive land areas, and are hence termed "continental." These continental islands consist in every case of a variety of stratified rocks of various ages, thus corresponding closely with the usual structure of continents; although many of the islands are small like Jersey or the Shetland Islands, or far from continental land like the Falkland Islands or New Zealand. They all contain indigenous mammalia or batrachia, and generally a much greater variety of birds, reptiles, insects, and plants, than do the oceanic islands. From these various characteristics we conclude that they have all once formed parts of continents, or at all events of much larger land areas, and have become isolated, either by subsidence of the intervening land or by the effects of long-continued marine denudation.

Now, if we trace the thousand-fathom line around all our existing continents we find that, with only two exceptions, every island which can be classed as "continental" falls within this line, while all that lie beyond it have the undoubted characteristics of "oceanic" islands. We, therefore, conclude that the thousand-fathom line marks out, approximately, the "continental area,"—that is, the limits within which continental development and change throughout known geological time have gone on. There may, of course, have been some extensions of land beyond this limit, while some areas within it may always have been ocean; but so far as we have any direct evidence, this line may be taken to mark out, approximately, the most probable boundary between the "continental area," which has always consisted of land and shallow sea in varying proportions, and the great oceanic basins, within the limits of which volcanic activity has been building up numerous islands, but whose profound depths have apparently undergone little change.

Madagascar and New Zealand.

The two exceptions just referred to are Madagascar and New Zealand, and all the evidence goes to show that in these cases the land connection with the nearest continental area was very remote in time. The extraordinary isolation of the productions of Madagascar—almost all the most characteristic forms of mammalia, birds, and reptiles of Africa being absent from it—renders it certain that it must have been separated from that continent very early in the Tertiary, if not as far back as the latter part of the Secondary period; and this extreme antiquity is indicated by a depth of considerably more than a thousand fathoms in the Mozambique Channel, though this deep portion is less than a hundred miles wide between the Comoro Islands and the mainland.[1] Madagascar is the only island on the globe with a fairly rich mammalian fauna which is separated from a continent by a depth greater than a thousand fathoms; and no other island presents so many peculiarities in these animals, or has preserved so many lowly organised and archaic forms. The exceptional character of its productions agrees exactly with its exceptional isolation by means of a very deep arm of the sea.

New Zealand possesses no known mammals and only a single species of batrachian; but its geological structure is perfectly continental. There is also much evidence that it does possess one mammal, although no specimens have been yet obtained.[2] Its reptiles and birds are highly peculiar and more numerous than in any truly oceanic island. Now the sea which directly separates New Zealand from Australia is more than 2000 fathoms deep, but in a north-west direction there is an extensive bank under 1000 fathoms, extending to and including Lord Howe's Island, while north of this are other banks of the same depth, approaching towards a submarine extension of Queensland on the one hand, and New Caledonia on the other, and altogether suggestive of a land union with Australia at some very remote period. Now the peculiar relations of the New Zealand fauna and flora with those of Australia and of the tropical Pacific Islands to the northward indicate such a connection, probably during the Cretaceous period; and here, again, we have the exceptional depth of the dividing sea and the form of the ocean bottom according well with the altogether exceptional isolation of New Zealand, an isolation which has been held by some naturalists to be great enough to justify its claim to be one of the primary Zoological Regions.

The Teachings of the Thousand-Fathom Line.

If now we accept the annexed map as showing us approximately how far beyond their present limits our continents may

[1] For a full account of the peculiarities of the Madagascar fauna, see my *Island Life*, chap. xix.

[2] See *Island Life*, p. 446, and the whole of chaps. xxi. xxii. More recent soundings have shown that the Map at p. 443, as well as that of the Madagascar group at p. 387, are erroneous, the ocean around Norfolk Island and in the Straits of Mozambique being more than 1000 fathoms deep. The general argument is, however, unaffected.

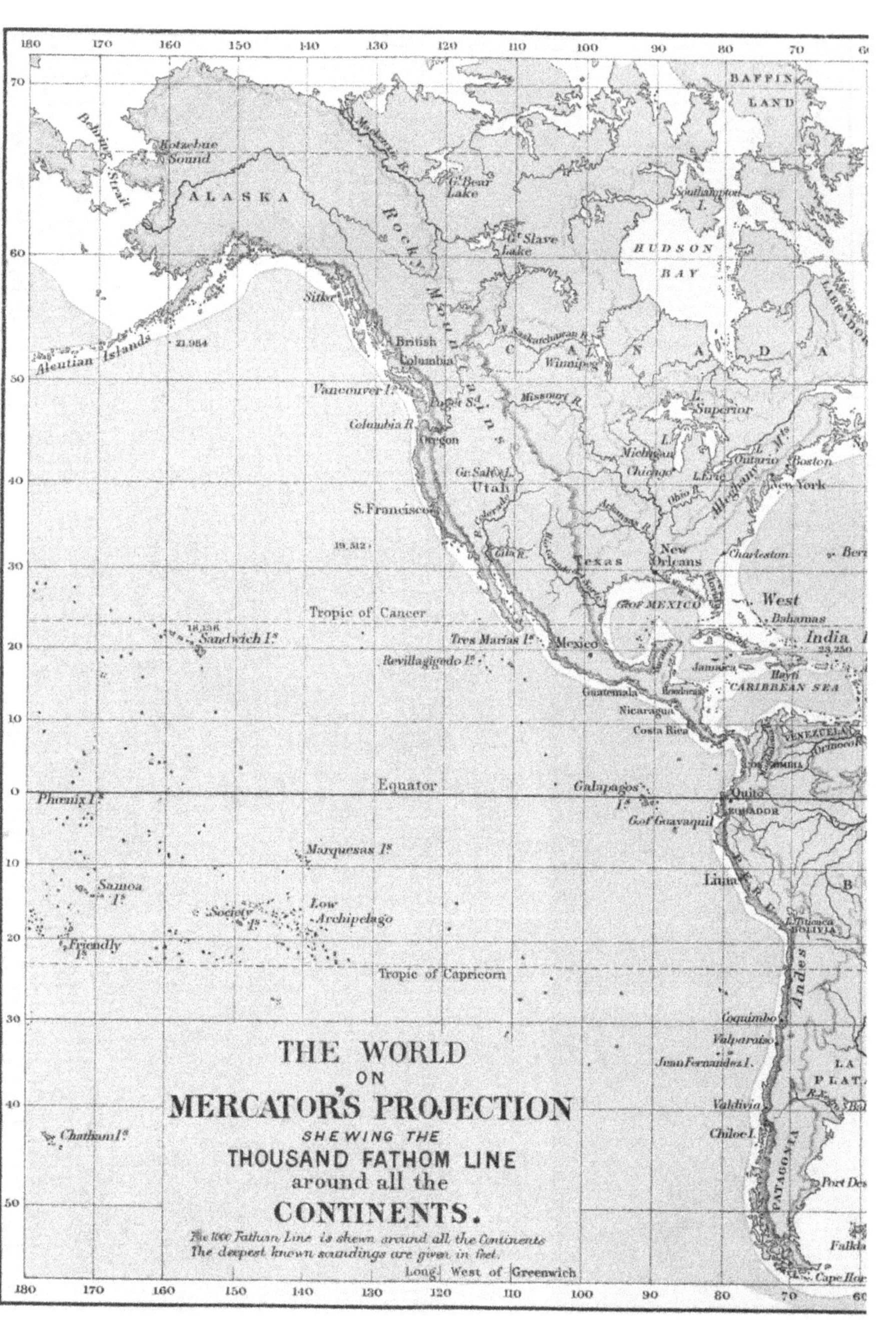
THE WORLD
ON
MERCATOR'S PROJECTION
SHEWING THE
THOUSAND FATHOM LINE
around all the
CONTINENTS.
The 1000 Fathom Line is shewn around all the Continents
The deepest known soundings are given in feet.
Long. West of Greenwich
ALASKA
Behring Strait
Kotzebue Sound
Aleutian Islands
Sitka
Rocky Mountains
British Columbia
Vancouver I.
Columbia R.
Oregon
Gr. Salt L.
Utah
S. Francisco
Gt. Bear Lake
Gt. Slave Lake
HUDSON BAY
BAFFIN LAND
Southampton I.
CANADA
Winnipeg
Missouri R.
L. Superior
Michigan
Chicago
Ontario
Boston
New York
Alleghany Mts.
Texas
New Orleans
Charleston
G. of MEXICO
Mexico
Tres Marias Is.
Revillagigedo Is.
West India Is.
Bahamas
Jamaica
Hayti
CARIBBEAN SEA
Guatemala
Honduras
Nicaragua
Costa Rica
VENEZUELA
Orinoco R.
COLOMBIA
Quito
ECUADOR
Galapagos Is.
G. of Guayaquil
Lima
PERU
BOLIVIA
L. Titicaca
Andes
Coquimbo
Valparaiso
Juan Fernandez I.
Valdivia
Chiloe I.
PATAGONIA
LA PLATA
Port Desire
Falkland
Cape Horn
Tropic of Cancer
Equator
Tropic of Capricorn
Sandwich Is.
Phœnix I.
Marquesas Is.
Samoa Is.
Society Is.
Low Archipelago
Friendly Is.
Chatham Is.
21,984
19,312
18,136
23,250

London: Macmillan & Co.

S I B E R I A
R. Oby
R. Lena
Yakutsk
Okhotsk
OKHOTSK SEA
KAMCHATKA SEA
Kamchatka
L. Baikal
Dauria
Amur R.
L. Balkhash
Gobi or Shamo Desert
Manchuria
Kurile Islands
Yesso
Hakodadi
JAPAN SEA
Japan Isles
Niphon
Yedo
Pekin
MONGOLIA
Corea
YELLOW SEA
TIBET
Cashmere
Moupin
Shanghai
Himalayas
Nepal
Delhi
Assam
CHINA
Bonin Is.
Loo Choo Is.
Amoy
Formosa I.
Calcutta
BURMA
Hong Kong
Bombay
BENGAL SEA
Hainan I.
CHINA SEA
SIAM
Philippine Is.
Marianne Is.
Madras
Andaman Is.
Ceylon
Nicobar Is.
Malay Penin.
Marshall Is.
Maldive Is.
Borneo
Celebes
Gilolo
New Ireland
Sumatra
Ceram
New Guinea
Solomon Is.
Chagos Is.
Java
Lombok
Bally
Timor
Torres St.
C. York
Keeling Is.
N. AUSTRALIA
New Hebrides
Fiji Is.
Rodriguez I.
QUEENSLAND
New Caledonia
W. AUSTRALIA
Norfolk I.
S. AUSTRALIA
N. S. WALES
Sydney
Ld. Howe Is.
Murray
Melbourne
Amsterdam I.
St. Paul I.
New Zealand
Tasmania
Kerguelen I.
Auckland Is.
Macquarie I.
Long. East of Greenwich
24,232
27,930
26,850
Stanford's Geographical Estab.t London.

have extended during any portion of the Tertiary and Secondary periods, we shall obtain a foundation of inestimable value for our inquiries into those migrations of animals and plants during past ages which have resulted in their present peculiarities of distribution. We see, for instance, that the South American and African continents have always been separated by nearly as wide an ocean as at present, and that whatever similarities there may be in their productions must be due to the similar forms having been derived from a common origin in one of the great northern continents. The radical difference between the higher forms of life of the two continents accords perfectly with their permanent separation. If there had been any direct connection between them during Tertiary times, we should hardly have found the deep-seated differences between the Quadrumana of the two regions—no family even being common to both; nor the peculiar Insectivora of the one continent, and the equally peculiar Edentata of the other. The very numerous families of birds quite peculiar to one or other of these continents, many of which, by their structural isolation and varied development of generic and specific forms, indicate a high antiquity, equally suggest that there has been no near approach to a land connection during the same epoch.

Looking to the two great northern continents, we see indications of a possible connection between them both in the North Atlantic and the North Pacific oceans; and when we remember that from middle Tertiary times backward—so far as we know continuously to the earliest Palæozoic epoch—a temperate and equable climate, with abundant woody vegetation, prevailed up to and within the arctic circle, we see what facilities may have been afforded for migration from one continent to the other, sometimes between America and Europe, sometimes between America and Asia. Admitting these highly probable connections, no bridging of the Atlantic in more southern latitudes (of which there is not a particle of evidence) will have been necessary to account for all the intermigration that has occurred between the two continents. If, on the other hand, we remember how long must have been the route, and how diverse must always have been the conditions between the more northern and the more southern portions of the American and Euro-Asiatic continents, we shall not be surprised that many widespread forms in either continent have not crossed into the other; and that while the skunks (Mephitis), the pouched rats (Saccomyidæ), and the turkeys (Meleagris) are confined to America, the pigs and the hedgehogs, the true flycatchers and the pheasants are found only in the Euro-Asiatic continent. But, just as there have been periods which facilitated intermigration between America and the Old World, there have almost certainly been periods, perhaps of long duration even geologically, when these continents have been separated by seas as wide as, or even wider than, those of the present day; and thus may be explained such curious anomalies as the origination of the camel-tribe in America, and its entrance into Asia in comparatively recent Tertiary times, while the introduction of oxen and bears into America from the Euro-Asiatic continent appears to have been equally recent.[1]

We shall find on examination that this view of the general permanence of the oceanic and continental areas, with constant minor fluctuations of land and sea over the whole extent of the latter, enables us to understand, and offer a rational explanation of, most of the difficult problems of geographical distribution; and further, that our power of doing this is in direct proportion to our acquaintance with the distribution of fossil forms of life during the Tertiary period. We must, also, take due note of many other facts of almost equal importance for a due appreciation of the problems presented for solution, the most essential being, the various powers of dispersal possessed by the different groups of animals and plants, the geological antiquity of the species and genera, and the width and depth of the seas which separate the countries they inhabit. A few illustrations will now be given of the way in which these branches of knowledge enable us to deal with the difficulties and anomalies that present themselves.

The Distribution of Marsupials.

This singular and lowly organised type of mammals constitutes almost the sole representative of the class in Australia

[1] For some details of these migrations, see the author's *Geographical Distribution of Animals*, vol. i. p. 140; also Heilprin's *Geographical and Geological Distribution of Animals.*

and New Guinea, while it is entirely unknown in Asia, Africa, or Europe. It reappears in America, where several species of opossums are found; and it was long thought necessary to postulate a direct southern connection of these distant countries, in order to account for this curious fact of distribution. When, however, we look to what is known of the geological history of the marsupials the difficulty vanishes. In the Upper Eocene deposits of Western Europe the remains of several animals closely allied to the American opossums have been found; and as, at this period, a very mild climate prevailed far up into the arctic regions, there is no difficulty in supposing that the ancestors of the group entered America from Europe or Northern Asia during early Tertiary times.

But we must go much further back for the origin of the Australian marsupials. All the chief types of the higher mammalia were in existence in the Eocene, if not in the preceding Cretaceous period, and as we find none of these in Australia, that country must have been finally separated from the Asiatic continent during the Secondary or Mesozoic period. Now during that period, in the Upper and the Lower Oolite and in the still older Trias, the jaw-bones of numerous small mammalia have been found, forming eight distinct genera, which are believed to have been either marsupials or some allied lowly forms. In North America also, in beds of the Jurassic and Triassic formations, the remains of an equally great variety of these small mammalia have been discovered; and from the examination of more than sixty specimens, belonging to at least six distinct genera, Professor Marsh is of opinion that they represent a generalised type, from which the more specialised marsupials and insectivora were developed.

From the fact that very similar mammals occur both in Europe and America at corresponding periods, and in beds which represent a long succession of geological time, and that during the whole of this time no fragments of any higher forms have been discovered, it seems probable that both the northern continents (or the larger portion of their area) were then inhabited by no other mammalia than these, with perhaps other equally low types. It was, probably, not later than the Jurassic age when some of these primitive marsupials were able to enter Australia, where they have since remained almost completely isolated; and, being free from the competition of higher forms, they have developed into the great variety of types we now behold there. These occupy the place, and have to some extent acquired the form and structure of distinct orders of the higher mammals—the rodents, the insectivora, and the carnivora,—while still preserving the essential characteristics and lowly organisation of the marsupials. At a much later period—probably in late Tertiary times—the ancestors of the various species of rats and mice which now abound in Australia, and which, with the aerial bats, constitute its only forms of placental mammals, entered the country from some of the adjacent islands. For this purpose a land connection was not necessary, as these small creatures might easily be conveyed among the branches or in the crevices of trees uprooted by floods and carried down to the sea, and then floated to a shore many miles distant. That no actual land connection with, or very close approximation to, an Asiatic island has occurred in recent times, is sufficiently proved by the fact that no squirrel, pig, civet, or other widespread mammal of the Eastern hemisphere has been able to reach the Australian continent.

The Distribution of Tapirs.

These curious animals form one of the puzzles of geographical distribution, being now confined to two very remote regions of the globe—the Malay Peninsula and adjacent islands of Sumatra and Borneo, inhabited by one species, and tropical America, where there are three or four species, ranging from Brazil to Ecuador and Guatemala. If we considered these living forms only, we should be obliged to speculate on enormous changes of land and sea in order that these tropical animals might have passed from one country to the other. But geological discoveries have rendered all such hypothetical changes unnecessary. During Miocene and Pliocene times tapirs abounded over the whole of Europe and Asia, their remains having been found in the tertiary deposits of France, India, Burmah, and China. In both North and South America fossil remains of tapirs occur only in caves and deposits of Post-Pliocene age, showing that they are comparatively recent immigrants into that continent. They perhaps entered by the route of Kamchatka and Alaska, where the climate, even now so much milder and more equable than on the north-east of America, might have been warm enough in late Pliocene times to have allowed the migration of these animals. In Asia they were driven southwards by the competition of numerous higher and more powerful forms, but have found a last resting-place in the swampy forests of the Malay region.

What these Facts Prove.

Now these two cases, of the marsupials and the tapirs, are in the highest degree instructive, because they show us that, without any hypothetical bridging of deep oceans, and with only such changes of sea and land as are indicated by the extent of the comparatively shallow seas surrounding and connecting the existing continents, we are able to account for the anomaly of allied forms occurring only in remote and widely separated areas. These examples really constitute crucial tests, because, of all classes of animals, mammalia are least able to surmount physical barriers. They are obviously unable to pass over wide arms of the sea, while the necessity for constant supplies of food and water renders sandy deserts or snow-clad plains equally impassable. Then, again, the peculiar kinds of food on which alone many of them can subsist, and their liability to the attacks of other animals, put a further check upon their migrations. In these respects almost all other organisms have great advantages over mammals. Birds can often fly long distances, and can thus cross arms of the sea, deserts, or mountain ranges; insects not only fly, but are frequently carried great distances by gales of wind, as shown by the numerous cases of their visits to ships hundreds of miles from land. Reptiles, though slow of movement, have advantages in their greater capacity for enduring hunger or thirst, their power of resisting cold or drought in a state of torpidity, and they have also some facilities for migration across the sea by means of their eggs, which may be conveyed in crevices of timber or among masses of floating vegetable matter. And when we come to the vegetable kingdom, the means of transport are at their maximum, numbers of seeds having special adaptations for being carried by mammalia or birds, and for floating in the water, or through the air, while many are so small and so light that there is practically no limit to the distances they may be carried by gales and hurricanes.

We may, therefore, feel quite certain that the means of distribution that have enabled the larger mammalia to reach the most remote regions from a common starting-point, will be at least as efficacious, and usually far more efficacious, with all other land animals and plants; and if in every case the existing distribution of this class can be explained on the theory of oceanic and continental permanence, with the limited changes of sea and land already referred to, no valid objections can be taken against this theory founded on anomalies of distribution in other orders. Yet nothing is more common than for students of this or that group to assert that the theory of oceanic permanence is quite inconsistent with the distribution of its various species and genera. Because a few Indian genera and closely allied species of birds are found in Madagascar, a land termed "Lemuria" has been supposed to have united the two countries during a comparatively recent geological epoch; while the similarity of fossil plants and reptiles, from the Permian and Miocene formations of India and South Africa, has been adduced as further evidence of this connection. But there are also genera of snakes, of insects, and of plants, common to Madagascar and South America only, which have been held to necessitate a direct land connection between these countries. These views evidently refute themselves, because any such land connections must have led to a far greater similarity in the productions of the several countries than actually exists, and would besides render altogether inexplicable the absence of all the chief types of African and Indian mammalia from Madagascar, and its marvellous individuality in every department of the organic world.[1]

Powers of Dispersal as illustrated by Insular Organisms.

Having arrived at the conclusion that our existing oceans have remained practically unaltered throughout the Tertiary and Secondary periods of geology, and that the distribution of the

[1] For a full discussion of this question, see *Island Life*, pp. 390-420.

mammalia is such as might have been brought about by their known powers of dispersal, and by such changes of land and sea as have probably or certainly occurred, we are, of course, restricted to similar causes to explain the much wider and sometimes more eccentric distribution of other classes of animals and of plants. In doing so, we have to rely partly on direct evidence of dispersal, afforded by the land organisms that have been observed far out at sea, or which have taken refuge on ships, as well as by the periodical visitants to remote islands; but very largely on indirect evidence, afforded by the frequent presence of certain groups on remote oceanic islands, which some ancestral forms must, therefore, have reached by transmission across the ocean from distant lands.

Birds.

These vary much in their powers of flight, and their capability of traversing wide seas and oceans. Many swimming and wading birds can continue long on the wing, fly swiftly, and have, besides, the power of resting safely on the surface of the water. These would hardly be limited by any width of ocean, except for the need of food; and many of them, as the gulls, petrels, and divers, find abundance of food on the surface of the sea itself. These groups have a wide distribution *across* the oceans; while waders—especially plovers, sandpipers, snipes, and herons—are equally cosmopolitan, travelling *along* the coasts of all the continents, and across the narrow seas which separate them. Many of these birds seem unaffected by climate, and as the organisms on which they feed are equally abundant on arctic, temperate, and tropical shores, there is hardly any limit to the range even of some of the species.

Land-birds are much more restricted in their range, owing to their usually limited powers of flight, their inability to rest on the surface of the sea or to obtain food from it, and their greater specialisation, which renders them less able to maintain themselves in the new countries they may occasionally reach. Many of them are adapted to live only in woods, or in marshes, or in deserts; they need particular kinds of food or a limited range of temperature; and they are adapted to cope only with the special enemies or the particular group of competitors among which they have been developed. Such birds as these may pass again and again to a new country, but are never able to establish themselves in it; and it is this organic barrier, as it is termed, rather than any physical barrier, which, in many cases, determines the presence of a species in one area and its absence from another. We must always remember, therefore, that, although the presence of a species in a remote oceanic island clearly proves that its ancestors must at one time have found their way there, the absence of a species does not prove the contrary, since it also may have reached the island, but have been unable to maintain itself, owing to the inorganic or organic conditions not being suitable to it. This general principle applies to all classes of organisms, and there are many striking illustrations of it. In the Azores there are eighteen species of land-birds which are permanent residents, but there are also several others which reach the islands almost every year after great storms, but have never been able to establish themselves. In Bermuda the facts are still more striking, since there are only ten species of resident birds, while no less than twenty other species of land-birds and more than a hundred species of waders and aquatics are frequent visitors, often in great numbers, but are never able to establish themselves. On the same principle we account for the fact that, of the many continental insects and birds that have been let loose, or have escaped from confinement, in this country, hardly one has been able to maintain itself, and the same phenomenon is still more striking in the case of plants. Of the thousands of hardy plants which grow easily in our gardens, very few have ever run wild, and when the experiment is purposely tried it invariably fails. Thus A. de Candolle informs us that several botanists of Paris, Geneva, and especially of Montpellier, have sown the seeds of many hundreds of species of exotic hardy plants, in what appeared to be the most favourable situations, but that in hardly a single case has any one of them become naturalised.[1] Still more, then, in plants than in animals the absence of a species does not prove that it has never reached the locality, but merely that it has not been able to maintain itself in com-

[1] *Géographie Botanique*, p. 798.

petition with the native productions. In other cases, as we have seen, facts of an exactly opposite nature occur. The rat, the pig, and the rabbit, the water-cress, the clover, and many other plants, when introduced into New Zealand, flourish exceedingly, and even exterminate their native competitors; so that in these cases we may feel sure that the species in question did not exist in New Zealand simply because they had been unable to reach that country by their natural means of dispersal. I will now give a few cases, in addition to those recorded in my previous works, of birds and insects which have been observed far from any land.

Birds and Insects at Sea.

Captain D. Fullarton of the ship *Timaru* recorded in his log the occurrence of a great number of small land-birds about the ship on 15th March 1886, when in Lat. 48° 31′ N., Long. 8° 16′ W. He says: "A great many small land-birds about us; put about sixty into a coop, evidently tired out." And two days later, 17th March, "Over fifty of the birds cooped on 15th died, though fed. Sparrows, finches, water-wagtails, two small birds, name unknown, one kind like a linnet, and a large bird like a starling. In all there have been on board over seventy birds, besides some that hovered about us for some time and then fell into the sea exhausted." Easterly winds and severe weather were experienced at the time.[1] The spot where this remarkable flight of birds was met with is about 160 miles due west of Brest, and this is the least distance the birds must have been carried. It is interesting to note that the position of the ship is nearly in the line from the English and French coasts to the Azores, where, after great storms, so many bird stragglers arrive annually. These birds were probably blown out to sea during their spring migration along the south coast of England to Wales and Ireland. During the autumnal migration, however, great flocks of birds—especially starlings, thrushes, and fieldfares—have been observed every year flying out to sea from the west coast of Ireland, almost the whole of which must perish. At the Nash Lighthouse, in the Bristol Channel on the coast of Glamorganshire, an enormous number of small birds were observed on 3d September, includ-

[1] *Nature*, 1st April 1886.

ing nightjars, buntings, white-throats, willow-wrens, cuckoos, house-sparrows, robins, wheatears, and blackbirds. These had probably crossed from Somersetshire, and had they been caught by a storm the larger portion of them must have been blown out to sea.[1]

These facts enable us to account sufficiently well for the birds of oceanic islands, the number and variety of which are seen to be proportionate to their facilities for reaching the island and maintaining themselves in it. Thus, though more birds yearly reach Bermuda than the Azores, the number of residents in the latter islands is much larger, due to the greater extent of the islands, their number, and their more varied surface. In the Galapagos the land-birds are still more numerous, due in part to their larger area and greater proximity to the continent, but chiefly to the absence of storms, so that the birds which originally reached the islands have remained long isolated and have developed into many closely allied species adapted to the special conditions. All the species of the Galapagos but one are peculiar to the islands, while the Azores possess only one peculiar species, and Bermuda none—a fact which is clearly due to the continual immigration of fresh individuals keeping up the purity of the breed by intercrossing. In the Sandwich Islands, which are extremely isolated, being more than 2000 miles from any continent or large island, we have a condition of things similar to what prevails in the Galapagos, the land-birds, eighteen in number, being all peculiar, and belonging, except one, to peculiar genera. These birds have probably all descended from three or four original types which reached the islands at some remote period, probably by means of intervening islets that have since disappeared. In St. Helena we have a degree of permanent isolation which has prevented any land-birds from reaching the island; for although its distance from the continent, 1100 miles, is not so great as in the case of the Sandwich Islands, it is situated in an ocean almost entirely destitute of small islands, while its position within the tropics renders it free from violent storms. Neither is there, on the nearest part of the coast of Africa, a perpetual stream of migrating birds like that which

[1] Report of the Brit. Assoc. Committee on Migration of Birds during 1886.

supplies the innumerable stragglers which every year reach Bermuda and the Azores.

Insects.

Winged insects have been mainly dispersed in the same way as birds, by their power of flight, aided by violent or long-continued winds. Being so small, and of such low specific gravity, they are occasionally carried to still greater distances; and thus no islands, however remote, are altogether without them. The eggs of insects, being often deposited in borings or in crevices of timber, may have been conveyed long distances by floating trees, as may the larvæ of those species which feed on wood. Several cases have been published of insects coming on board ships at great distances from land; and Darwin records having caught a large grasshopper when the ship was 370 miles from the coast of Africa, whence the insect had probably come.

In the *Entomologists' Monthly Magazine* for June 1885, Mr. MacLachlan has recorded the occurrence of a swarm of moths in the Atlantic ocean, from the log of the ship *Pleione*. The vessel was homeward bound from New Zealand, and in Lat. 6° 47′ N., Long. 32° 50′ W., hundreds of moths appeared about the ship, settling in numbers on the spars and rigging. The wind for four days previously had been very light from north, north-west, or north-east, and sometimes calm. The north-east trade wind occasionally extends to the ship's position at that time of year. The captain adds that "frequently, in that part of the ocean, he has had moths and butterflies come on board." The position is 960 miles south-west of the Cape Verde Islands, and about 440 north-east of the South American coast. The specimen preserved is Deiopeia pulchella, a very common species in dry localities in the Eastern tropics, and rarely found in Britain, but, Mr. MacLachlan thinks, not found in South America. They must have come, therefore, from the Cape Verde Islands, or from some parts of the African coast, and must have traversed about a thousand miles of ocean with the assistance, no doubt, of a strong north-east trade wind for a great part of the distance. In the British Museum collection there is a specimen of the same moth caught at sea during the voyage of the *Rattlesnake*, in Lat. 6° N., Long. 22½° W., being between the former position and Sierra Leone, thus rendering it probable that the moths came from that part of the African coast, in which case the swarm encountered by the *Pleione* must have travelled more than 1200 miles.

A similar case was recorded by Mr. F. A. Lucas in the American periodical *Science* of 8th April 1887. He states that in 1870 he met with numerous moths of many species while at sea in the South Atlantic (Lat. 25° S., Long. 24° W.), about 1000 miles from the coast of Brazil. As this position is just beyond the south-east trades, the insects may have been brought from the land by a westerly gale. In the *Zoologist* (1864, p. 8920) is the record of a small longicorn beetle which flew on board a ship 500 miles off the west coast of Africa. Numerous other cases are recorded of insects at less distances from land, and, taken in connection with those already given, they are sufficient to show that great numbers must be continually carried out to sea, and that occasionally they are able to reach enormous distances. But the reproductive powers of insects are so great that all we require, in order to stock a remote island, is that some few specimens shall reach it even once in a century, or once in a thousand years.

Insects at great Altitudes.

Equally important is the proof we possess that insects are often carried to great altitudes by upward currents of air. Humboldt noticed them up to heights of 15,000 and 18,000 feet in South America, and Mr. Albert Müller has collected many interesting cases of the same character in Europe.[1] A moth (Plusia gamma) has been found on the summit of Mont Blanc; small hymenoptera and moths have been seen on the Pyrenees at a height of 11,000 feet, while numerous flies and beetles, some of considerable size, have been caught on the glaciers and snow-fields of various parts of the Alps. Upward currents of air, whirlwinds and tornadoes, occur in all parts of the world, and large numbers of insects are thus carried up into the higher regions of the atmosphere, where they are liable to be caught by strong winds, and thus conveyed enormous distances over seas or continents. With such

[1] *Trans. Ent. Soc.*, 1871, p. 184.

powerful means of dispersal the distribution of insects over the entire globe, and their presence in the most remote oceanic islands, offer no difficulties.

The Dispersal of Plants.

The dispersal of seeds is effected in a greater variety of ways than are available in the case of any animals. Some fruits or seed-vessels, and some seeds, will float for many weeks, and after immersion in salt water for that period the seeds will often germinate. Extreme cases are the double cocoa-nut of the Seychelles, which has been found on the coast of Sumatra, about 3000 miles distant; the fruits of the Sapindus saponaria (soap-berry), which has been brought to Bermuda by the Gulf Stream from the West Indies, and has grown after a journey in the sea of about 1500 miles; and the West Indian bean, Entada scandens, which reached the Azores from the West Indies, a distance of full 3000 miles, and afterwards germinated at Kew. By these means we can account for the similarity in the shore flora of the Malay Archipelago and most of the islands of the Pacific; and from an examination of the fruits and seeds, collected among drift during the voyage of the *Challenger*, Mr. Hemsley has compiled a list of 121 species which are probably widely dispersed by this means.

A still larger number of species owe their dispersal to birds in several distinct ways. An immense number of fruits in all parts of the world are devoured by birds, and have been attractively coloured (as we have seen), in order to be so devoured, because the seeds pass through the birds' bodies and germinate where they fall. We have seen how frequently birds are forced by gales of wind across a wide expanse of ocean, and thus seeds must be occasionally carried. It is a very suggestive fact, that all the trees and shrubs in the Azores bear berries or small fruits which are eaten by birds; while all those which bear larger fruits, or are eaten chiefly by mammals—such as oaks, beeches, hazels, crabs, etc.—are entirely wanting. Game-birds and waders often have portions of mud attached to their feet, and Mr. Darwin has proved by experiment that such mud frequently contains seeds. One partridge had such a quantity of mud attached to its foot as to contain seeds from which eighty-two plants germinated; this proves that a very small portion of mud may serve to convey seeds, and such an occurrence repeated even at long intervals may greatly aid in stocking remote islands with vegetation. Many seeds also adhere to the feathers of birds, and thus, again, may be conveyed as far as birds are ever carried. Dr. Guppy found a small hard seed in the gizzard of a Cape Petrel, taken about 550 miles east of Tristan da Cunha.

Dispersal of Seeds by the Wind.

In the preceding cases we have been able to obtain direct evidence of transportal; but although we know that many seeds are specially adapted to be dispersed by the wind, we cannot obtain direct proof that they are so carried for hundreds or thousands of miles across the sea, owing to the difficulty of detecting single objects which are so small and inconspicuous. It is probable, however, that the wind as an agent of dispersal is really more effective than any of those we have hitherto considered, because a very large number of plants have seeds which are very small and light, and are often of such a form as to facilitate aerial carriage for enormous distances. It is evident that such seeds are especially liable to be transported by violent winds, because they become ripe in autumn at the time when storms are most prevalent, while they either lie upon the surface of the ground, or are disposed in dry capsules on the plant ready to be blown away. If inorganic particles comparable in weight, size, or form with such seeds are carried for great distances, we may be sure that seeds will also be occasionally carried in the same way. It will, therefore, be necessary to give a few examples of wind-carriage of small objects.

On 27th July 1875 a remarkable shower of small pieces of hay occurred at Monkstown, near Dublin. They appeared floating slowly down from a great height, as if falling from a dark cloud which hung overhead. The pieces picked up were wet, and varied from single blades of grass to tufts weighing one or two ounces. A similar shower occurred a few days earlier in Denbighshire, and was observed to travel in a direction contrary to that of the wind in the lower atmosphere.[1] There is no evidence of the distance from which the hay was

[1] *Nature* (1875), vol. xii. pp. 279, 298.

brought, but as it had been carried to a great height, it was in a position to be conveyed to almost any distance by a violent wind, had such occurred at the time.

Mineral Matter carried by the Wind.

The numerous cases of sand and volcanic dust being carried enormous distances through the atmosphere sufficiently prove the importance of wind as a carrier of solid matter, but unfortunately the matter collected has not been hitherto examined with a view to determine the maximum size and weight of the particles. A few facts, however, have been kindly furnished me by Professor Judd, F.R.S. Some dust which fell at Genoa on 15th October 1885, and was believed to have been brought from the African desert, consisted of quartz, hornblende, and other minerals, and contained particles having a diameter of $\frac{1}{500}$ inch, each weighing $\frac{1}{200,000}$ grain. This dust had probably travelled over 600 miles. In the dust from Krakatoa, which fell at Batavia, about 100 miles distant, during the great eruption, there are many solid particles even larger than those mentioned above. Some of this dust was given me by Professor Judd, and I found in it several ovoid particles of a much larger size, being $\frac{1}{50}$ inch long, and $\frac{1}{70}$ wide and deep. The dust from the same eruption, which fell on board the ship *Arabella*, 970 miles from the volcano, also contained solid particles $\frac{1}{500}$ inch diameter. Mr. John Murray of the *Challenger* Expedition writes to me that he finds in the deep sea deposits 500 and even 700 miles west of the coast of Africa, rounded particles of quartz, having a diameter of $\frac{1}{250}$ inch, and similar particles are found at equally great distances from the south-west coasts of Australia; and he considers these to be atmospheric dust carried to that distance by the wind. Taking the sp. gr. of quartz at 2·6, these particles would weigh about $\frac{1}{25,000}$ grain each. These interesting facts can, however, by no means be taken as indicating the extreme limits of the power of wind in carrying solid particles. During the Krakatoa eruption no gale of special violence occurred, and the region is one of comparative calms. The grains of quartz found by Mr. Murray more nearly indicate the limit, but the very small portions of matter brought up by the dredge, as compared with the enormous areas of sea-bottom, over which the atmospheric dust must have been scattered, render it in the highest degree improbable that the maximum limit either of size of particles, or of distance from land has been reached.

Let us, however, assume that the quartz grains, found by Mr. Murray in the deep-sea ooze 700 miles from land, give us the extreme limit of the power of the atmosphere as a carrier of solid particles, and let us compare with these the weights of some seeds. From a small collection of the seeds of thirty

No.	Species.	Approximate No. of Seeds in one Grain.	Approximate Dimensions.	Remarks.
			in. in. in.	
1	Draba verna	1,800	$\frac{1}{60} \times \frac{1}{90} \times \frac{1}{150}$	Oval, flat.
2	Hypericum perforatum	520	$\frac{1}{30} \times \frac{1}{80}$	Cylindrical.
3	Astilbe rivularis	4,500	$\frac{1}{50} \times \frac{1}{100}$	Elongate, flat, tailed, wavy.
4	Saxifraga coriophylla	750	$\frac{1}{40} \times \frac{1}{75}$	Surface rough, adhere to the dry capsules.
5	Œnothera rosea	640	$\frac{1}{40} \times \frac{1}{80}$	Ovate.
6	Hypericum hirsutum	700	$\frac{1}{30} \times \frac{1}{100}$	Cylindrical, rough.
7	Mimulus luteus	2,900	$\frac{1}{60} \times \frac{1}{100}$	Oval, minute.
8	Penthorum sedoides	8,000 *	$\frac{1}{70} \times \frac{1}{150}$	Flattened, very minute.
9	Sagina procumbens	12,000 *	$\frac{1}{120}$	Sub-triangular, flat.
10	Orchis maculata	15,000 *	...	Margined, flat, very minute.
11	Gentiana purpurea	35	$\frac{1}{25}$	Wavy, rough, with this coriaceous margins.
12	Silene alpina	...	$\frac{1}{30}$	Flat, with fringed margins.
13	Adenophora communis	...	$\frac{1}{20} \times \frac{1}{40}$	Very thin, wavy, light.
	Quartz grains	25,000	$\frac{1}{250}$	Deep sea . . 700 miles.
	Do.	200,000	$\frac{1}{500}$	Genoa . . . 600 miles.

species of herbaceous plants sent me from Kew, those in the above table were selected, and small portions of eight of them carefully weighed in a chemical balance.[1] By counting these portions I was able to estimate the number of seeds weighing one grain. The three very minute species, whose numbers are marked with an asterisk (*), were estimated by the comparison of their sizes with those of the smaller weighed seeds.

If now we compare the seeds with the quartz grains, we

[1] I am indebted to Professor R. Meldola of the Finsbury Technical Institute, and Rev. T. D. Titmas of Charterhouse for furnishing me with the weights required.

find that several are from twice to three times the weight of the grains found by Mr. Murray, and others five times, eight times, and fifteen times as heavy; but they are proportionately very much larger, and, being usually irregular in shape or compressed, they expose a very much larger surface to the air. The surface is often rough, and several have dilated margins or tailed appendages, increasing friction and rendering the uniform rate of falling through still air immensely less than in the case of the smooth, rounded, solid quartz grains. With these advantages it is a moderate estimate that seeds ten times the weight of the quartz grains could be carried quite as far through the air by a violent gale and under the most favourable conditions. These limits will include five of the seeds here given, as well as hundreds of others which do not exceed them in weight; and to these we may add some larger seeds which have other favourable characteristics, as is the case with numbers 11-13, which, though very much larger than the rest, are so formed as in all probability to be still more easily carried great distances by a gale of wind. It appears, therefore, to be absolutely certain that every autumnal gale capable of conveying solid mineral particles to great distances, must also carry numbers of small seeds at least as far; and if this is so, the wind alone will form one of the most effective agents in the dispersal of plants.

Hitherto this mode of conveyance, as applying to the transmission of seeds for great distances across the ocean, has been rejected by botanists, for two reasons. In the first place, there is said to be no direct evidence of such conveyance; and, secondly, the peculiar plants of remote oceanic islands do not appear to have seeds specially adapted for aerial transmission. I will consider briefly each of these objections.

Objection to the Theory of Wind-Dispersal.

To obtain direct evidence of the transmission of such minute and perishable objects, which do not exist in great quantities, and are probably carried to the greatest distances but rarely and as single specimens, is extremely difficult. A bird or insect can be seen if it comes on board ship, but who would ever detect the seeds of Mimulus or Orchis even if a score of them fell on a ship's deck? Yet if but one such seed per century were carried to an oceanic island, that island might become rapidly overrun by the plant, if the conditions were favourable to its growth and reproduction. It is further objected that search has been made for such seeds, and they have not been found. Professor Kerner of Innsbruck examined the snow on the surface of glaciers, and assiduously collected all the seeds he could find, and these were all of plants which grew in the adjacent mountains or in the same district. In like manner, the plants growing on moraines were found to be those of the adjacent mountains, plateaux, or lowlands. Hence he concluded that the prevalent opinion that seeds may be carried through the air for very great distances "is not supported by fact."[1] The opinion is certainly not supported by Kerner's facts, but neither is it opposed by them. It is obvious that the seeds that would be carried by the wind to moraines or to the surface of glaciers would be, first and in the greatest abundance, those of the immediately surrounding district; then, very much more rarely, those from more remote mountains; and lastly, in extreme rarity, those from distant countries or altogether distinct mountain ranges. Let us suppose the first to be so abundant that a single seed could be found by industrious search on each square yard of the surface of the glacier; the second so scarce that only one could possibly be found in a hundred yards square; while to find one of the third class it would be necessary exhaustively to examine a square mile of surface. Should we expect that *one* ever to be found, and should the fact that it could not be found be taken as a proof that it was not there? Besides, a glacier is altogether in a bad position to receive such remote wanderers, since it is generally surrounded by lofty mountains, often range behind range, which would intercept the few air-borne seeds that might have been carried from a distant land. The conditions in an oceanic island, on the other hand, are the most favourable, since the land, especially if high, will intercept objects carried by the wind, and will thus cause more of the solid matter to fall on it than on an equal area of ocean. We know that winds at sea often blow violently for days together, and the rate of motion is indicated by the fact that 72 miles an hour was the average velocity

[1] See *Nature*, vol. vi. p. 164, for a summary of Kerner's paper.

of the wind observed during twelve hours at the Ben Nevis observatory, while the velocity sometimes rises to 120 miles an hour. A twelve hours' gale might, therefore, carry light seeds a thousand miles as easily and certainly as it could carry quartz-grains of much greater specific gravity, rotundity, and smoothness, 500 or even 100 miles; and it is difficult even to imagine a sufficient reason why they should not be so carried—perhaps very rarely and under exceptionally favourable conditions,—but this is all that is required.

As regards the second objection, it has been observed that orchideæ, which have often exceedingly small and light seeds, are remarkably absent from oceanic islands. This, however, may be very largely due to their extreme specialisation and dependence on insect agency for their fertilisation; while the fact that they do occur in such very remote islands as the Azores, Tahiti, and the Sandwich Islands, proves that they must have once reached these localities either by the agency of birds or by transmission through the air; and the facts I have given above render the latter mode at least as probable as the former. Sir Joseph Hooker remarks on the composite plant of Kerguelen Island (Cotula plumosa) being found also on Lord Auckland and MacQuarrie Islands, and yet having no pappus, while other species of the genus possess it. This is certainly remarkable, and proves that the plant must have, or once have had, some other means of dispersal across wide oceans.[1] One of the most widely dispersed species in the whole world (Sonchus oleraceus) possesses pappus, as do four out of five of the species which are common to Europe and New Zealand, all of which have a very wide distribution. The same author remarks on the limited area occupied by most species of Compositæ, notwithstanding their facilities for dispersal by means of their feathered seeds; but it has been

[1] It seems quite possible that the absence of pappus in this case is a recent adaptation, and that it has been brought about by causes similar to those which have reduced or aborted the wings of insects in oceanic islands. For when a plant has once reached one of the storm-swept islands of the southern ocean, the pappus will be injurious for the same reason that the wings of insects are injurious, since it will lead to the seeds being blown out to sea and destroyed. The seeds which are heaviest and have least pappus will have the best chance of falling on the ground and remaining there to germinate, and this process of selection might rapidly lead to the entire disappearance of the pappus.

already shown that limitations of area are almost always due to the competition of allied forms, facilities for dispersal being only one of many factors in determining the wide range of species. It is, however, a specially important factor in the case of the inhabitants of remote oceanic islands, since, whether they are peculiar species or not, they or their remote ancestors must at some time or other have reached their present position by natural means.

I have already shown elsewhere, that the flora of the Azores strikingly supports the view of the species having been introduced by aerial transmission only, that is, by the agency of birds and the wind, because all plants that could not possibly have been carried by these means are absent.[1] In the same way we may account for the extreme rarity of Leguminosæ in all oceanic islands. Mr. Hemsley, in his Report on Insular Floras, says that they "are wanting in a large number of oceanic islands where there is no true littoral flora," as St. Helena, Juan Fernandez, and all the islands of the South Atlantic and South Indian Oceans. Even in the tropical islands, such as Mauritius and Bourbon, there are no endemic species, and very few in the Galapagos and the remoter Pacific Islands. All these facts are quite in accordance with the absence of facilities for transmission through the air, either by birds or the wind, owing to the comparatively large size and weight of the seeds; and an additional proof is thus afforded of the extreme rarity of the successful floating of seeds for great distances across the ocean.[2]

Explanation of North Temperate Plants in the Southern Hemisphere.

If we now admit that many seeds which are either minute in size, of thin texture or wavy form, or so fringed or margined as to afford a good hold to the air, are capable of being carried for many hundreds of miles by exceptionally

[1] See *Island Life*, p. 251.

[2] Mr. Hemsley suggests that it is not so much the difficulty of transmission by floating, as the bad conditions the seeds are usually exposed to when they reach land. Many, even if they germinate, are destroyed by the waves, as Burchell noticed at St. Helena; while even a flat and sheltered shore would be an unsuitable position for many inland plants. Air-borne seeds, on the other hand, may be carried far inland, and so scattered that some of them are likely to reach suitable stations.

violent and long-continued gales of wind, we shall not only be better able to account for the floras of some of the remotest oceanic islands, but shall also find in the fact a sufficient explanation of the wide diffusion of many genera, and even species, of arctic and north temperate plants in the southern hemisphere or on the summits of tropical mountains. Nearly fifty of the flowering plants of Tierra-del-Fuego are found also in North America or Europe, but in no intermediate country; while fifty-eight species are common to New Zealand and Northern Europe; thirty-eight to Australia, Northern Europe, and Asia; and no less than seventy-seven common to New Zealand, Australia, and South America.[1] On lofty mountains far removed from each other, identical or closely allied plants often occur. Thus the fine Primula imperialis of a single mountain peak in Java has been found (or a closely allied species) in the Himalayas; and many other plants of the high mountains of Java, Ceylon, and North India are either identical or closely allied forms. So, in Africa, some species, found on the summits of the Cameroons and Fernando Po in West Africa, are closely allied to species in the Abyssinian highlands and in Temperate Europe; while other Abyssinian and Cameroons species have recently been found on the mountains of Madagascar. Some peculiar Australian forms have been found represented on the summit of Kini Balu in Borneo. Again, on the summit of the Organ mountains in Brazil there are species allied to those of the Andes, but not found in the intervening lowlands.

No Proof of Recent Lower Temperature in the Tropics.

Now all these facts, and numerous others of like character, were supposed by Mr. Darwin to be due to a lowering of temperature during glacial epochs, which allowed these temperate forms to migrate across the intervening tropical lowlands. But any such change within the epoch of existing species is almost inconceivable. In the first place, it would necessitate the extinction of much of the tropical flora (and with it of the insect life), because without such extinction alpine herbaceous plants could certainly never spread over tropical forest low-

[1] For fuller particulars, see Sir J. Hooker's *Introduction to Floras of New Zealand and Australia*, and a summary in my *Island Life*, chaps. xxii. xxiii.

lands; and, in the next place, there is not a particle of direct evidence that any such lowering of temperature in intertropical lowlands ever took place. The only alleged evidence of the kind is that adduced by the late Professor Agassiz and Mr. Hartt; but I am informed by my friend, Mr. J. C. Branner (now State Geologist of Arkansas, U.S.), who succeeded Mr. Hartt, and spent several years completing the geological survey of Brazil, that the supposed moraines and glaciated granite rocks near Rio Janeiro and elsewhere, as well as the so-called boulder-clay of the same region, are entirely explicable as the results of sub-aerial denudation and weathering, and that there is no proof whatever of glaciation in any part of Brazil.

Lower Temperature not needed to Explain the Facts.

But any such vast physical change as that suggested by Darwin, involving as it does such tremendous issues as regards its effects on the tropical fauna and flora of the whole world, is really quite uncalled for, because the facts to be explained are of the same essential nature as those presented by remote oceanic islands, between which and the nearest continents no temperate land connection is postulated. In proportion to their limited area and extreme isolation, the Azores, St. Helena, the Galapagos, and the Sandwich Islands, each possess a fairly rich—the last a very rich—indigenous flora; and the means which sufficed to stock them with a great variety of plants would probably suffice to transmit others from mountain-top to mountain-top in various parts of the globe. In the case of the Azores, we have large numbers of species identical with those of Europe, and others closely allied, forming an exactly parallel case to the species found on the various mountain summits which have been referred to. The distances from Madagascar to the South African mountains and to Kilimandjaro, and from the latter to Abyssinia, are no greater than from Spain to the Azores, while there are other equatorial mountains forming stepping-stones at about an equal distance to the Cameroons. Between Java and the Himalayas we have the lofty mountains of Sumatra and of North-western Burma, forming steps at about the same distance apart; while between Kini Balu and the Australian Alps we have the unexplored snow mountains of New Guinea, the Bellenden Ker mountains in Queensland, and the New England and Blue Mountains of New South Wales. Between Brazil and Bolivia the distances are no greater; while the unbroken range of mountains from Arctic America to Tierra-del-Fuego offers the greatest facilities for transmission, the partial gap between the lofty peak of Chiriqui and the high Andes of New Grenada being far less than from Spain to the Azores. Thus, whatever means have sufficed for stocking oceanic islands must have been to some extent effective in transmitting northern forms from mountain to mountain, across the equator, to the southern hemisphere; while for this latter form of dispersal there are special facilities, in the abundance of fresh and unoccupied surfaces always occurring in mountain regions, owing to avalanches, torrents, mountain-slides, and rock-falls, thus affording stations on which air-borne seeds may germinate and find a temporary home till driven out by the inroads of the indigenous vegetation. These temporary stations may be at much lower altitudes than the original habitat of the species, if other conditions are favourable. Alpine plants often descend into the valleys on glacial moraines, while some arctic species grow equally well on mountain summits and on the seashore. The distances above referred to between the loftier mountains may thus be greatly reduced by the occurrence of suitable conditions at lower altitudes, and the facilities for transmission by means of aerial currents proportionally increased.[1]

Facts Explained by the Wind-Carriage of Seeds.

But if we altogether reject aerial transmission of seeds for great distances, except by the agency of birds, it will be difficult, if not impossible, to account for the presence of so many identical species of plants on remote mountain summits, or for that "continuous current of vegetation" described by Sir Joseph Hooker as having apparently long existed from the northern to the southern hemisphere. It may be admitted that we can, possibly, account for the greater portion of the floras of remote oceanic islands by the agency of birds alone; because, when blown out to sea land-birds must reach some island

[1] For a fuller discussion of this subject, see my *Island Life*, chap. xxiii.

or perish, and all which come within sight of an island will struggle to reach it as their only refuge. But, with mountain summits the case is altogether different, because, being surrounded by land instead of by sea, no bird would need to fly, or to be carried by the wind, for several hundred miles at a stretch to another mountain summit, but would find a refuge in the surrounding uplands, ridges, valleys, or plains. As a rule the birds that frequent lofty mountain tops are peculiar species, allied to those of the surrounding district; and there is no indication whatever of the passage of birds from one remote mountain to another in any way comparable with the flights of birds which are known to reach the Azores annually, or even with the few regular migrants from Australia to New Zealand. It is almost impossible to conceive that the seeds of the Himalayan primula should have been thus carried to Java; but, by means of gales of wind, and intermediate stations from fifty to a few hundred miles apart, where the seeds might vegetate for a year or two and produce fresh seed to be again carried on in the same manner, the transmission might, after many failures, be at last effected.

A very important consideration is the vastly larger scale on which wind-carriage of seeds must act, as compared with bird-carriage. It can only be a few birds which carry seeds attached to their feathers or feet. A very small proportion of these would carry the seeds of Alpine plants; while an almost infinitesimal fraction of these latter would convey the few seeds attached to them safely to an oceanic island or remote mountain. But winds, in the form of whirlwinds or tornadoes, gales or hurricanes, are perpetually at work over large areas of land and sea. Insects and light particles of matter are often carried up to the tops of high mountains; and, from the very nature and origin of winds, they usually consist of ascending or descending currents, the former capable of suspending such small and light objects as are many seeds long enough for them to be carried enormous distances. For each single seed carried away by external attachment to the feet or feathers of a bird, countless millions are probably carried away by violent winds; and the chance of conveyance to a great distance and in a definite direction must be many times greater by the latter mode than by the former.[1] We have seen that inorganic particles of much greater specific gravity than seeds, and nearly as heavy as the smallest kinds, are carried to great distances through the air, and we can therefore hardly doubt that some seeds are carried as far. The direct agency of the wind, as a supplement to bird-transport, will help to explain the presence in oceanic islands of plants growing in dry or rocky places whose small seeds are not likely to become attached to birds; while it seems to be the only effective agency possible in the dispersal of those species of alpine or sub-alpine plants found on the summits of distant mountains, or still more widely separated in the temperate zones of the northern and southern hemispheres.

Concluding Remarks.

On the general principles that have been now laid down, it will be found that all the chief facts of the geographical distribution of animals and plants can be sufficiently understood. There will, of course, be many cases of difficulty and some seeming anomalies, but these can usually be seen to depend on our ignorance of some of the essential factors of the problem. Either we do not know the distribution of the group in recent geological times, or we are still ignorant of the special methods by which the organisms are able to cross the sea. The latter difficulty applies especially to the lizard tribe, which are found

[1] A very remarkable case of wind conveyance of seeds on a large scale is described in a letter from Mr. Thomas Hanbury to his brother, the late Daniel Hanbury, which has been kindly communicated to me by Mr. Hemsley of Kew. The letter is dated "Shanghai, 1st May 1856," and the passage referred to is as follows:—

"For the past three days we have had very warm weather for this time of year, in fact almost as warm as the middle of summer. Last evening the wind suddenly changed round to the north and blew all night with considerable violence, making a great change in the atmosphere.

"This morning, myriads of small white particles are floating about in the air; there is not a single cloud and no mist, yet the sun is quite obscured by this substance, and it looks like a white fog in England. I enclose thee a sample, thinking it may interest. It is evidently a vegetable production; I think, apparently, some kind of seed."

Mr. Hemsley adds, that this substance proves to be the plumose seeds of a poplar or willow. In order to produce the effects described—*quite obscuring the sun like a white fog*,—the seeds must have filled the air to a very great height; and they must have been brought from some district where there were extensive tracts covered with the tree which produced them.

in almost all the tropical oceanic islands; but the particular mode in which they are able to traverse a wide expanse of ocean, which is a perfect barrier to batrachia and almost so to snakes, has not yet been discovered. Lizards are found in all the larger Pacific Islands as far as Tahiti, while snakes do not extend beyond the Fiji Islands; and the latter are also absent from Mauritius and Bourbon, where lizards of seven or eight species abound. Naturalists resident in the Pacific Islands would make a valuable contribution to our science by studying the life-history of the native lizards, and endeavouring to ascertain the special facilities they possess for crossing over wide spaces of ocean.

CHAPTER XIII

THE GEOLOGICAL EVIDENCES OF EVOLUTION

What we may expect—The number of known species of extinct animals—Causes of the imperfection of the geological record—Geological evidences of evolution—Shells—Crocodiles—The rhinoceros tribe—The pedigree of the horse tribe—Development of deer's horns—Brain development—Local relations of fossil and living animals—Cause of extinction of large animals—Indications of general progress in plants and animals—The progressive development of plants—Possible cause of sudden late appearance of exogens—Geological distribution of insects—Geological succession of vertebrata—Concluding remarks.

THE theory of evolution in the organic world necessarily implies that the forms of animals and plants have, broadly speaking, progressed from a more generalised to a more specialised structure, and from simpler to more complex forms. We know, however, that this progression has been by no means regular, but has been accompanied by repeated degradation and degeneration; while extinction on an enormous scale has again and again stopped all progress in certain directions, and has often compelled a fresh start in development from some comparatively low and imperfect type.

The enormous extension of geological research in recent times has made us acquainted with a vast number of extinct organisms, so vast that in some important groups—such as the mollusca—the fossil are more numerous than the living species; while in the mammalia they are not much less numerous, the preponderance of living species being chiefly in the smaller and in the arboreal forms which have not been so well preserved as the members of the larger groups. With such a wealth of material to illustrate the successive stages through which animals have passed, it will naturally be expected that we should find important evidence of evolution. We should hope to learn the steps by which some isolated forms have been connected with their nearest allies, and in many cases to have the gaps filled up which now separate genus from genus, or species from species. In some cases these expectations are fulfilled, but in many other cases we seek in vain for evidence of the kind we desire; and this absence of evidence with such an apparent wealth of material is held by many persons to throw doubt on the theory of evolution itself. They urge, with much appearance of reason, that all the arguments we have hitherto adduced fall short of demonstration, and that the crucial test consists in being able to show, in a great number of cases, those connecting links which we say must have existed. Many of the gaps that still remain are so vast that it seems incredible to these writers that they could ever have been filled up by a close succession of species, since these must have spread over so many ages, and have existed in such numbers, that it seems impossible to account for their total absence from deposits in which great numbers of species belonging to other groups are preserved and have been discovered. In order to appreciate the force, or weakness, of these objections, we must inquire into the character and completeness of that record of the past life of the earth which geology has unfolded, and ascertain the nature and amount of the evidence which, under actual conditions, we may expect to find.

The Number of known Species of Extinct Animals.

When we state that the known fossil mollusca are considerably more numerous than those which now live on the earth, it appears at first sight that our knowledge is very complete, but this is far from being the case. The species have been continually changing throughout geological time, and at each period have probably been as numerous as they are now. If we divide the fossiliferous strata into twelve great divisions —the Pliocene, Miocene, Eocene, Cretaceous, Oolite, Lias, Trias, Permian, Carboniferous, Devonian, Silurian, and Cambrian,—we find not only that each has a very distinct and characteristic molluscan fauna, but that the different subdivisions often present a widely different series of species; so that although a certain number of species are common to two or more of the great divisions, the totality of the species that have lived upon the earth must be very much more than twelve times—perhaps even thirty or forty times—the number now living. In like manner, although the species of fossil mammals now recognised by more or less fragmentary fossil remains may not be much less numerous than the living species, yet the duration of existence of these was comparatively so short that they were almost completely changed, perhaps six or seven times, during the Tertiary period; and this is certainly only a fragment of the geological time during which mammalia existed on the globe.

There is also reason to believe that the higher animals were much more abundant in species during past geological epochs than now, owing to the greater equability of the climate which rendered even the arctic regions as habitable as the temperate zones are in our time.

The same equable climate would probably cause a more uniform distribution of moisture, and render what are now desert regions capable of supporting abundance of animal life. This is indicated by the number and variety of the species of large animals that have been found fossil in very limited areas which they evidently inhabited at one period. M. Albert Gaudry found, in the deposits of a mountain stream at Pikermi in Greece, an abundance of large mammalia such as are nowhere to be found living together at the present time. Among them were two species of Mastodon, two different rhinoceroses, a gigantic wild boar, a camel and a giraffe larger than those now living, several monkeys, carnivora ranging from martens and civets to lions and hyænas of the largest size, numerous antelopes of at least five distinct genera, and besides these many forms altogether extinct. Such were the great herds of Hipparion, an ancestral form of horse; the Helladotherium, a huge animal bigger than the giraffe; the Ancylotherium, one of the Edentata; the huge Dinotherium; the Aceratherium, allied to the rhinoceros; and the monstrous Chalicotherium, allied to the swine and ruminants, but as large as a rhinoceros; and to prey upon these, the great Machairodus or sabre-toothed tiger. And all these remains were found in a space 300 paces long by 60 paces broad, many of the species existing in enormous quantities.

The Pikermi fossils belong to the Upper Miocene formation, but an equally rich deposit of Upper Eocene age has been discovered in South-Western France at Quercy, where M. Filhol has determined the presence of no less than forty-two species of beasts of prey alone. Equally remarkable are the various discoveries of mammalian fossils in North America, especially in the old lake bottoms now forming what are called the "bad lands" of Dakota and Nebraska, belonging to the Miocene period. Here are found an enormous assemblage of remains, often perfect skeletons, of herbivora and carnivora, as varied and interesting as those from the localities already referred to in Europe; but altogether distinct, and far exceeding, in number and variety of species of the larger animals, the whole existing fauna of North America. Very similar phenomena occur in South America and in Australia, leading us to the conclusion that the earth at the present time is impoverished as regards the larger animals, and that at each successive period of Tertiary time, at all events, it contained a far greater number of species than now inhabit it. The very richness and abundance of the remains which we find in limited areas, serve to convince us how imperfect and fragmentary must be our knowledge of the earth's fauna at any one past epoch; since we cannot believe that all, or nearly all, of the animals which inhabited any district were entombed in a single lake, or overwhelmed by the floods of a single river.

But the spots where such rich deposits occur are exceedingly few and far between when compared with the vast areas of continental land, and we have every reason to believe that in past ages, as now, numbers of curious species were rare or local, the commoner and more abundant species giving a very imperfect idea of the existing series of animal forms. Yet more important, as showing the imperfection of our knowledge, is the enormous lapse of time between the several formations in which we find organic remains in any abundance, so vast that in many cases we find ourselves almost in a new world, all the species and most of the genera of the higher animals having undergone a complete change.

Causes of the Imperfection of the Geological Record.

These facts are quite in accordance with the conclusions of geologists as to the necessary imperfection of the geological record, since it requires the concurrence of a number of favourable conditions to preserve any adequate representation of the life of a given epoch. In the first place, the animals to be preserved must not die a natural death by disease, or old age, or by being the prey of other animals, but must be destroyed by some accident which shall lead to their being embedded in the soil. They must be either carried away by floods, sink into bogs or quicksands, or be enveloped in the mud or ashes of a volcanic eruption; and when thus embedded they must remain undisturbed amid all the future changes of the earth's surface.

But the chances against this are enormous, because denudation is always going on, and the rocks we now find at the earth's surface are only a small fragment of those which were originally laid down. The alternations of marine and freshwater deposits, and the frequent unconformability of strata with those which overlie them, tell us plainly of repeated elevations and depressions of the surface, and of denudation on an enormous scale. Almost every mountain range, with its peaks, ridges, and valleys, is but the remnant of some vast plateau eaten away by sub-aerial agencies; every range of sea-cliffs tell us of long slopes of land destroyed by the waves; while almost all the older rocks which now form the surface of the earth have been once covered with newer deposits which have long since disappeared. Nowhere are the evidences of this denudation more apparent than in North and South America, where granitic or metamorphic rocks cover an area hardly less than that of all Europe. The same rocks are largely developed in Central Africa and Eastern Asia; while, besides those portions that appear exposed on the surface, areas of unknown extent are buried under strata which rest on them uncomformably, and could not, therefore, constitute the original capping under which the whole of these rocks must once have been deeply buried; because granite can only be formed, and metamorphism can only go on, deep down in the crust of the earth. What an overwhelming idea does this give us of the destruction of whole piles of rock, miles in thickness and covering areas comparable with those of continents; and how great must have been the loss of the innumerable fossil forms which those rocks contained! In view of such destruction we are forced to conclude that our palæontological collections, rich though they may appear, are really but small and random samples, giving no adequate idea of the mighty series of organism which have lived upon the earth.[1]

Admitting, however, the extreme imperfection of the geological record as a whole, it may be urged that certain limited portions of it are fairly complete—as, for example, the various Miocene deposits of India, Europe, and North America,—and that in these we ought to find many examples of species and genera linked together by intermediate forms. It may be replied that in several cases this really occurs; and the reason why it does not occur more often is, that the theory of evolution requires that distinct genera should be linked together, not by a direct passage, but by the descent of both from a common ancestor, which may have lived in some much earlier age the record of which is either wanting or very incomplete. An illustration given by Mr. Darwin will make this more clear to those who have not studied the subject. The fantail and pouter pigeons are two very distinct and unlike breeds, which we yet know to have been both derived from the common wild rock-pigeon. Now, if we had every variety of living pigeon before us, or even all those which have lived during the present century, we should find no intermediate types between these two—none combining in any degree the characters of the pouter with that of the fantail. Neither should we ever find such an intermediate form, even had there been preserved a specimen of every breed of pigeon since the ancestral rock-pigeon was first tamed by man—a period of probably several thousand years. We thus see that a complete passage from one very distinct species to another could not be expected even had we a complete record of the life of any one period. What we require is a complete

[1] The reader who desires to understand this subject more fully, should study chap. x. of the *Origin of Species,* and chap. xiv. of Sir Charles Lyell's *Principles of Geology.*

record of all the species that have existed since the two forms began to diverge from their common ancestor, and this the known imperfection of the record renders it almost impossible that we should ever attain. All that we have a right to expect is, that, as we multiply the fossil forms in any group, the gaps that at first existed in that group shall become less wide and less numerous; and also that, in some cases, a tolerably direct series shall be found, by which the more specialised forms of the present day shall be connected with more generalised ancestral types. We might also expect that when a country is now characterised by special groups of animals, the fossil forms that immediately preceded them shall, for the most part, belong to the same groups; and further, that, comparing the more ancient with the more modern types, we should find indications of progression, the earlier forms being, on the whole, lower in organisation, and less specialised in structure than the later. Now evidence of evolution of these varied kinds is what we do find, and almost every fresh discovery adds to their number and cogency. In order, therefore, to show that the testimony given by geology is entirely in favour of the theory of descent with modification, some of the more striking of the facts will now be given.

Geological Evidences of Evolution.

In an article in *Nature* (vol. xiv. p. 275), Professor Judd calls attention to some recent discoveries in the Hungarian plains, of fossil lacustrine shells, and their careful study by Dr. Neumayr and M. Paul of the Austrian Geological Survey. The beds in which they occur have accumulated to the thickness of 2000 feet, containing throughout abundance of fossils, and divisible into eight zones, each of which exhibits a well-marked and characteristic fauna. Professor Judd then describes the bearing of these discoveries as follows—

"The group of shells which affords the most interesting evidence of the origin of new forms through descent with modification is that of the genus Vivipara or Paludina, which occurs in prodigious abundance throughout the whole series of fresh-water strata. We shall not, of course, attempt in this place to enter into any details concerning the forty distinct *forms* of this genus (Dr. Neumayr very properly hesitates to call them all *species*), which are named and described in this monograph, and between which, as the authors show, so many connecting links, clearly illustrating the derivation of the newer from the older types, have been detected. On the minds of those who carefully examine the admirably engraved figures given in the plates accompanying this valuable memoir, or still better, the very large series of specimens from among which the subjects of these figures are selected, and which are now in the museum of the Reichsanstalt of Vienna, but little doubt will, we suspect, remain that the authors have fully made out their case, and have demonstrated that, beyond all controversy, the series with highly complicated ornamentation were variously derived by descent—the lines of which are in most cases perfectly clear and obvious—from the simple and unornamented Vivipara achatinoides of the Congerien-Schichten (the lower division of the series of strata). It is interesting to notice that a large portion of these unquestionably derived forms depart so widely from the type of the genus Vivipara, that they have been separated on so high an authority as that of Sandberger, as a new genus, under the name of Tulotoma. And hence we are led to the conclusion that a vast number of forms, certainly exhibiting specific distinctions, and according to some naturalists, differences even entitled to be regarded of generic value, have all a common ancestry."

It is, as Professor Judd remarks, owing to the exceptionally favourable circumstances of a long-continued and unbroken series of deposits being formed under physical conditions either identical or very slowly changing, that we owe so complete a record of the process of organic change. Usually, some disturbing elements, such as a sudden change of physical conditions, or the immigration of new sets of forms from other areas and the consequent retreat or partial extinction of the older fauna, interferes with the continuity of organic development, and produces those puzzling discordances so generally met with in geological formations of marine origin. While a case of the kind now described affords evidence of the origin of species complete and conclusive, though on a necessarily very limited scale, the very rarity of the conditions which are essential to such completeness serves to explain why it is that in most cases the direct evidence of evolution is not to be obtained.

Another illustration of the filling up of gaps between existing groups is afforded by Professor Huxley's researches on fossil crocodiles. The gap between the existing crocodiles and the lizards is very wide, but as we go back in geological time we meet with fossil forms which are to some extent intermediate and form a connected series. The three living genera—Crocodilus, Alligator, and Gavialis—are found in the Eocene formation, and allied forms of another genus, Holops, in the Chalk. From the Chalk backward to the Lias another group of genera occurs, having anatomical characteristics intermediate between the living crocodiles and the most ancient forms. These, forming two genera Belodon and Stagonolepis, are found in a still older formation, the Trias. They have characters resembling some lizards, especially the remarkable Hatteria of New Zealand, and have also some resemblances to the Dinosaurians—reptiles which in some respects approach birds. Considering how comparatively few are the remains of this group of animals, the evidence which it affords of progressive development is remarkably clear.[1]

Among the higher animals the rhinoceros, the horse, and the deer afford good evidence of advance in organisation and of the filling up of the gaps which separate the living forms from their nearest allies. The earliest ancestral forms of the rhinoceroses occur in the Middle Eocene of the United States, and were to some extent intermediate between the rhinoceros and tapir families, having like the latter four toes to the front feet, and three to those behind. These are followed in the Upper Eocene by the genus Amynodon, in which the skull assumes more distinctly the rhinocerotic type. Following this in the Lower Miocene we have the Aceratherium, like the last in its feet, but still more decidedly a rhinoceros in its general structure. From this there are two diverging lines—one in the Old World, the other in the New. In the former, to which the Aceratherium is supposed to have migrated in early Miocene times, when a mild climate and luxuriant vegetation prevailed far within the arctic circle, it gave rise to the Ceratorhinus and the various horned rhinoceroses of late Tertiary times and of those now living. In America a

[1] On "Stagonolepis Robertsoni and on the Evolution of the Crocodilia," in *Q. J. of Geological Society*, 1875; and abstract in *Nature*, vol. xii. p. 38.

number of large hornless rhinoceroses were developed—they are found in the Upper Miocene, Pliocene, and Post-Pliocene formations—and then became extinct. The true rhinoceroses have three toes on all the feet.[1]

The Pedigree of the Horse Tribe.

Yet more remarkable is the evidence afforded by the ancestral forms of the horse tribe which have been discovered in the American tertiaries. The family Equidæ, comprising the living horse, asses, and zebras, differ widely from all other mammals in the peculiar structure of the feet, all of which terminate in a single large toe forming the hoof. They have forty teeth, the molars being formed of hard and soft material in crescentic folds, so as to be a powerful agent in grinding up hard grasses and other vegetable food. The former peculiarities depend upon modifications of the skeleton, which have been thus described by Professor Huxley :—

"Let us turn in the first place to the fore-limb. In most quadrupeds, as in ourselves, the fore-arm contains distinct bones, called the radius and the ulna. The corresponding region in the horse seems at first to possess but one bone. Careful observation, however, enables us to distinguish in this bone a part which clearly answers to the upper end of the ulna. This is closely united with the chief mass of the bone which represents the radius, and runs out into a slender shaft, which may be traced for some distance downwards upon the back of the radius, and then in most cases thins out and vanishes. It takes still more trouble to make sure of what is nevertheless the fact, that a small part of the lower end of the bone of a horse's fore-arm, which is only distinct in a very young foal, is really the lower extremity of the ulna.

"What is commonly called the knee of a horse is its wrist. The 'cannon bone' answers to the middle bone of the five metacarpal bones which support the palm of the hand in ourselves. The pastern, coronary, and coffin bones of veterinarians answer to the joints of our middle fingers, while the hoof is simply a greatly enlarged and thickened nail. But if

[1] From a paper by Messrs. Scott and Osborne, "On the Origin and Development of the Rhinoceros Group," read before the British Association in 1883.

what lies below the horse's 'knee' thus corresponds to the middle finger in ourselves, what has become of the four other fingers or digits? We find in the places of the second and fourth digits only two slender splintlike bones, about two-thirds as long as the cannon bone, which gradually taper to their lower ends and bear no finger joints, or, as they are termed, phalanges. Sometimes, small bony or gristly nodules are to be found at the bases of these two metacarpal splints, and it is probable that these represent rudiments of the first and fifth toes. Thus, the part of the horse's skeleton which corresponds with that of the human hand, contains one overgrown middle digit, and at least two imperfect lateral digits; and these answer, respectively, to the third, the second, and the fourth fingers in man.

"Corresponding modifications are found in the hind limb. In ourselves, and in most quadrupeds, the leg contains two distinct bones, a large bone, the tibia, and a smaller and more slender bone, the fibula. But, in the horse, the fibula seems, at first, to be reduced to its upper end; a short slender bone united with the tibia, and ending in a point below, occupying its place. Examination of the lower end of a young foal's shin-bone, however, shows a distinct portion of osseous matter which is the lower end of the fibula; so that the, apparently single, lower end of the shin-bone is really made up of the coalesced ends of the tibia and fibula, just as the, apparently single, lower end of the forearm bone is composed of the coalesced radius and ulna.

"The heel of the horse is the part commonly known as the hock. The hinder cannon bone answers to the middle metatarsal bone of the human foot, the pastern, coronary, and coffin bones, to the middle toe bones; the hind hoof to the nail; as in the forefoot. And, as in the forefoot, there are merely two splints to represent the second and the fourth toes. Sometimes a rudiment of a fifth toe appears to be traceable.

"The teeth of a horse are not less peculiar than its limbs. The living engine, like all others, must be well stoked if it is to do its work; and the horse, if it is to make good its wear and tear, and to exert the enormous amount of force required for its propulsion, must be well and rapidly fed. To this end, good cutting instruments and powerful and lasting crushers are needful. Accordingly, the twelve cutting teeth of a horse are close-set and concentrated in the forepart of its mouth, like so many adzes or chisels. The grinders or molars are large, and have an extremely complicated structure, being composed of a number of different substances of unequal hardness. The consequence of this is that they wear away at different rates; and, hence, the surface of each grinder is always as uneven as that of a good millstone."[1]

We thus see that the Equidæ differ very widely in structure from most other mammals. Assuming the truth of the theory of evolution, we should expect to find traces among extinct animals of the steps by which this great modification has been effected; and we do really find traces of these steps, imperfectly among European fossils, but far more completely among those of America.

It is a singular fact that, although no horse inhabited America when discovered by Europeans, yet abundance of remains of extinct horses have been found both in North and South America in Post-Tertiary and Upper Pliocene deposits; and from these an almost continuous series of modified forms can be traced in the Tertiary formation, till we reach, at the very base of the series, a primitive form so unlike our perfected animal, that, had we not the intermediate links, few persons would believe that the one was the ancestor of the other. The tracing out of this marvellous history we owe chiefly to Professor Marsh of Yale College, who has himself discovered no less than thirty species of fossil Equidæ; and we will allow him to tell the story of the development of the horse from a humble progenitor in his own words.

"The oldest representative of the horse at present known is the diminutive Eohippus from the Lower Eocene. Several species have been found, all about the size of a fox. Like most of the early mammals, these ungulates had forty-four teeth, the molars with short crowns and quite distinct in form from the premolars. The ulna and fibula were entire and distinct, and there were four well-developed toes and a rudiment of another on the forefeet, and three toes behind. In the structure of the feet and teeth, the Eohippus unmistak-

[1] American Addresses, pp. 73-76.

ably indicates that the direct ancestral line to the modern horse has already separated from the other perissodactyles, or odd-toed ungulates.

"In the next higher division of the Eocene another genus, Orohippus, makes its appearance, replacing Eohippus, and showing a greater, though still distant, resemblance to the equine type. The rudimentary first digit of the forefoot has disappeared, and the last premolar has gone over to the molar series. Orohippus was but little larger than Eohippus, and in most other respects very similar. Several species have been found, but none occur later than the Upper Eocene.

"Near the base of the Miocene, we find a third closely allied genus, Mesohippus, which is about as large as a sheep, and one stage nearer the horse. There are only three toes and a rudimentary splint on the forefeet, and three toes behind. Two of the premolar teeth are quite like the molars. The ulna is no longer distinct or the fibula entire, and other characters show clearly that the transition is advancing.

"In the Upper Miocene Mesohippus is not found, but in its place a fourth form, Miohippus, continues the line. This genus is near the Anchitherium of Europe, but presents several important differences. The three toes in each foot are more nearly of a size, and a rudiment of the fifth metacarpal bone is retained. All the known species of this genus are larger than those of Mesohippus, and none of them pass above the Miocene formation.

"The genus Protohippus of the Lower Pliocene is yet more equine, and some of its species equalled the ass in size. There are still three toes on each foot, but only the middle one, corresponding to the single toe of the horse, comes to the ground. This genus resembles most nearly the Hipparion of Europe.

"In the Pliocene we have the last stage of the series before reaching the horse, in the genus Pliohippus, which has lost the small hooflets, and in other respects is very equine. Only in the Upper Pliocene does the true Equus appear and complete the genealogy of the horse, which in the Post-Tertiary roamed over the whole of North and South America, and soon after became extinct. This occurred long before the discovery of the continent by Europeans, and no satisfactory

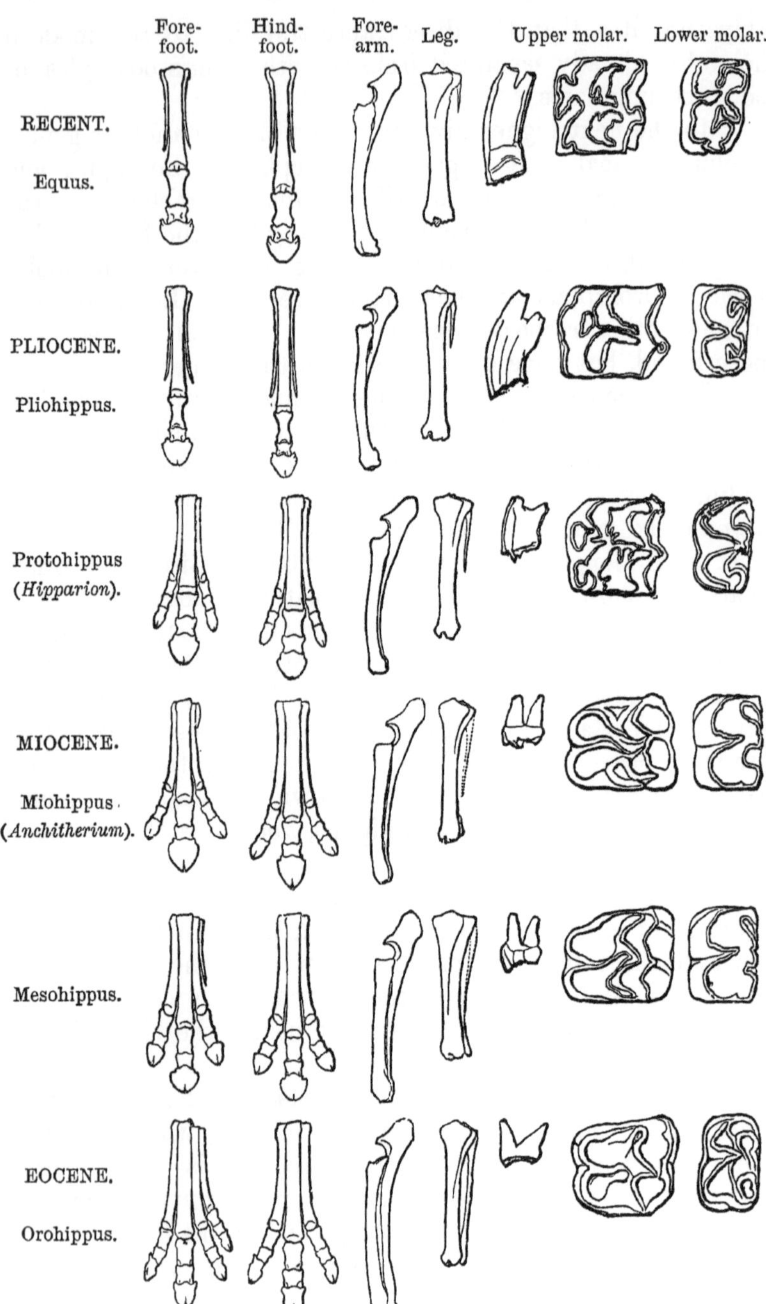

FIG. 33.—Geological development of the horse tribe (Eohippus since discovered).

reason for the extinction has yet been given. Besides the characters I have mentioned, there are many others in the skeleton, skull, teeth, and brain of the forty or more intermediate species, which show that the transition from the Eocene Eohippus to the modern Equus has taken place in the order indicated"[1] (see Fig. 33).

Well may Professor Huxley say that this is demonstrative evidence of evolution; the doctrine resting upon exactly as secure a foundation as did the Copernican theory of the motions of the heavenly bodies at the time of its promulgation. Both have the same basis—the coincidence of the observed facts with the theoretical requirements.

Development of Deer's Horns.

Another clear and unmistakable proof of evolution is afforded by one of the highest and latest developed tribes of mammals—the true deer. These differ from all other ruminants in possessing solid deciduous horns which are always more or less branched. They first appear in the Middle Miocene formation, and continue down to our time; and their development has been carefully traced by Professor Boyd Dawkins, who thus summarises his results:—

"In the middle stage of the Miocene the cervine antler consists merely of a simple forked crown (as in Cervus dicroceros), which increases in size in the Upper Miocene, although it still remains small and erect, like that of the roe. In Cervus Matheroni it measures 11·4 inches, and throws off not more than four tines, all small. The deer living in Auvergne in the succeeding or Pliocene age, present us with another stage in the history of antler development. There, for the first time, we see antlers of the Axis and Rusa type, larger and longer, and more branching than any antlers were before, and possessing three or more well-developed tines. Deer of this type abounded in Pliocene Europe. They belong to the Oriental division of the Cervidæ, and their presence in Europe confirms the evidence of the flora, brought forward by the Comte de Saporta, that the Pliocene climate was warm. They have probably disappeared from Europe in consequence

[1] Lecture on the Introduction and Succession of Vertebrate Life in America, *Nature*, vol. xvi. p. 471.

of the lowering of the temperature in the Pleistocene age, while their descendants have found a congenial home in the warmer regions of Eastern Asia.

"In the latest stage of the Pliocene—the Upper Pliocene of the Val d'Arno—the Cervus dicranios of Nesti presents us with antlers much smaller than those of the Irish elk, but very complicated in their branching. This animal survived into the succeeding age, and is found in the pre-glacial forest bed of Norfolk, being described by Dr. Falconer under the name of Sedgwick's deer. The Irish elk, moose, stag, reindeer, and fallow deer appear in Europe in the Pleistocene age, all with highly complicated antlers in the adult, and the first possessing the largest antlers yet known. Of these the Irish elk disappeared in the Prehistoric age, after having lived in countless herds in Ireland, while the rest have lived on into our own times in Euro-Asia, and, with the exception of the last, also in North America.

"From this survey it is obvious that the cervine antlers have increased in size and complexity from the Mid-Miocene to the Pleistocene age, and that their successive changes are analogous to those which are observed in the development of antlers in the living deer, which begin with a simple point, and increase in number of tines till their limit of growth be reached. In other words, the development of antlers indicated at successive and widely-separated pages of the geological record is the same as that observed in the history of a single living species. It is also obvious that the progressive diminution of size and complexity in the antlers, from the present time back into the early Tertiary age, shows that we are approaching the zero of antler development in the Mid-Miocene. No trace of any antler-bearing ruminant has been met with in the lower Miocenes, either of Europe or the United States."[1]

Progressive Brain-Development.

The three illustrations now given sufficiently prove that, whenever the geological record approaches to completeness, we have evidence of the progressive change of species in definite directions, and from less developed to more de-

[1] *Nature*, vol. xxv. p. 84.

veloped types—exactly such a change as we may expect to find if the evolution theory be the true one. Many other illustrations of a similar change could be given, but the animal groups in which they occur being less familiar, the details would be less interesting, and perhaps hardly intelligible. There is, however, one very remarkable proof of development that must be briefly noticed—that afforded by the steady increase in the size of the brain. This may be best stated in the words of Professor Marsh :—

"The real progress of mammalian life in America, from the beginning of the Tertiary to the present, is well illustrated by the brain-growth, in which we have the key to many other changes. The earliest known Tertiary mammals all had very small brains, and in some forms this organ was proportionally less than in certain reptiles. There was a gradual increase in the size of the brain during this period, and it is interesting to find that this growth was mainly confined to the cerebral hemispheres, or higher portion of the brain. In most groups of mammals the brain has gradually become more convoluted, and thus increased in quality as well as quantity. In some also the cerebellum and olfactory lobes, the lower parts of the brain, have even diminished in size. In the long struggle for existence during Tertiary time the big brains won, then as now; and the increasing power thus gained rendered useless many structures inherited from primitive ancestors, but no longer adapted to new conditions."

This remarkable proof of development in the organ of the mental faculties, forms a fitting climax to the evidence already adduced of the progressive evolution of the general structure of the body, as illustrated by the bony skeleton. We now pass on to another class of facts equally suggestive of evolution.

The Local Relations of Fossil and Living Animals.

If all existing animals have been produced from ancestral forms—mostly extinct—under the law of variation and natural selection, we may expect to find in most cases a close relation between the living forms of each country and those which inhabited it in the immediately preceding epoch. But if species have originated in some quite different way, either by any kind of special creation, or by sudden advances of organisation in the offspring of preceding types, such close relationship would not be found; and facts of this kind become, therefore, to some extent a test of evolution under natural selection or some other law of gradual change. Of course the relationship will not appear when extensive migration has occurred, by which the inhabitants of one region have been able to take possession of another region, and destroy or drive out its original inhabitants, as has sometimes happened. But such cases are comparatively rare, except where great changes of climate are known to have occurred; and we usually do find a remarkable continuity between the existing fauna and flora of a country and those of the immediately preceding age. A few of the more remarkable of these cases will now be briefly noticed.

The mammalian fauna of Australia consists, as is well known, wholly of the lowest forms—the Marsupials and Monotremata—except only a few species of mice. This is accounted for by the complete isolation of the country from the Asiatic continent during the whole period of the development of the higher animals. At some earlier epoch the ancestral marsupials, which abounded both in Europe and North America in the middle of the Secondary period, entered the country, and have since remained there, free from the competition of higher forms, and have undergone a special development in accordance with the peculiar conditions of a limited area. While in the large continents higher forms of mammalia have been developed, which have almost or wholly exterminated the less perfect marsupials, in Australia these latter have become modified into such varied forms as the leaping kangaroos, the burrowing wombats, the arboreal phalangers, the insectivorous bandicoots, and the carnivorous Dasyuridæ or native cats, culminating in the Thylacinus or "tiger-wolf" of Tasmania—animals as unlike each other as our sheep, rabbits, squirrels, and dogs, but all retaining the characteristic features of the marsupial type.

Now in the caves and late Tertiary or Post-Tertiary deposits of Australia the remains of many extinct mammalia have been found, but all are marsupials. There are many kangaroos, some larger than any living species, and others more allied to the tree-kangaroos of New Guinea; a large wombat as large as a tapir; the Diprotodon, a thick-limbed kangaroo the size of a rhinoceros or small elephant; and a quite different animal, the Nototherium, nearly as large. The carnivorous Thylacinus of Tasmania is also found fossil; and a huge phalanger, Thylacoleo, the size of a lion, believed by Professor Owen and by Professor Oscar Schmidt to have been equally carnivorous and destructive.[1] Besides these, there are many other species more resembling the living forms both in size and structure, of which they may be, in some cases, the direct ancestors. Two species of extinct Echidna, belonging to the very low Monotremata, have also been found in New South Wales.

Next to Australia, South America possesses the most remarkable assemblage of peculiar mammals, in its numerous Edentata—the sloths, ant-eaters, and armadillos; its rodents, such as the cavies and chinchillas; its marsupial opossums, and its quadrumana of the family Cebidæ. Remains of extinct species of all these have been found in the caves of Brazil, of Post-Pliocene age; while in the earlier Pliocene deposits of the pampas many distinct genera of these groups have been found, some of gigantic size and extraordinary form. There are armadillos of many types, some being as large as elephants; gigantic sloths of the genera Megatherium, Megalonyx, Mylodon, Lestodon, and many others; rodents belonging to the American families Cavidæ and Chinchillidæ; and ungulates allied to the llama; besides many other extinct forms of intermediate types or of uncertain affinities.[2] The extinct Moas of New Zealand —huge wingless birds allied to the living Apteryx—illustrate the same general law.

The examples now quoted, besides illustrating and enforcing the general fact of evolution, throw some light on the usual character of the modification and progression of animal forms. In the cases where the geological record is tolerably complete, we find a continuous development of some kind—either in complexity of ornamentation, as in the fossil Paludinas of the Hungarian lake-basins; in size and in the specialisation of the

[1] See *The Mammalia in their Relation to Primeval Times*, p. 102.

[2] For a brief enumeration and description of these fossils, see the author's *Geographical Distribution of Animals*, vol. i. p. 146.

feet and teeth, as in the American fossil horses; or in the increased development of the branching horns, as in the true deer. In each of these cases specialisation and adaptation to the conditions of the environment appear to have reached their limits, and any change of these conditions, especially if it be at all rapid or accompanied by the competition of less developed but more adaptable forms, is liable to cause the extinction of the most highly developed groups. Such we know was the case with the horse tribe in America, which totally disappeared in that continent at an epoch so recent that we cannot be sure that the disappearance was not witnessed, perhaps caused, by man; while even in the Eastern hemisphere it is the smaller species—the asses and the zebras—that have persisted, while the larger and more highly developed true horses have almost, if not quite, disappeared in a state of nature. So we find, both in Australia and South America, that in a quite recent period many of the largest and most specialised forms have become extinct, while only the smaller types have survived to our day; and a similar fact is to be observed in many of the earlier geological epochs, a group progressing and reaching a maximum of size or complexity and then dying out, or leaving at most but few and pigmy representatives.

Cause of Extinction of Large Animals.

Now there are several reasons for the repeated extinction of large rather than of small animals. In the first place, animals of great bulk require a proportionate supply of food, and any adverse change of conditions would affect them more seriously than it would smaller animals. In the next place, the extreme specialisation of many of these large animals would render it less easy for them to be modified in any new direction suited to changed conditions. Still more important, perhaps, is the fact that very large animals always increase slowly as compared with small ones—the elephant producing a single young one every three years, while a rabbit may have a litter of seven or eight young two or three times a year. Now the probability of favourable variations will be in direct proportion to the population of the species, and as the smaller animals are not only many hundred times more numerous than the largest, but also increase perhaps a hundred times as rapidly, they are able to become quickly modified by variation and natural selection in harmony with changed conditions, while the large and bulky species, being unable to vary quickly enough, are obliged to succumb in the struggle for existence. As Professor Marsh well observes: "In every vigorous primitive type which was destined to survive many geological changes, there seems to have been a tendency to throw off lateral branches, which became highly specialised and soon died out, because they were unable to adapt themselves to new conditions." And he goes on to show how the whole narrow path of the persistent Suilline type, throughout the entire series of the American tertiaries, is strewed with the remains of such ambitious offshoots, many of them attaining the size of a rhinoceros; "while the typical pig, with an obstinacy never lost, has held on in spite of catastrophes and evolution, and still lives in America to-day."

Indications of General Progression in Plants and Animals.

One of the most powerful arguments formerly adduced against evolution was, that geology afforded no evidence of the gradual development of organic forms, but that whole tribes and classes appeared suddenly at definite epochs, and often in great variety and exhibiting a very perfect organisation. The mammalia, for example, were long thought to have first appeared in Tertiary times, where they are represented in some of the earlier deposits by all the great divisions of the class fully developed—carnivora, rodents, insectivora, marsupials, and even the perissodactyle and artiodactyle divisions of the ungulata—as clearly defined as at the present day. The discovery in 1818 of a single lower jaw in the Stonesfield Slate of Oxfordshire hardly threw doubt on the generalisation, since either its mammalian character was denied, or the geological position of the strata, in which it was found, was held to have been erroneously determined. But since then, at intervals of many years, other remains of mammalia have been discovered in the Secondary strata, ranging from the Upper Oolite to the Upper Trias both in Europe and the United States, and one even (Tritylodon) in the Trias of South Africa. All these are either marsupials, or of some still lower type of mammalia; but they consist of many distinct forms classed in about twenty genera. Nevertheless, a great gap still exists between these mammals and those of the Tertiary strata, since no mammal of any kind has been found in any part of the Cretaceous formation, although in several of its subdivisions abundance of land plants, freshwater shells, and air-breathing reptiles have been discovered. So with fishes. In the last century none had been obtained lower than the Carboniferous formation; thirty years later they were found to be very abundant in the Devonian rocks, and later still they were discovered in the Upper Ludlow and Lower Ludlow beds of the Silurian formation.

We thus see that such sudden appearances are deceptive, and are, in fact, only what we ought to expect from the known imperfection of the geological record. The conditions favourable to the fossilisation of any group of animals occur comparatively rarely, and only in very limited areas; while the conditions essential for their permanent preservation in the rocks, amid all the destruction caused by denudation or metamorphism, are still more exceptional. And when they are thus preserved to our day, the particular part of the rocks in which they lie hidden may not be on the surface but buried down deep under other strata, and may thus, except in the case of mineral-bearing deposits, be altogether out of our reach. Then, again, how large a proportion of the earth consists of wild and uncivilised regions in which no exploration of the rocks has been yet made, so that whether we shall find the fossilised remains of any particular group of animals which lived during a limited period of the earth's history, and in a limited area, depends upon at least a fivefold combination of chances. Now, if we take each of these chances separately as only ten to one against us (and some are certainly more than this), then the actual chance against our finding the fossil remains, say of any one order of mammalia, or of land plants, at any particular geological horizon, will be about a hundred thousand to one.

It may be said, if the chances are so great, how is it that we find such immense numbers of fossil species exceeding in number, in some groups, all those that are now living? But this is exactly what we should expect, because the number of species of organisms that have ever lived upon the earth, since the earliest geological times, will probably be many hundred times greater than those now existing of which we have any knowledge; and hence the enormous gaps and chasms in the geological record of extinct forms is not to be wondered at. Yet, notwithstanding these chasms in our knowledge, if evolution is true, there ought to have been, on the whole, progression in all the chief types of life. The higher and more specialised forms should have come into existence later than the lower and more generalised forms; and however fragmentary the portions we possess of the whole tree of life upon the earth, they ought to show us broadly that such a progressive evolution has taken place. We have seen that in some special groups, already referred to, such a progression is clearly visible, and we will now cast a hasty glance over the entire series of fossil forms, in order to see if a similar progression is manifested by them as a whole.

The Progressive Development of Plants.

Ever since fossil plants have been collected and studied, the broad fact has been apparent that the early plants—those of the Coal formation—were mainly cryptogamous, while in the Tertiary deposits the higher flowering plants prevailed. In the intermediate secondary epoch the gymnosperms—cycads and coniferæ—formed a prominent part of the vegetation, and as these have usually been held to be a kind of transition form between the flowerless and flowering plants, the geological succession has always, broadly speaking, been in accordance with the theory of evolution. Beyond this, however, the facts were very puzzling. The highest cryptogams—ferns, lycopods, and equisetaceæ—appeared suddenly, and in immense profusion in the Coal formation, at which period they attained a development they have never since surpassed or even equalled; while the highest plants—the dicotyledonous and monocotyledonous angiosperms—which now form the bulk of the vegetation of the world, and exhibit the most wonderful modifications of form and structure, were almost unknown till the Tertiary period, when they suddenly appeared in full development, and, for the most part, under the same generic forms as now exist.

During the latter half of the present century, however, great additions have been made to our knowledge of fossil plants; and although there are still indications of vast gaps in our knowledge, due, no doubt, to the very exceptional conditions required for the preservation of plant remains, we now possess evidence of a more continuous development of the various types of vegetation. According to Mr. Lester F. Ward, between 8000 and 9000 species of fossil plants have been described or indicated; and, owing to the careful study of the nervation of leaves, a large number of these are referable to their proper orders or genera, and therefore give us some notion—which, though very imperfect, is probably accurate in its main outlines—of the progressive development of vegetation on the earth.[1] The following is a summary of the facts as given by Mr. Ward:—

The lowest forms of vegetable life—the cellular plants—have been found in Lower Silurian deposits in the form of three species of marine algæ; and in the whole Silurian formation fifty species have been recognised. We cannot for a moment suppose, however, that this indicates the first appearance of vegetable life upon the earth, for in these same Lower Silurian beds the more highly organised vascular cryptogams appear in the form of rhizocarps—plants allied to Marsilea and Azolla,—and a very little higher, ferns, lycopods, and even conifers appear. We have indications, however, of a still more ancient vegetation, in the carbonaceous shales and thick beds of graphite far down in the Middle Laurentian, since there is no other known agency than the vegetable cell by means of which carbon can be extracted from the atmo-

[1] Sketch of Palæobotany in Fifth Annual Report of U.S. Geological Survey, 1883-84, pp. 363-452, with diagrams. Sir J. William Dawson, speaking of the value of leaves for the determination of fossil plants, says: "In my own experience I have often found determinations of the leaves of trees confirmed by the discovery of their fruits or of the structure of their stems. Thus, in the rich cretaceous plant-beds of the Dunvegan series, we have beech-nuts associated in the same bed with leaves referred to *Fagus*. In the Laramie beds I determined many years ago nuts of the *Trapa* or water-chestnut, and subsequently Lesquereux found in beds in the United States leaves which he referred to the same genus. Later, I found in collections made on the Red Deer River of Canada my fruits and Lesquereux's leaves on the same slab. The presence of trees of the genera *Carya* and *Juglans* in the same formation was inferred from their leaves, and specimens have since been obtained of silicified wood with the microscopic structure of the modern butternut. Still we are willing to admit that determinations from leaves alone are liable to doubt."—*The Geological History of Plants*, p. 196.

sphere and fixed in the solid state. These great beds of graphite, therefore, imply the existence of abundance of vegetable life at the very commencement of the era of which we have any geological record.[1]

Ferns, as already stated, begin in the Middle Silurian formation with the Eopteris Morrieri. In the Devonian, we have 79 species, in the Carboniferous 627, and in the Permian 186 species; after which fossil ferns diminish greatly, though they are found in every formation; and the fact that fully 3000 living species are known, while the richest portion of the Tertiary in fossil plants—the Miocene—has only produced 87 species, will serve to indicate the extreme imperfection of the geological record.

The Equisetaceæ (horsetails) which also first appear in the Silurian and reach their maximum development in the Coal formation, are, in all succeeding formations, far less numerous than ferns, and only thirty living species are known. Lycopodiaceæ, though still more abundant in the Coal formation, are very rarely found in any succeeding deposit, though the living species are tolerably numerous, about 500 having been described. As we cannot suppose them to have really diminished and then increased again in this extraordinary manner, we have another indication of the exceptional nature of plant preservation and the extreme and erratic character of the imperfection of the record.

Passing now to the next higher division of plants—the gymnosperms—we find Coniferæ appearing in the Upper Silurian, becoming tolerably abundant in the Devonian, and reaching a maximum in the Carboniferous, from which formation more than 300 species are known, equal to the number recorded as now living. They occur in all succeeding formations, being abundant in the Oolite, and excessively so in the Miocene, from which 250 species have been described. The allied family of gymnosperms, the Cycadaceæ, first appear in the Carboniferous era, but very scantily; are most abundant in the Oolite, from which formation 116 species are known, and then steadily diminish to the Tertiary, although there are seventy-five living species.

We now come to the true flowering plants, and we first

[1] Sir J. William Dawson's *Geological History of Plants*, p. 18.

meet with monocotyledons in the Carboniferous and Permian formations. The character of these fossils was long disputed, but is now believed to be well established; and the sub-class continues to be present in small numbers in all succeeding deposits, becoming rather plentiful in the Upper Cretaceous, and very abundant in the Eocene and Miocene. In the latter formation 272 species have been discovered; but the 116 species in the Eocene form a larger proportion of the total vegetation of the period.

True dicotyledons appear very much later, in the Cretaceous period, and only in its upper division, if we except a single species from the Urgonian beds of Greenland. The remarkable thing is that we here find the sub-class fully developed and in great luxuriance of types, all the three divisions—Apetalæ, Polypetalæ, and Gamopetalæ—being represented, with a total of no less than 770 species. Among them are such familiar forms as the poplar, the birch, the beech, the sycamore, and the oak; as well as the fig, the true laurel, the sassafras, the persimmon, the maple, the walnut, the magnolia, and even the apple and the plum tribes. Passing on to the Tertiary period the numbers increase, till they reach their maximum in the Miocene, where more than 2000 species of dicotyledons have been discovered. Among these the proportionate number of the higher gamopetalæ has slightly increased, but is considerably less than at the present day.

Possible Cause of sudden late Appearance of Exogens.

The sudden appearance of fully developed exogenous flowering plants in the Cretaceous period is very analogous to the equally sudden appearance of all the chief types of placental mammalia in the Eocene; and in both cases we must feel sure that this suddenness is only apparent, due to unknown conditions which have prevented their preservation (or their discovery) in earlier formations. The case of the dicotyledonous plants is in some respects the most extraordinary, because in the earlier Mesozoic formations we appear to have a fair representation of the flora of the period, including such varied forms as ferns, equisetums, cycads, conifers, and monocotyledons. The only hint at an explanation of this anomaly has been given by Mr. Ball, who supposes that all these groups inhabited the lowlands, where there was not only excessive heat and moisture, but also a superabundance of carbonic acid in the atmosphere—conditions under which these groups had been developed, but which were prejudicial to the dicotyledons. These latter are supposed to have originated on the high table-lands and mountain ranges, in a rarer and drier atmosphere in which the quantity of carbonic acid gas was much less; and any deposits formed in lake beds at high altitudes and at such a remote epoch have been destroyed by denudation, and hence we have no record of their existence.[1]

During a few weeks spent recently in the Rocky Mountains, I was struck by the great scarcity of monocotyledons and ferns in comparison with dicotyledons—a scarcity due apparently to the dryness and rarity of the atmosphere favouring the higher groups. If we compare Coulter's *Rocky Mountain Botany* with Gray's *Botany of the Northern* (*East*) *United States,* we have two areas which differ chiefly in the points of altitude and atmospheric moisture. Unfortunately, in neither of these works are the species consecutively numbered; but by taking the pages occupied by the two divisions of dicotyledons on the one hand, monocotyledons and ferns on the other, we can obtain a good approximation. In this way we find that in the flora of the North-Eastern States the monocotyledons and ferns are to the dicotyledons in the proportion of 45 to 100; in the Rocky Mountains they are in the proportion of only 34 to 100; while if we take an exclusively Alpine flora, as given by Mr. Ball, there are not one-fifth as many monocotyledons as dicotyledons. These facts show that even at the present day elevated plateaux and mountains are more favourable to dicotyledons than to monocotyledons, and we may, therefore, well suppose that the former originated within such elevated areas, and were for long ages confined to them. It is interesting to note that their richest early remains have been found in the central regions of the North American continent, where they now, proportionally, most abound, and where the conditions of altitude and a dry atmosphere were probably present at a very early period.

[1] "On the Origin of the Flora of the European Alps," *Proc. of Roy. Geog. Society*, vol. i. (1879), pp. 564-588.

The diagram (Fig. 34), slightly modified from one given

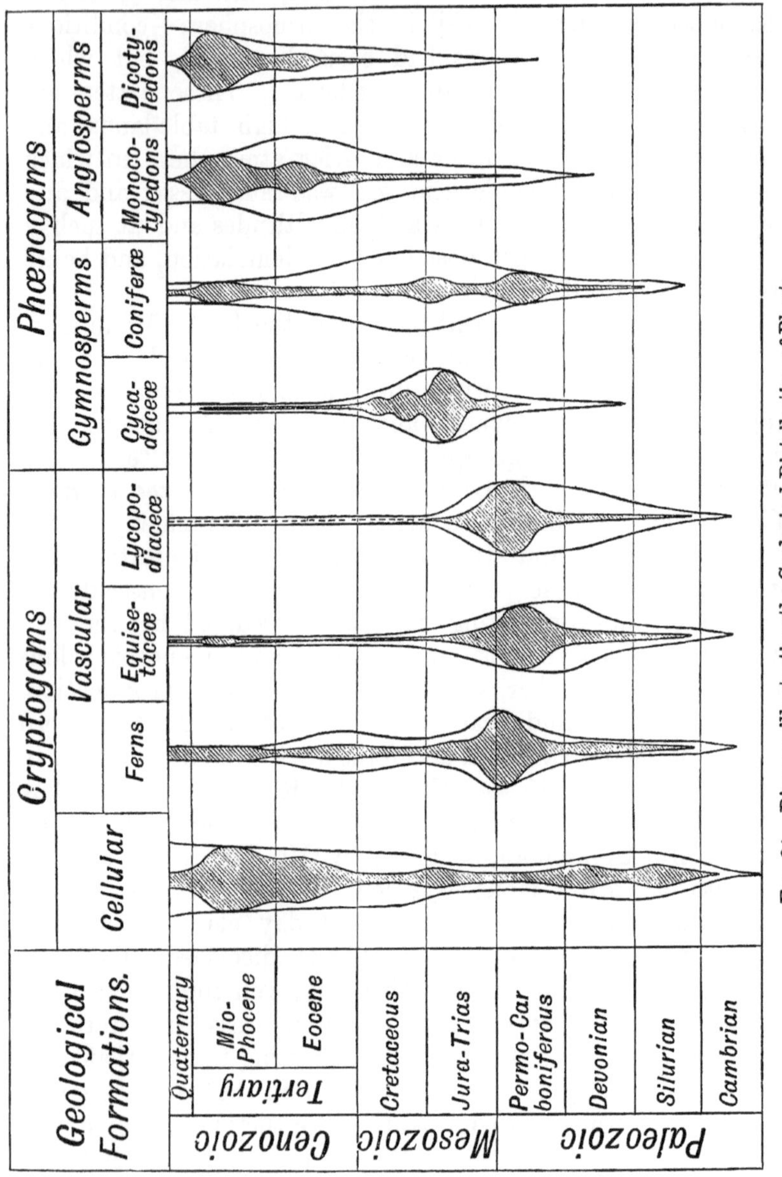

FIG. 34.—Diagram illustrating the Geological Distribution of Plants.

by Mr. Ward, will illustrate our present knowledge of the development of the vegetable kingdom in geological time.

The shaded vertical bands exhibit the proportions of the fossil forms actually discovered, while the outline extensions are intended to show what we may fairly presume to have been the approximate periods of origin, and progressive increase of the number of species, of the chief divisions of the vegetable kingdom. These seem to accord fairly well with their respective grades of development, and thus offer no obstacle to the acceptance of the belief in their progressive evolution.

Geological Distribution of Insects.

The marvellous development of insects into such an endless variety of forms, their extreme specialisation, and their adaptation to almost every possible condition of life, would almost necessarily imply an extreme antiquity. Owing, however, to their small size, their lightness, and their usually aerial habits, no class of animals has been so scantily preserved in the rocks; and it is only recently that the whole of the scattered material relating to fossil insects and their allies have been brought together by Mr. Samuel H. Scudder of Boston, and we have thus learned their bearing on the theory of evolution.[1]

The most striking fact which presents itself on a glance at the distribution of fossil insects, is the completeness of the representation of all the chief types far back in the Secondary period, at which time many of the existing families appear to have been perfectly differentiated. Thus in the Lias we find dragonflies "apparently as highly specialised as to-day, no less than four tribes being present." Of beetles we have undoubted Curculionidæ from the Lias and Trias; Chrysomelidæ in the same deposits; Cerambycidæ in the Oolites; Scarabæidæ in the Lias; Buprestidæ in the Trias; Elateridæ, Trogositidæ, and Nitidulidæ in the Lias; Staphylinidæ in the English Purbecks; while Hydrophilidæ, Gyrinidæ, and Carabidæ occur in the Lias. All these forms are well represented, but there are many other families doubtfully identified in equally ancient rocks. Diptera of the families Empidæ, Asilidæ, and Tipulidæ have been found as far back as the Lias. Of Lepidoptera, Sphingidæ and Tineidæ have been found

[1] Systematic Review of our Present Knowledge of Fossil Insects, including Myriapods and Arachnids (*Bull. of U. S. Geol. Survey*, No. 31, Washington, 1886).

in the Oolite; while ants, representing the highly specialised Hymenoptera, have occurred in the Purbeck and Lias.

This remarkable identity of the families of very ancient with those of existing insects is quite comparable with the apparently sudden appearance of existing genera of trees in the Cretaceous epoch. In both cases we feel certain that we must go very much farther back in order to find the ancestral forms from which they were developed, and that at any moment some fresh discovery may revolutionise our ideas as to the antiquity of certain groups. Such a discovery was made while Mr. Scudder's work was passing through the press. Up to that date all the existing orders of true insects appeared to have originated in the Trias, the alleged moth and beetle of the Coal formation having been incorrectly determined. But now, undoubted remains of beetles have been found in the Coal measures of Silesia, thus supporting the interpretation of the borings in carboniferous trees as having been made by insects of this order, and carrying back this highly specialised form of insect life well into Palæozoic times. Such a discovery renders all speculation as to the origin of true insects premature, because we may feel sure that all the other orders of insects, except perhaps hymenoptera and lepidoptera, were contemporaneous with the highly specialised beetles.

The less highly organised terrestrial arthropoda—the Arachnida and Myriapoda—are, as might be expected, much more ancient. A fossil spider has been found in the Carboniferous, and scorpions in the Upper Silurian rocks of Scotland, Sweden, and the United States. Myriapoda have been found abundantly in the Carboniferous and Devonian formations; but all are of extinct orders, exhibiting a more generalised structure than living forms.

Much more extraordinary, however, is the presence in the Palæozoic formations of ancestral forms of true insects, termed by Mr. Scudder Palæodictyoptera. They consist of generalised cockroaches and walking-stick insects (Orthopteroidea); ancient mayflies and allied forms, of which there are six families and more than thirty genera (Neuropteroidea); three genera of Hemipteroidea resembling various Homoptera and Hemiptera, mostly from the Carboniferous formation, a few from the Devonian, and one ancestral cockroach (Palæoblattina) from the Middle Silurian sandstone of France. If this occurrence of a true hexapod insect from the Middle Silurian be really established, taken in connection with the well-defined Coleoptera from the Carboniferous, the origin of the entire group of terrestrial arthropoda is necessarily thrown back into the Cambrian epoch, if not earlier. And this cannot be considered improbable in view of the highly differentiated land plants—ferns, equisetums, and lycopods—in the Middle or Lower Silurian, and even a conifer (Cordaites Robbii) in the Upper Silurian; while the beds of graphite in the Laurentian were probably formed from terrestrial vegetation.

On the whole, then, we may affirm that, although the geological record of the insect life of the earth is exceptionally imperfect, it yet decidedly supports the evolution hypothesis. The most specialised order, Lepidoptera, is the most recent, only dating back to the Oolite; the Hymenoptera, Diptera, and Homoptera go as far as the Lias; while the Orthoptera and Neuroptera extend to the Trias. The recent discovery of Coleoptera in the Carboniferous shows, however, that the preceding limits are not absolute, and will probably soon be overpassed. Only the more generalised ancestral forms of winged insects have been traced back to Silurian time, and along with them the less highly organised scorpions; facts which serve to show us the extreme imperfection of our knowledge, and indicate possibilities of a world of terrestrial life in the remotest Palæozoic times.

Geological Succession of Vertebrata.

The lowest forms of vertebrates are the fishes, and these appear first in the geological record in the Upper Silurian formation. The most ancient known fish is a Pteraspis, one of the bucklered ganoids or plated fishes—by no means a very low type—allied to the sturgeon (Accipenser) and alligator-gar (Lepidosteus), but, as a group, now nearly extinct. Almost equally ancient are the sharks, which under various forms still abound in our seas. We cannot suppose these to be nearly the earliest fishes, especially as the two lowest orders, now represented by the Amphioxus or lancelet and the lampreys, have not yet been found fossil. The ganoids were greatly developed in the Devonian era, and continued till the

Cretaceous, when they gave way to the true osseous fishes, which had first appeared in the Jurassic period, and have continued to increase till the present day. This much later appearance of the higher osseous fishes is quite in accordance with evolution, although some of the very lowest forms, the lancelet and the lampreys, together with the archaic ceratodus, have survived to our time.

The Amphibia, represented by the extinct labyrinthodons, appear first in the Carboniferous rocks, and these peculiar forms became extinct early in the Secondary period. The labyrinthodons were, however, highly specialised, and do not at all indicate the origin of the class, which may be as ancient as the lower forms of fishes. Hardly any recognisable remains of our existing groups—the frogs, toads, and salamanders—are found before the Tertiary period, a fact which indicates the extreme imperfection of the record as regards this class of animals.

True reptiles have not been found till we reach the Permian where Prohatteria and Proterosaurus occur, the former closely allied to the lizard-like Sphenodon of New Zealand, the latter having its nearest allies in the same group of reptiles—Rhyncocephala, other forms of which occur in the Trias. In this last-named formation the earliest crocodiles—Phytosaurus (Belodon) and Stagonolepis occur, as well as the earliest tortoises—Chelytherium, Proganochelys, and Psephoderma.[1] Fossil serpents have been first found in the Cretaceous formation, but the conditions for the preservation of these forms have evidently been unfavourable, and the record is correspondingly incomplete. The marine Plesiosauri and Ichthyosauri, the flying Pterodactyles, the terrestrial Iguanodon of Europe, and the huge Atlantosaurus of Colorado—the largest land animal that has ever lived upon the earth[2]—all belong to special developments of the reptilian type which flourished during the Secondary epoch, and then became extinct.

[1] For the facts as to the early appearance of the above named groups of reptiles I am indebted to Mr. R. Lydekker of the Geological Department of the Natural History Museum.

[2] According to Professor Marsh this creature was 50 or 60 feet long, and when erect, at least 30 feet in height. It fed upon the foliage of the mountain forests of the Cretaceous epoch, the remains of which are preserved with it.

Birds are among the rarest of fossils, due, no doubt, to their aerial habits removing them from the ordinary dangers of flood, bog, or ice which overwhelm mammals and reptiles, and also to their small specific gravity which keeps them floating on the surface of water till devoured. Their remains were long confined to Tertiary deposits, where many living genera and a few extinct forms have been found. The only birds yet known from the older rocks are the toothed birds (Odontornithes) of the Cretaceous beds of the United States, belonging to two distinct families and many genera; a penguin-like form (Enaliornis) from the Upper Greensand of Cambridge; and the well-known long-tailed Archæopteryx from the Upper Oolite of Bavaria. The record is thus imperfect and fragmentary in the extreme; but it yet shows us, in the few birds discovered in the older rocks, more primitive and generalised types, while the Tertiary birds had already become specialised like those living, and had lost both the teeth and the long vertebral tail, which indicate reptilian affinities in the earlier ages.

Mammalia have been found, as already stated, as far back as the Trias formation, in Europe in the United States and in South Africa, all being very small, and belonging either to the Marsupial order, or to some still lower and more generalised type, out of which both Marsupials and Insectivora were developed. Other allied forms have been found in the Lower and Upper Oolite both of Europe and the United States. But there is then a great gap in the whole Cretaceous formation, from which no mammal has been obtained, although both in the Wealden and the Upper Chalk in Europe, and in the Upper Cretaceous deposits of the United States an abundant and well-preserved terrestrial flora has been discovered. Why no mammals have left their remains here it is impossible to say. We can only suppose that the limited areas in which land plants have been so abundantly preserved, did not present the conditions which are needed for the fossilisation and preservation of mammalian remains.

When we come to the Tertiary formation, we find mammals in abundance; but a wonderful change has taken place. The obscure early types have disappeared, and we discover in their place a whole series of forms belonging to existing orders, and even sometimes to existing families. Thus, in the Eocene we have remains of the opossum family; bats apparently belonging to living genera; rodents allied to the South American cavies and to dormice and squirrels; hoofed animals belonging to the odd-toed and even-toed groups; and ancestral forms of cats, civets, dogs, with a number of more generalised forms of carnivora. Besides these there are whales, lemurs, and many strange ancestral forms of proboscidea.[1]

The great diversity of forms and structures at so remote an epoch would require for their development an amount of time, which, judging by the changes that have occurred in other groups, would carry us back far into the Mesozoic period. In order to understand why we have no record of these changes in any part of the world, we must fall back upon some such supposition as we made in the case of the dicotyledonous plants. Perhaps, indeed, the two cases are really connected, and the upland regions of the primeval world, which saw the development of our higher vegetation, may have also afforded the theatre for the gradual development of the varied mammalian types which surprise us by their sudden appearance in Tertiary times.

Notwithstanding these irregularities and gaps in the record, the accompanying table, summarising our actual knowledge of the geological distribution of the five classes of vertebrata,

GEOLOGICAL DISTRIBUTION OF MAMMALIA.

	Silurian.	Devonian.	Carboniferous.	Permian.	Trias.	Lias.	Oolite.	Cretaceous.	Tertiary.
Fishes . . .	━	━	━	━	━	━	━	━	━
Amphibia . .			━	━	━	━	━	━	━
Reptiles . .				━	━	━	━	━	━
Birds . . .							━	━	━
Mammalia . .					━	━	━	━	━

[1] For fuller details, see the author's *Geographical Distribution of Animals*, and Heilprin's *Geographical and Geological Distribution of Animals.*

exhibits a steady progression from lower to higher types, excepting only the deficiency in the bird record which is easily explained. The comparative perfection of type in which each of these classes first appears, renders it certain that the origin of each and all of them must be sought much farther back than any records which have yet been discovered. The researches of palæontologists and embryologists indicate a reptilian origin for birds and mammals, while reptiles and amphibia arose, perhaps independently, from fishes.

Concluding Remarks.

The brief review we have now taken of the more suggestive facts presented by the geological succession of organic forms, is sufficient to show that most, if not all, of the supposed difficulties which it presents in the way of evolution, are due either to imperfections in the geological record itself, or to our still very incomplete knowledge of what is really recorded in the earth's crust. We learn, however, that just as discovery progresses, gaps are filled up and difficulties disappear; while, in the case of many individual groups, we have already obtained all the evidence of progressive development that can reasonably be expected. We conclude, therefore, that the geological difficulty has now disappeared; and that this noble science, when properly understood, affords clear and weighty evidence of evolution.

CHAPTER XIV

FUNDAMENTAL PROBLEMS IN RELATION TO VARIATION AND HEREDITY

Fundamental difficulties and objections—Mr. Herbert Spencer's factors of organic evolution—Disuse and effects of withdrawal of natural selection—Supposed effects of disuse among wild animals—Difficulty as to co-adaptation of parts by variation and selection—Direct action of the environment—The American school of evolutionists—Origin of the feet of the ungulates—Supposed action of animal intelligence—Semper on the direct influence of the environment—Professor Geddes's theory of variation in plants—Objections to the theory—On the origin of spines—Variation and selection overpower the effects of use and disuse—Supposed action of the environment in imitating variations—Weismann's theory of heredity—The cause of variation—The non-heredity of acquired characters—The theory of instinct—Concluding remarks.

HAVING now set forth and illustrated at some length the most important of the applications of the development hypothesis in the explanation of the broader and more generally interesting phenomena presented by the organic world, we propose to discuss some of the more fundamental problems and difficulties which have recently been adduced by eminent naturalists. It is the more necessary to do this, because there is now a tendency to minimise the action of natural selection in the production of organic forms, and to set up in its place certain fundamental principles of variation or laws of growth, which it is urged are the real originators of the several lines of development, and of most of the variety of form and structure in the vegetable and animal kingdoms. These views have, moreover, been seized upon by popular writers to throw doubt and discredit on the whole theory of evolution, and especially on Darwin's presentation of that theory, to the bewilderment of the general public, who are quite unable to decide how far the new views, even if well established, tend to subvert the Darwinian theory, or whether they are really more than subsidiary parts of it, and quite powerless without it to produce any effect whatever.

The writers whose special views we now propose to consider are: (1) Mr. Herbert Spencer, on modification of structures arising from modification of functions, as set forth in his *Factors of Organic Evolution.* (2) Dr. E. D. Cope, who advocates similar views in detail, in his work entitled *The Origin of the Fittest,* and may be considered the head of a school of American naturalists who minimise the agency of natural selection. (3) Dr. Karl Semper, who has especially studied the direct influence of the environment in the whole animal kingdom, and has set forth his views in a volume on *The Natural Conditions of Existence as they Affect Animal Life.* (4) Mr. Patrick Geddes, who urges that fundamental laws of growth, and the antagonism of vegetative and reproductive forces, account for much that has been imputed to natural selection.

We will now endeavour to ascertain what are the more important facts and arguments adduced by each of the above writers, and how far they offer a substitute for the action of natural selection; having done which, a brief account will be given of the views of Dr. Aug. Weismann, whose theory of heredity will, if established, strike at the very root of the arguments of the first three of the writers above referred to.

Mr. Herbert Spencer's Factors of Organic Evolution.

Mr. Spencer, while fully recognising the importance and wide range of the principle of natural selection, thinks that sufficient weight has not been given to the effects of use and disuse as a factor in evolution, or to the direct action of the environment in determining or modifying organic structures. As examples of the former class of actions, he adduces the decreased size of the jaws in the civilised races of mankind, the inheritance of nervous disease produced by overwork, the great and inherited development of the udders in cows and goats, and the shortened legs, jaws, and snout in improved races of pigs—the two latter examples being quoted from Mr. Darwin,—and other cases of like nature. As examples of the latter, Mr. Darwin is again quoted as admitting that there are many cases in which the action of similar conditions appears to have produced corresponding changes in different species; and we have a very elaborate discussion of the direct action of the medium in modifying the protoplasm of simple organisms, so as to bring about the difference between the outer surface and the inner part that characterises the cells or other units of which they are formed.

Now, although this essay did little more than bring together facts which had been already adduced by Mr. Darwin or by Mr. Spencer himself, and lay stress upon their importance, its publication in a popular review was immediately seized upon as "an avowed and definite declaration against some of the leading ideas on which the Mechanical Philosophy depends," and as being "fatal to the adequacy of the Mechanical Philosophy as any explanation of organic evolution,"[1]—an expression of opinion which would be repudiated by every Darwinian. For, even admitting the interpretation which Mr. Spencer puts on the facts he adduces, they are all included in the causes which Darwin himself recognised as having acted in bringing about the infinitude of forms in the organic world. In the concluding chapter of the *Origin of Species* he says: "I have now recapitulated the facts and considerations which have thoroughly convinced me that species have been modified during a long course of descent. This has been effected chiefly through the natural selection of numerous successive, slight, favourable variations; aided in an important manner by the inherited effects of the use and disuse of parts; and in an unimportant manner—that is, in relation to adaptive structures whether past or present, by the direct action of external conditions, and by variations which seem to us, in our ignorance, to arise spontaneously." This passage, summarising Darwin's whole inquiry, and explaining his final point of view, shows how very inaccurate may be the popular notion, as expressed by the Duke of Argyll, of any supposed additions to the causes of change of species as recognised by Darwin.

[1] See the Duke of Argyll's letter in *Nature*, vol. xxxiv. p. 336.

But, as we shall see presently, there is now much reason to believe that the supposed inheritance of acquired modifications—that is, of the effects of use and disuse, or of the direct influence of the environment—is not a fact; and if so, the very foundation is taken away from the whole class of objections on which so much stress is now laid. It therefore becomes important to inquire whether the facts adduced by Darwin, Spencer, and others, do really necessitate such inheritance, or whether any other interpretation of them is possible. I believe there is such an interpretation; and we will first consider the cases of disuse on which Mr. Spencer lays most stress.

The cases Mr. Spencer adduces as demonstrating the effects of disuse in diminishing the size and strength of organs are, the diminished size of the jaws in the races of civilised men, and the diminution of the muscles used in closing the jaws in the case of pet-dogs fed for generations on soft food. He argues that the minute reduction in any one generation could not possibly have been useful, and, therefore, not the subject of natural selection; and against the theory of correlation of the diminished jaw with increased brain in man, he urges that there are cases of large brain development, accompanied by jaws above the average size. Against the theory of economy of nutrition in the case of the pet-dogs, he places the abundant food of these animals which would render such economy needless.

But neither he nor Mr. Darwin has considered the effects of the withdrawal of the action of natural selection in keeping up the parts in question to their full dimensions, which, of itself, seems to me quite adequate to produce the results observed. Recurring to the evidence, adduced in Chapter III, of the constant variation occurring in all parts of the organism, while selection is constantly acting on these variations in eliminating all that fall below the best working standard, and preserving only those that are fully up to it; and, remembering further, that, of the whole number of the increase produced annually, only a small percentage of the best adapted can be preserved, we shall see that every useful organ will be kept up nearly to its higher limit of size and efficiency. Now Mr. Galton has proved experimentally that, when any part has thus been increased (or diminished) by selection, there is in the offspring a strong tendency to revert to a mean or average size, which tends to check further increase. And this mean appears to be, not the mean of the actual existing individuals but a lower mean, or that from which they had been recently raised by selection.[1] He calls this the law of "Regression towards Mediocrity," and it has been proved by experiments with vegetables and by observations on mankind. This regression, in every generation, takes place even when both parents have been selected for their high development of the organ in question; but when there is no such selection, and crosses are allowed among individuals of every grade of development, the deterioration will be very rapid; and after a time not only will the average size of the part be greatly reduced, but the instances of full development will become very rare. Thus what Weismann terms "panmixia," or free intercrossing, will co-operate with Galton's law of "regression towards mediocrity," and the result will be that, whenever selection ceases to act on any part or organ which has heretofore been kept up to a maximum of size and efficiency, the organ in question will rapidly decrease till it reaches a mean value considerably below the mean of the progeny that has usually been produced each year, and very greatly below the mean of that portion which has survived annually; and this will take place by the general law of heredity, and quite irrespective of any *use* or *disuse* of the part in question. Now, no observations have been adduced by Mr. Spencer or others, showing that the average amount of change supposed to be due to *disuse* is greater than that due to the law of regression towards mediocrity; while even if it were somewhat greater, we can see many possible contributory causes to its production. In the case of civilised man's diminished jaw, there may well be some correlation between the jaw and the brain, seeing that increased mental activity would lead to the withdrawal of blood and of nervous energy from adjacent parts, and might thus lead to diminished growth of those parts in the individual. And in the case of pet-dogs, the selection of small or short-headed individuals would imply the unconscious selection of those with less massive temporal muscles, and thus lead to the concomitant

[1] *Journal of the Anthropological Institute*, vol. xv. pp. 246-260.

reduction of those muscles. The amount of reduction observed by Darwin in the wing-bones of domestic ducks and poultry, and in the hind legs of tame rabbits, is very small, and is certainly no greater than the above causes will well account for; while so many of the external characters of all our domestic animals have been subject to long-continued artificial selection, and we are so ignorant of the possible correlations of different parts, that the phenomena presented by them seem sufficiently explained without recurrence to the assumption that any changes in the individual, due to disuse, are inherited by the offspring.

Supposed Effects of Disuse among Wild Animals.

It may be urged, however, that among wild animals we have many undoubted results of disuse much more pronounced than those among domestic kinds, results which cannot be explained by the causes already adduced. Such are the reduced size of the wings of many birds on oceanic islands; the abortion of the eyes in many cave animals, and in some which live underground; and the loss of the hind limbs in whales and in some lizards. These cases differ greatly in the amount of the reduction of parts which has taken place, and may be due to different causes. It is remarkable that in some of the birds of oceanic islands the reduction is little if any greater than in domestic birds, as in the water-hen of Tristan d'Acunha. Now if the reduction of wing were due to the hereditary effects of disuse, we should expect a very much greater effect in a bird inhabiting an oceanic island than in a domestic bird, where the disuse has been in action for an indefinitely shorter period. In the case of many other birds, however—as some of the New Zealand rails and the extinct dodo of Mauritius—the wings have been reduced to a much more rudimentary condition, though it is still obvious that they were once organs of flight; and in these cases we certainly require some other causes than those which have reduced the wings of our domestic fowls. One such cause may have been of the same nature as that which has been so efficient in reducing the wings of the insects of oceanic islands—the destruction of those which, during the occasional use of their wings, were carried out to sea. This form of natural selection may well have acted in the case of birds whose powers of flight were already somewhat reduced, and to whom, there being no enemies to escape from, their use was only a source of danger. We may thus, perhaps, account for the fact that many of these birds retain small but useless wings with which they never fly; for, the wings having been reduced to this functionless condition, no power could reduce them further except correlation of growth or economy of nutrition, causes which only rarely come into play.

The complete loss of eyes in some cave animals may, perhaps, be explained in a somewhat similar way. Whenever, owing to the total darkness, they became useless, they might also become injurious, on account of their delicacy of organisation and liability to accidents and disease; in which case natural selection would begin to act to reduce, and finally abort them; and this explains why, in some cases, the rudimentary eye remains, although completely covered by a protective outer skin. Whales, like moas and cassowaries, carry us back to a remote past, of whose conditions we know too little for safe speculation. We are quite ignorant of the ancestral forms of either of these groups, and are therefore without the materials needful for determining the steps by which the change took place, or the causes which brought it about.[1]

On a review of the various examples that have been given by Mr. Darwin and others of organs that have been reduced or aborted, there seems too much diversity in the results for all to be due to so direct and uniform a cause as the individual effects of disuse accumulated by heredity. For if that were the only or chief efficient cause, and a cause capable of producing a decided effect during the comparatively short period

[1] The idea of the non-heredity of acquired variations was suggested by the summary of Professor Weismann's views, in *Nature*, referred to later on. But since this chapter was written I have, through the kindness of Mr. E. B. Poulton, seen some of the proofs of the forthcoming translation of Weismann's Essays on Heredity, in which he sets forth an explanation very similar to that here given. On the difficult question of the almost entire disappearance of organs, as in the limbs of snakes and of some lizards, he adduces "a certain form of correlation, which Roux calls 'the struggle of the parts in the organism,'" as playing an important part. Atrophy following disuse is nearly always attended by the corresponding increase of other organs: blind animals possess more developed organs of touch, hearing, and smell; the loss of power in the wings is accompanied by increased strength of the legs, etc. Now as these latter characters, being useful, will be selected, it is easy to understand that a congenital increase of these will be accompanied by a cor-

of the existence of animals in a state of domestication, we should expect to find that, in wild species, all unused parts or organs had been reduced to the smallest rudiments, or had wholly disappeared. Instead of this we find various grades of reduction, indicating the probable result of several distinct causes; sometimes acting separately, sometimes in combination, such as those we have already pointed out.

And if we find no positive evidence of *disuse*, acting by its direct effect on the individual, being transmitted to the offspring, still less can we find such evidence in the case of the *use* of organs. For here the very fact of *use*, in a wild state, implies *utility*, and utility is the constant subject for the action of natural selection; while among domestic animals those parts which are exceptionally used are so used in the service of man, and have thus become the subjects of artificial selection. Thus "the great and inherited development of the udders in cows and goats," quoted by Spencer from Darwin, really affords no proof of inheritance of the increase due to use, because, from the earliest period of the domestication of these animals, abundant milk-production has been highly esteemed, and has thus been the subject of selection; while there are no cases among wild animals that may not be better explained by variation and natural selection.

Difficulty as to Co-adaptation of Parts by Variation and Selection.

Mr. Spencer again brings forward this difficulty, as he did in his *Principles of Biology* twenty-five years ago, and urges that all the adjustments of bones, muscles, blood-vessels, and nerves which would be required during, for example, the development of the neck and fore-limbs of the giraffe, could

responding congenital diminution of the unused organ; and in cases where the means of nutrition are deficient, every diminution of these useless parts will be a gain to the whole organism, and thus their complete disappearance will, in some cases, be brought about directly by natural selection. This corresponds with what we know of these rudimentary organs.

It must, however, be pointed out that the non-heredity of acquired characters was maintained by Mr. Francis Galton more than twelve years ago, on theoretical considerations almost identical with those urged by Professor Weismann; while the insufficiency of the evidence for their hereditary transmission was shown, by similar arguments to those used above and in the work of Professor Weismann already referred to (see "A Theory of Heredity," in *Journ. Anthrop. Instit.*, vol. v. pp. 343-345).

not have been effected by "simultaneous fortunate spontaneous variations." But this difficulty is fully disposed of by the facts of simultaneous variation adduced in our third chapter, and has also been specially considered in Chapter VI, p. 127. The best answer to this objection may, perhaps, be found in the fact that the very thing said to be impossible by variation and natural selection has been again and again effected by variation and artificial selection. During the process of formation of such breeds as the greyhound or the bull-dog, of the race-horse and cart-horse, of the fantail pigeon or the otter-sheep, many co-ordinate adjustments have been produced; and no difficulty has occurred, whether the change has been effected by a single variation—as in the last case named—or by slow steps, as in all the others. It seems to be forgotten that most animals have such a surplus of vitality and strength for all the ordinary occasions of life that any slight superiority in one part can be at once utilised; while the moment any want of balance occurs, variations in the insufficiently developed parts will be selected to bring back the harmony of the whole organisation. The fact that, in all domestic animals, variations do occur, rendering them swifter or stronger, larger or smaller, stouter or slenderer, and that such variations can be separately selected and accumulated for man's purposes, is sufficient to render it certain that similar or even greater changes may be effected by natural selection, which, as Darwin well remarks, "acts on every internal organ, on every shade of constitutional difference, on the whole machinery of life." The difficulty as to co-adaptation of parts by variation and natural selection appears to me, therefore, to be a wholly imaginary difficulty which has no place whatever in the operations of nature.

Direct Action of the Environment.

Mr. Spencer's last objection to the wide scope given by Darwinians to the agency of natural selection is, that organisms are acted upon by the environment, which produces in them definite changes, and that these changes in the individual are transmitted by inheritance, and thus become increased in successive generations. That such changes are produced in the individual there is ample evidence, but that they are inherited independently of any form of selection or of reversion is exceedingly doubtful, and Darwin nowhere expresses himself as satisfied with the evidence. The two very strongest cases he mentions are the twenty-nine species of American trees which all differed in a corresponding way from their nearest European allies; and the American maize which became changed after three generations in Europe. But in the case of the trees the differences alleged may be partly due to correlation with constitutional peculiarities dependent on climate, especially as regards the deeper tint of the fading leaves and the smaller size of the buds and seeds in America than in Europe; while the less deeply toothed or serrated leaves in the American species are, in our present complete ignorance of the causes and uses of serration, quite as likely to be due to some form of adaptation as to any direct action of the climate. Again, we are not told how many of the allied species do not vary in this particular manner, and this is certainly an important factor in any conclusion we may form on the question.

In the case of the maize it appears that one of the more remarkable and highly selected American varieties was cultivated in Germany, and in three years nearly all resemblance to the original parent was lost; and in the sixth year it closely resembled a common European variety, but was of somewhat more vigorous growth. In this case no selection appears to have been practised, and the effects may have been due to that "reversion to mediocrity" which invariably occurs, and is more especially marked in the case of varieties which have been rapidly produced by artificial selection. It may be considered as a partial reversion to the wild or unimproved stock; and the same thing would probably have occurred, though perhaps less rapidly, in America itself. As this is stated by Darwin to be the most remarkable case known to him "of the direct and prompt action of climate on a plant," we must conclude that such direct effects have not been proved to be accumulated by inheritance, independently of reversion or selection.

The remaining part of Mr. Spencer's essay is devoted to a consideration of the hypothetical action of the environment on the lower organisms which consist of simple cells or formless masses of protoplasm; and he shows with great elaboration that the outer and inner parts of these are necessarily subject to different conditions; and that the outer actions of air or water lead to the formation of integuments, and sometimes to other definite modifications of the surface, whence arise permanent differences of structure. Although in these cases also it is very difficult to determine how much is due to direct modification by external agencies transmitted and accumulated by inheritance, and how much to spontaneous variations accumulated by natural selection, the probabilities in favour of the former mode of action are here greater, because there is no differentiation of nutritive and reproductive cells in these simple organisms; and it can be readily seen that any change produced in the latter will almost certainly affect the next generation.[1] We are thus carried back almost to the origin of life, and can only vaguely speculate on what took place under conditions of which we know so little.

The American School of Evolutionists.

The tentative views of Mr. Spencer which we have just discussed, are carried much further, and attempts have been made to work them out in great detail, by many American naturalists, whose best representative is Dr. E. D. Cope of Philadelphia.[2] This school endeavours to explain all the chief modifications of form in the animal kingdom by fundamental laws of growth and the inherited effects of use and effort, returning, in fact, to the teachings of Lamarck as being at least equally important with those of Darwin.

The following extract will serve to show the high position claimed by this school as original discoverers, and as having made important additions to the theory of evolution:—

"Wallace and Darwin have propounded as the cause of modification in descent their law of natural selection. This law has been epitomised by Spencer as the 'survival of the fittest.' This neat expression no doubt covers the case, but it leaves the origin of the fittest entirely untouched. Darwin assumes a 'tendency to variation' in nature, and it is plainly

[1] This explanation is derived from Weismann's Theory of the Continuity of the Germ-Plasm as summarised in *Nature*.

[2] See a collection of his essays under the title, *The Origin of the Fittest: Essays on Evolution*. D. Appleton and Co. New York. 1887.

necessary to do this, in order that materials for the exercise of a selection should exist. Darwin and Wallace's law is then only restrictive, directive, conservative, or destructive of something already created. I propose, then, to seek for the originative laws by which these subjects are furnished; in other words, for the causes of the origin of the fittest."[1]

Mr. Cope lays great stress on the existence of a special developmental force termed "bathmism" or growth-force, which acts by means of retardation and acceleration "without any reference to fitness at all;" that "instead of being controlled by fitness it is the controller of fitness." He argues that "all the characteristics of generalised groups from genera up (excepting, perhaps, families) have been evolved under the law of acceleration and retardation," combined with some intervention of natural selection; and that specific characters, or species, have been evolved by natural selection with some assistance from the higher law. He, therefore, makes species and genera two absolutely distinct things, the latter not developed out of the former; generic characters and specific characters are, in his opinion, fundamentally different, and have had different origins, and whole groups of species have been simultaneously modified, so as to belong to another genus; whence he thinks it "highly probable that the same specific form has existed through a succession of genera, and perhaps in different epochs of geologic time."

Useful characters, he concludes, have been produced by the special location of growth-force by use; useless ones have been produced by location of growth-force without the influence of use. Another element which determines the direction of growth-force, and which precedes use, is effort; and "it is thought that effort becomes incorporated into the metaphysical acquisitions of the parent, and is inherited with other metaphysical qualities by the young, which, during the period of growth, is much more susceptible to modifying influences, and is likely to exhibit structural change in consequence."[2]

From these few examples of their teachings, it is clear that

[1] *Origin of the Fittest*, p. 174.

[2] *Ibid.* p. 29. It may be here noted that Darwin found these theories unintelligible. In a letter to Professor E. T. Morse in 1877, he writes: "There is one point which I regret you did not make clear in your Ad-

these American evolutionists have departed very widely from the views of Mr. Darwin, and in place of the well-established causes and admitted laws to which he appeals have introduced theoretical conceptions which have not yet been tested by experiments or facts, as well as metaphysical conceptions which are incapable of proof. And when they come to illustrate these views by an appeal to palæontology or morphology, we find that a far simpler and more complete explanation of the facts is afforded by the established principles of variation and natural selection. The confidence with which these new ideas are enunciated, and the repeated assertion that without them Darwinism is powerless to explain the origin of organic forms, renders it necessary to bestow a little more time on the explanations they give us of well-known phenomena with which, they assert, other theories are incompetent to grapple.

As examples of use producing structural change, Mr. Cope adduces the hooked and toothed beaks of the falcons and the butcher-birds, and he argues that the fact of these birds belonging to widely different groups proves that similarity of use has produced a similar structural result. But no attempt is made to show any direct causal connection between the use of a bill to cut or tear flesh and the development of a tooth on the mandible. Such use might conceivably strengthen the bill or increase its size, but not cause a special tooth-like outgrowth which was not present in the ancestral thrush-like forms of the butcher-bird. On the other hand, it is clear that any variations of the bill tending towards a hook or tooth would give the possessor some advantage in seizing and tearing its prey, and would thus be preserved and increased by natural selection. Again, Mr. Cope urges the effects of a supposed "law of polar or centrifugal growth" to counteract a tendency to unsymmetrical growth, where one side of the body is used more than the other. But the undoubted hurtfulness of want of symmetry in many important actions or functions would rapidly eliminate any such tendency. When, however, it has

dress, namely, what is the meaning and importance of Professors Cope and Hyatt's views on acceleration and retardation? I have endeavoured, and given up in despair, the attempt to grasp their meaning" (*Life and Letters*, vol. iii. p. 233).

become useful, as in the case of the single enlarged claw of many crustacea, it has been preserved by natural selection.

Origin of the Feet of the Ungulates.

Perhaps the most original and suggestive of Mr. Cope's applications of the theory of use and effort in modifying structure are, his chapters "On the Origin of the Foot-Structure of the Ungulates;" and that "On the Effect of Impacts and Strains on the Feet of Mammalia;" and they will serve also to show the comparative merits of this theory and that of natural selection in explaining a difficult case of modification, especially as it is an explanation claimed as new and original when first enunciated in 1881. Let us, then, see how he deals with the problem.

The remarkable progressive change of a four or five-toed ancestor into the one-toed horse, and the equally remarkable division of the whole group of ungulate animals into the odd-toed and even-toed divisions, Mr. Cope attempts to explain by the effects of impact and use among animals which frequented hard or swampy ground respectively. On hard ground, it is urged, the long middle toe would be most used and subjected to the greatest strains, and would therefore acquire both strength and development. It would then be still more exclusively used, and the extra nourishment required by it would be drawn from the adjacent less-used toes, which would accordingly diminish in size, till, after a long series of changes, the records of which are so well preserved in the American tertiary rocks, the true one-toed horse was developed. In soft or swampy ground, on the other hand, the tendency would be to spread out the foot so that there were two toes on each side. The two middle toes would thus be most used and most subject to strains, and would, therefore, increase at the expense of the lateral toes. There would be, no doubt, an advantage in these two functional toes being of equal size, so as to prevent twisting of the foot while walking; and variations tending to bring this about would be advantageous, and would therefore be preserved. Thus, by a parallel series of changes in another direction, adapted to a distinct set of conditions, we should arrive at the symmetrical divided hoofs of our deer and cattle. The fact that sheep and goats are specially mountain and rock-loving animals may be explained by their being a later modification, since the divided hoof once formed is evidently well adapted to secure a firm footing on rugged and precipitous ground, although it could hardly have been first developed in such localities. Mr. Cope thus concludes: "Certain it is that the length of the bones in the feet of the ungulate orders has a direct relation to the dryness of the ground they inhabit, and the possibility of speed which their habit permits them or necessarily imposes on them."[1]

If there is any truth in the explanation here briefly summarised, it must entirely depend on the fact of individual modifications thus produced being hereditary, and we yet await the proof of this. In the meantime it is clear that the very same results could have been brought about by variation and natural selection. For the toes, like all other organs, vary in size and proportions, and in their degree of union or separation; and if in one group of animals it was beneficial to have the middle toe larger and longer, and in another set to have the two middle toes of the same size, nothing can be more certain than that these particular modifications would be continuously preserved, and the very results we see ultimately produced.

The oft-repeated objections that the cause of variations is unknown, that there must be something to determine variations in the right direction; that "natural selection includes no actively progressive principle, but must wait for the development of variation, and then, after securing the survival of the best, wait again for the best to project its own variations for selection," we have already sufficiently answered by showing that variation—in abundant or typical species—is always present in ample amount; that it exists in all parts and organs; that these vary, for the most part, independently, so that any required combination of variations can be secured; and finally, that all variation is necessarily either in excess or defect of the mean condition, and that, consequently, the right or favourable variations are so frequently present that the unerring power of natural selection never wants materials to work upon.

[1] *Origin of the Fittest*, p. 374.

Supposed Action of Animal Intelligence.

The following passage briefly summarises Mr. Cope's position: "Intelligence is a conservative principle, and will always direct effort and use into lines which will be beneficial to its possessor. Here we have the source of the fittest, *i.e.* addition of parts by increase and location of growth-force, directed by the influence of various kinds of compulsion in the lower, and intelligent option among higher animals. Thus intelligent choice, taking advantage of the successive evolution of physical conditions, may be regarded as the *originator of the fittest*, while natural selection is the tribunal to which all results of accelerated growth are submitted. This preserves or destroys them, and determines the new points of departure on which accelerated growth shall build."[1]

This notion of "intelligence"—the intelligence of the animal itself—determining its own variation, is so evidently a very partial theory, inapplicable to the whole vegetable kingdom, and almost so to all the lower forms of animals, amongst which, nevertheless, there is the very same adaptation and co-ordination of parts and functions as among the highest, that it is strange to see it put forward with such confidence as necessary for the completion of Darwin's theory. If "the various kinds of compulsion"—by which are apparently meant the laws of variation, growth, and reproduction, the struggle for existence, and the actions necessary to preserve life under the conditions of the animal's environment—are sufficient to have developed the varied forms of the lower animals and of plants, we can see no reason why the same "compulsion" should not have carried on the development of the higher animals also. The action of this "intelligent option" is altogether unproved; while the acknowledgment that natural selection is the tribunal which either preserves or destroys the variations submitted to it, seems quite inconsistent with the statement that intelligent choice is the "orginator of the fittest," since whatever is really "the fittest" can never be destroyed by natural selection, which is but another name for the survival of the fittest. If "the fittest" is always definitely

[1] *Origin of the Fittest*, p. 40.

produced by some other power, then natural selection is not wanted. If, on the other hand, both fit and unfit are produced, and natural selection decides between them, that is pure Darwinism, and Mr. Cope's theories have added nothing to it.

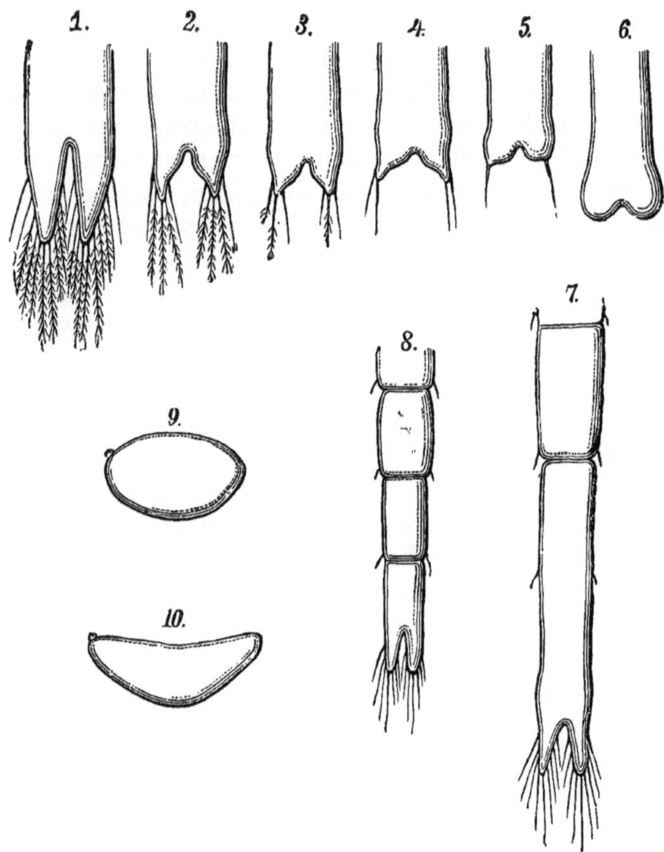

FIG. 35.—Transformation of Artemia salina to A. Milhausenii; 1, tail-lobe of A. salina, and its transition through 2, 3, 4, 5, to 6, into that of A. Milhausenii; 7, post-abdomen of A. salina; 8, post-abdomen of a form bred in brackish water; 9, gill of A. Milhausenii; 10, gill of A. salina. (From Schmankewitsch.)

Semper on the Direct Influence of the Environment.

Another eminent naturalist, Professor Karl Semper of Würzburg, also adopts the view of the direct transforming power of the environment, and has brought together an immense body of interesting facts showing the influence of food, of light, of temperature, of still water and moving water, of the atmosphere and its currents, of gravitation, and of other organisms, in modifying the forms and other characteristics of animals.[1] He believes that these various influences produce a direct and important effect, and that this effect is accumulated by inheritance; yet he acknowledges that we have no direct evidence of this, and there is hardly a single case adduced in the book which is not equally well explained by adaptation, brought about by the survival of beneficial variations. Perhaps the most remarkable case he has brought forward is that of the transformation of species of crustaceans by a change in the saltness of the water (see Fig. 35). Artemia salina lives in brackish water, while A. Milhausenii inhabits water which is much salter. They differ greatly in the form of the tail-lobes, and in the presence or absence of spines upon the tail, and had always been considered perfectly distinct species. Yet either was transformed into the other in a few generations, during which the saltness of the water was gradually altered. Yet more, A. salina was gradually accustomed to fresher water, and in the course of a few generations, when the water had become perfectly fresh, the species was changed into Branchipus stagnalis, which had always been considered to belong to a different genus on account of differences in the form of the antennæ and of the posterior segments of the body (see Fig. 36). This certainly appears to be a proof of change of conditions producing a change of form independently of selection, and of that change of form, while remaining under the same conditions, being inherited. Yet there is this peculiarity in the case, that there is a chemical change in the water, and that this water permeates the whole body, and must be absorbed by the tissues, and thus affect the ova and even the reproductive

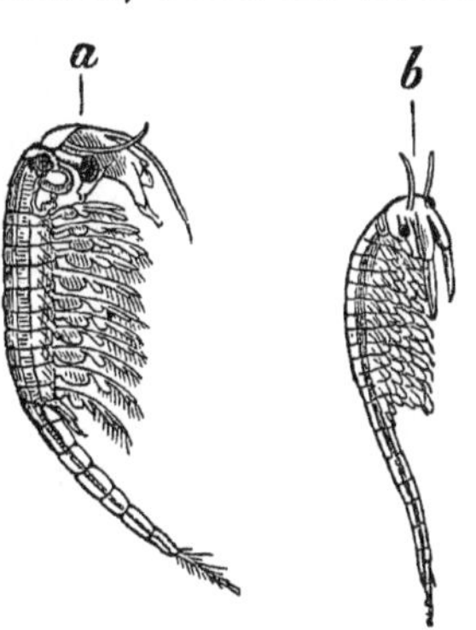

Fig. 36.
a. Branchipus stagnalis.
b. Artemia salina.

[1] *The Natural Conditions of Existence as they Affect Animal Life.* London, 1883.

elements, and in this way may profoundly modify the whole organisation. Why and how the external effects are limited to special details of the structure we do not know; but it does not seem as if any far-reaching conclusions as to the cumulative effect of external conditions on the higher terrestrial animals and plants, can be drawn from such an exceptional phenomenon. It seems rather analogous to those effects of external influences on the very lowest organisms in which the vegetative and reproductive organs are hardly differentiated, in which case such effects are doubtless inherited.[1]

Professor Geddes's Theory of Variation in Plants.

In a paper read before the Edinburgh Botanical Society in 1886 Mr. Patrick Geddes laid down the outlines of a fundamental theory of plant variation, which he has further extended in the article "Variation and Selection" in the *Encyclopædia Britannica*, and in a paper read before the Linnæan Society but not yet published.

A theory of variation should deal alike with the origin of specific distinctions and with those vaster differences which characterise the larger groups, and he thinks it should answer such questions as—How an axis comes to be arrested to form a flower? how the various forms of inflorescence were evolved? how did perigynous or epigynous flowers arise from hypogynous flowers? and many others equally fundamental. Natural selection acting upon numerous accidental variations will not, he urges, account for such general facts as these, which must depend on some constant law of variation. This law he believes to be the well-known antagonism of vegetative and reproductive growth acting throughout the whole course of plant development; and he uses it to explain many of the most characteristic features of the structure of flowers and fruits.

[1] In Dr. Weismann's essay on "Heredity," already referred to, he considers it not improbable that changes in organisms produced by climatic influences may be inherited, because, as these changes do not affect the external parts of an organism only, but often, as in the case of warmth or moisture permeate the whole structure, they may possibly modify the germ-plasm itself, and thus induce variations in the next generation. In this way, he thinks, may possibly be explained the climatic varieties of certain butterflies, and some other changes which seem to be effected by change of climate in a few generations.

Commencing with the origin of the flower, which all botanists agree in regarding as a shortened branch, he explains this shortening as an inevitable physiological fact, since the cost of the development of the reproductive elements is so great as necessarily to check vegetative growth. In the same manner the shortening of the inflorescence from raceme to spike or umbel, and thence to the capitulum or dense flower-head of the composite plants is brought about. This shortening, carried still further, produces the flattened leaf-like receptacle of Dorstenia, and further still the deeply hollowed fruity receptacle of the fig.

The flower itself undergoes a parallel modification due to a similar cause. It is formed by a series of modified leaves arranged round a shortened axis. In its earlier stages the number of these modified leaves is indefinite, as in many Ranunculaceæ; and the axis itself is not greatly shortened, as in Myosurus. The first advance is to a definite number of parts and a permanently shortened axis, in the arrangement termed hypogynous, in which all the whorls are quite distinct from each other. In the next stage there is a further shortening of the central axis, leaving the outer portion as a ring on which the petals are inserted, producing the arrangement termed perigynous. A still further advance is made by the contraction of the axis, so as to leave the central part forming the ovary quite below the flower, which is then termed epigynous.

These several modifications are said to be parallel and definite, and to be determined by the continuous checking of vegetation by reproduction along what is an absolute groove of progressive change. This being the case, the importance of natural selection is greatly diminished. Instead of selecting and accumulating spontaneous indefinite variations, its function is to retard them after the stage of maximum utility has been independently reached. The same simple conception is said to unlock innumerable problems of vegetable morphology, large and small alike. It explains the inevitable development of gymnosperm into angiosperm by the checked vegetative growth of the ovule-bearing leaf or carpel; while such minor adaptations as the splitting fruit of the geranium or the cupped stigma of the pansy, can be no longer looked upon as achievements of natural selection, but must be regarded as naturally traceable to the vegetative checking of their respective types of leaf organ. Again, a detailed examination of spiny plants practically excludes the hypothesis of mammalian selection altogether, and shows spines to arise as an expression of the diminishing vegetativeness—in fact, the ebbing vitality of a species.[1]

Objections to the Theory.

The theory here sketched out is enticing, and at first sight seems calculated to throw much light on the history of plant development; but on further consideration, it seems wanting in definiteness, while it is beset with difficulties at every step. Take first the shortening of the raceme into the umbel and the capitulum, said to be caused by arrest of vegetative growth, due to the antagonism of reproduction. If this were the whole explanation of the phenomenon, we should expect the quantity of seed to increase as this vegetative growth diminished, since the seed is the product of the reproductive energy of the plant, and its quantity the best measure of that energy. But is this the case? The ranunculus has comparatively few seeds, and the flowers are not numerous; while in the same order the larkspur and the columbine have far more seeds as well as more flowers, but there is no shortening of the raceme or diminution of the foliage, although the flowers are large and complex. So, the extremely shortened and compressed flower-heads of the compositæ produce comparatively few seeds—one only to each flower; while the foxglove, with its long spike of showy flowers, produces an enormous number.

Again, if the shortening of the central axis in the successive stages of hypogynous, perigynous, and epigynous flowers were an indication of preponderant reproduction and diminished vegetation, we should find everywhere some clear indications of this fact. The plants with hypogynous flowers should, as a rule, have less seed and more vigorous and abundant foliage than those at the other extreme with epigynous flowers. But the

[1] This brief indication of Professor Geddes's views is taken from the article "Variation and Selection" in the *Encyclopædia Britannica*, and a paper "On the Nature and Causes of Variation in Plants" in *Trans. and Proc. of the Edinburgh Botanical Society*, 1886; and is, for the most part, expressed in his own words.

hypogynous poppies, pinks, and St. John's worts have abundance of seed and rather scanty foliage; while the epigynous dogwoods and honeysuckles have few seeds and abundant foliage. If, instead of the number of the seeds, we take the size of the fruit as an indication of reproductive energy, we find this at a maximum in the gourd family, yet their rapid and luxuriant growth shows no diminution of vegetative power. So that the statement that plant modifications proceed "along an absolute groove of progressive change" is contradicted by innumerable facts indicating advance and regression, improvement or degradation, according as the ever-changing environment renders one form more advantageous than the other. As one instance I may mention the Anonaceæ or custard-apple tribe, which are certainly an advance from the Ranunculaceæ; yet in the genus Polyalthea the fruit consists of a number of separate carpels, each borne on a long stalk, as if reverting to the primitive stalked carpellary leaves.

On the Origin of Spines.

But perhaps the most extraordinary application of the theory is that which considers spines to be an indication of the "ebbing vitality of a species," and which excludes "mammalian selection altogether." If this were true, spines should occur mainly in feeble, rare, and dying-out species, instead of which we have the hawthorn, one of our most vigorous shrubs or trees, with abundant vitality and an extensive range over the whole Palæarctic region, showing that it is really a dominant species. In North America the numerous thorny species of Cratægus are equally vigorous, as are the false acacia (Robinia) and the honey-locust (Gleditschia). Neither have the numerous species of very spiny Acacias been noticed to be rarer or less vigorous than the unarmed kinds.

On the other point—that spines are not due to mammalian selection—we are able to adduce what must be considered direct and conclusive evidence. For if spines, admittedly produced by aborted branches, petioles, or peduncles, are due solely or mainly to diminished vegetativeness or ebbing vitality, they ought to occur in all countries alike, or at all events in all whose similar conditions tend to check vegetation; whereas, if they are, solely or mainly, developed as a protection against the attacks of herbivorous mammals, they ought to be most abundant where these are plentiful, and rare or absent where indigenous mammalia are wanting. Oceanic islands, as compared with continents, would thus furnish a crucial test of the two theories; and Mr. Hemsley of Kew, who has specially studied insular floras, has given me some valuable information on this point. He says: "There are no spiny or prickly plants in the indigenous element of the St. Helena flora. The relatively rich flora of the Sandwich Isles is not absolutely without a prickly plant, but almost so. All the endemic genera are unarmed, and the endemic species of almost every other genus. Even such genera as Zanthoxylon, Acacia, Xylosoma, Lycium, and Solanum, of which there are many armed species in other countries, are only represented by unarmed species. The two endemic Rubi have the prickles reduced to the setaceous condition, and the two palms are unarmed.

"The flora of the Galapagos includes a number of prickly plants, among them several cacti (these have not been investigated and may be American species), but I do not think one of the known endemic species of any family is prickly or spiny.

"Spiny and prickly plants are also rare in New Zealand, but there are the formidably armed species of wild Spaniard (Aciphylla), one species of Rubus, the pungent-leaved Epacrideæ and a few others."

Mr. J. G. Baker of Kew, who has specially studied the flora of Mauritius and the adjacent islands, also writes me on this point. He says: "Taking Mauritius alone, I do not call to mind a single species that is a spinose endemic tree or shrub. If you take the whole group of islands (Mauritius, Bourbon, Seychelles, and Rodriguez), there will be about a dozen species, but then nine of these are palms. Leaving out palms, the trees and shrubs of that part of the world are exceptionally non-spinose."

These are certainly remarkable facts, and quite inexplicable on the theory of spines being caused solely by checked vegetative growth, due to weakness of constitution or to an arid soil and climate. For the Galapagos and many parts of the Sandwich Islands are very arid, as is a considerable part of the North Island of New Zealand. Yet in our own moist climate and with our very limited number of trees and shrubs we have about eighteen spiny or prickly species, more, apparently, than in the whole endemic floras of the Mauritius, Sandwich Islands, and Galapagos, though these are all especially rich in shrubby and arboreal species. In New Zealand the prickly Rubus is a leafless trailing plant, and its prickles are probably a protection against the large snails of the country, several of which have shells from two to three and a half inches long.[1] The "wild Spaniards" are very spiny herbaceous Umbelliferæ, and may have gained their spines to preserve them from being trodden down or eaten by the Moas, which, for countless ages, took the place of mammals in New Zealand. The exact use or meaning of the spines in palms is more doubtful, though they are, no doubt, protective against some animals; but it is certainly an extraordinary fact that in the entire flora of the Mauritius, so largely consisting of trees and shrubs, not a single endemic species should be thorny or spiny.

If now we consider that every continental flora produces a considerable proportion of spiny and thorny species, and that these rise to a maximum in South Africa, where herbivorous mammalia were (before the settlement of the country), perhaps, more abundant and varied than in any other part of the world; while another district, remarkable for well-armed vegetation, is Chile, where the camel-like vicugnas, llamas, and alpacas, and an abundance of large rodents wage perpetual war against shrubby vegetation, we shall see the full significance of the almost total absence of thorny and spiny plants in the chief oceanic islands; and so far from "excluding the hypothesis of mammalian selection altogether," we shall find in this hypothesis the only satisfactory explanation of the facts.

From the brief consideration of Professor Geddes's theory now given, we conclude that, although the antagonism between vegetative and reproductive growth is a real agency, and must be taken account of in our endeavour to explain many of the fundamental facts in the structure and form of plants, yet it is so overpowered and directed at every step by the natural selection of favourable variations, that the results of its

[1] Placostylis bovinus, $3\frac{1}{2}$ inches long; Paryphanta Busbyi, 3 in. diam.; P. Hochstetteri, $2\frac{3}{4}$ in. diam.

exclusive and unmodified action are nowhere to be found in nature. It may be allowed to rank as one of those "laws of growth," of which so many have now been indicated, and which were always recognised by Darwin as underlying all variation; but unless we bear in mind that its action must always be subordinated to natural selection, and that it is continually checked, or diverted, or even reversed by the necessity of adaptation to the environment, we shall be liable to fall into such glaring errors as the imputing to "ebbing vitality" alone such a widespread phenomenon as the occurrence of spines and thorns, while ignoring altogether the influence of the organic environment in their production.[1]

The sketch now given of the chief attempts that have been made to prove that either the direct action of the environment or certain fundamental laws of variation are independent causes of modification of species, shows us that their authors have, in every case, failed to establish their contention. Any direct action of the environment, or any characters acquired by use or disuse, can have no effect whatever upon the race unless they are inherited; and that they are inherited in any case,

[1] The general arguments and objections here set forth will apply with equal force to Professor G. Henslow's theory of the origin of the various forms and structures of flowers as due to "the responsive actions of the protoplasm in consequence of the irritations set up by the weights, pressures, thrusts, tensions, etc., of the insect visitors" (*The Origin of Floral Structures through Insect and other Agencies*, p. 340). On the assumption that acquired characters are inherited, such irritations may have had something to do with the initiation of variations and with the production of certain details of structure, but they are clearly incompetent to have brought about the more important structural and functional modifications of flowers. Such are, the various adjustments of length and position of the stamens to bring the pollen to the insect and from the insect to the stigma; the various motions of stamens and styles at the right time and the right direction; the physiological adjustments bringing about fertility or sterility in heterostyled plants; the traps, springs, and complex movements of various parts of orchids; and innumerable other remarkable phenomena.

For the explanation of these we have no resource but variation and selection, to the effects of which, acting alternately with regression or degradation as above explained (p. 328) must be imputed the development of the countless floral structures we now behold. Even the primitive flowers, whose initiation may, perhaps, have been caused, or rendered possible, by the irritation set up by insects' visits, must, from their very origin, have been modified, in accordance with the supreme law of utility, by means of variation and survival of the fittest.

except when they directly affect the reproductive cells, has not been proved. On the other hand, as we shall presently show, there is much reason for believing that such acquired characters are in their nature non-heritable.

Variation and Selection Overpower the Effects of Use and Disuse.

But there is another objection to this theory arising from the very nature of the effects produced. In each generation the effects of use or disuse, or of effort, will certainly be very small, while of this small effect it is not maintained that the whole will be always inherited by the next generation. How small the effect is we have no means of determining, except in the case of disuse, which Mr. Darwin investigated carefully. He found that in twelve fancy breeds of pigeons, which are often kept in aviaries, or if free fly but little, the sternum had been reduced by about one-seventh or one-eighth of its entire length, and that of the scapula about one-ninth. In domestic ducks the weight of the wing-bones in proportion to that of the whole skeleton had decreased about one-tenth. In domestic rabbits the bones of the legs were found to have increased in weight in due proportion to the increased weight of the body, but those of the hind legs were rather less in proportion to those of the fore legs than in the wild animal, a difference which may be imputed to their being less used in rapid motion. The pigeons, therefore, afford the greatest amount of reduction by disuse—one-seventh of the length of the sternum. But the pigeon has certainly been domesticated four or five thousand years; and if the reduction of the wings by disuse has only been going on for the last thousand years, the amount of reduction in each generation would be absolutely imperceptible, and quite within the limits of the reduction due to the absence of selection, as already explained. But, as we have seen in Chapter III, the fortuitous variation of every part or organ usually amounts to one-tenth, and often to one-sixth of the average dimensions—that is, the fortuitous variation in one generation among a limited number of the individuals of a species is as great as the cumulative effects of disuse in a thousand generations! If we assume that the effects of use or of effort in the individual are equal to the effects of disuse, or even ten or a hundred times greater, they will even then not equal, in each generation, the amount of the fortuitous variations of the same part. If it be urged that the effects of use would modify all the individuals of a species, while the fortuitous variations to the amount named only apply to a portion of them, it may be replied, that that portion is sufficiently large to afford ample materials for selection, since it often equals the numbers that can annually survive; while the recurrence in each successive generation of a like amount of variation would render possible such a rapid adjustment to new conditions that the effects of use or disuse would be as nothing in comparison. It follows, that even admitting the modifying effects of the environment, and that such modifications are inherited, they would yet be entirely swamped by the greater effects of fortuitous variation, and the far more rapid cumulative results of the selection of such variations.

Supposed Action of the Environment in Initiating Variations.

It is, however, urged that the reaction of the environment initiates variations, which without it would never arise; such, for instance, as the origin of horns through the pressures and irritations caused by butting, or otherwise using the head as a weapon or for defence. Admitting, for the sake of argument, that this is so, all the evidence we possess shows that, from the very first appearance of the rudiment of such an organ, it would vary to a greater extent than the amount of growth directly produced by use; and these variations would be subject to selection, and would thus modify the organ in ways which use alone would never bring about. We have seen that this has been the case with the branching antlers of the stag, which have been modified by selection, so as to become useful in other ways than as a mere weapon; and the same has almost certainly been the case with the variously curved and twisted horns of antelopes. In like manner, every conceivable rudiment would, from its first appearance, be subject to the law of variation and selection, to which, thenceforth, the direct effect of the environment would be altogether subordinate.

A very similar mode of reasoning will apply to the other branch of the subject—the initiation of structures and organs by the action of the fundamental laws of growth. Admitting that such laws have determined some of the main divisions of the animal and vegetable kingdom, have originated certain important organs, and have been the fundamental cause of certain lines of development, yet at every step of the process these laws must have acted in entire subordination to the law of natural selection. No modification thus initiated could have advanced a single step, unless it were, on the whole, a useful modification; while its entire future course would be necessarily subject to the laws of variation and selection, by which it would be sometimes checked, sometimes hastened on, sometimes diverted to one purpose, sometimes to another, according as the needs of the organism, under the special conditions of its existence, required such modification. We need not deny that such laws and influences may have acted in the manner suggested, but what we do deny is that they could possibly escape from the ever-present and all-powerful modifying effects of variation and natural selection.[1]

Weismann's Theory of Heredity.

Professor August Weismann has put forth a new theory of heredity founded upon the "continuity of the germ-plasm," one of the logical consequences of which is, that acquired characters of whatever kind are not transmitted from parent to offspring. As this is a matter of vital importance to the theory of natural selection, and as, if well founded, it strikes away the foundations of most of the theories discussed in the present chapter, a brief outline of Weismann's views must be attempted,

[1] In an essay on "The Duration of Life," forming part of the translation of Dr. Weismann's papers already referred to, the author still further extends the sphere of natural selection by showing that the average duration of life in each species has been determined by it. A certain length of life is essential in order that the species may produce offspring sufficient to ensure its continuance under the most unfavourable conditions; and it is shown that the remarkable inequalities of longevity in different species and groups may be thus accounted for. Yet more, the occurrence of death in the higher organisms, in place of the continued survival of the unicellular organisms however much they may increase by subdivision, may be traced to the same great law of utility for the race and survival of the fittest. The whole essay is of exceeding interest, and will repay a careful perusal. A similar idea occurred to the present writer about twenty years back, and was briefly noted down at the time, but subsequently forgotten.

although it is very difficult to make them intelligible to persons unfamiliar with the main facts of modern embryology.[1]

The problem is thus stated by Weismann: "How is it that in the case of all higher animals and plants a single cell is able to separate itself from amongst the millions of most various kinds of which an organism is composed, and by division and complicated differentiation to reconstruct a new individual with marvellous likeness, unchanged in many cases even throughout whole geological periods?" Darwin attempted to solve the problem by his theory of "Pangenesis," which supposed that every individual cell in the body gave off gemmules or germs capable of reproducing themselves, and that portions of these germs of each of the almost infinite number of cells permeate the whole body and become collected in the generative cells, and are thus able to reproduce the whole organism. This theory is felt to be so ponderously complex and difficult that it has met with no general acceptance among physiologists.

The fact that the germ-cells *do* reproduce with wonderful accuracy not only the general characters of the species, but many of the individual characteristics of the parents or more remote ancestors, and that this process is continued from generation to generation, can be accounted for, Weismann thinks, only on two suppositions which are physiologically possible. Either the substance of the parent germ-cell, after passing through a cycle of changes required for the construction of a new individual, possesses the capability of producing anew germ-cells identical with those from which that individual was developed, or *the new germ-cells arise, as far as their essential and characteristic substance is concerned, not at all out of the body of the individual, but direct from the parent germ-cell.* This latter view Weismann holds to be the correct one, and, on this theory, heredity depends on the fact that a substance of special molecular composition passes over from one generation to another. This is the "germ-plasm," the power of which to develop itself into a perfect organism depends on the extraordinary complication of its minutest structure. At every new birth a portion

[1] The outline here given is derived from two articles in *Nature*, vol. xxxiii. p. 154, and vol. xxxiv. p. 629, in which Weismann's papers are summarised and partly translated.

of the specific germ-plasm, which the parent egg-cell contains, is not used up in producing the offspring, but is reserved unchanged to produce the germ-cells of the following generation. Thus the germ-cells—so far as regards their essential part the germ-plasm—are not a product of the body itself, but are related to one another in the same way as are a series of generations of unicellular organisms derived from one another by a continuous course of simple division. Thus the question of heredity is reduced to one of growth. A minute portion of the very same germ-plasm from which, first the germ-cell, and then the whole organism of the parent, were developed, becomes the starting-point of the growth of the child.

The Cause of Variation.

But if this were all, the offspring would reproduce the parent exactly, in every detail of form and structure; and here we see the importance of sex, for each new germ grows out of the united germ-plasms of two parents, whence arises a mingling of their characters in the offspring. This occurs in each generation; hence every individual is a complex result reproducing in ever-varying degrees the diverse characteristics of his two parents, four grandparents, eight great-grandparents, and other more remote ancestors; and that ever-present individual variation arises which furnishes the material for natural selection to act upon. Diversity of sex becomes, therefore, of primary importance as *the cause of variation.* Where asexual generation prevails, the characteristics of the individual alone are reproduced, and there are thus no means of effecting the change of form or structure required by changed conditions of existence. Under such changed conditions a complex organism, if only asexually propagated, would become extinct. But when a complex organism is sexually propagated, there is an ever-present cause of change which, though slight in any one generation, is cumulative, and under the influence of selection is sufficient to keep up the harmony between the organism and its slowly changing environment.[1]

[1] There are many indications that this explanation of the cause of variation is the true one. Mr. E. B. Poulton suggests one, in the fact that parthenogenetic reproduction only occurs in isolated species, not in groups of related species; as this shows that parthenogenesis cannot lead to the evolution of

The Non-Heredity of Acquired Characters.

Certain observations on the embryology of the lower animals are held to afford direct proof of this theory of heredity, but they are too technical to be made clear to ordinary readers. A logical result of the theory is the impossibility of the transmission of acquired characters, since the molecular structure of the germ-plasm is already determined within the embryo; and Weismann holds that there are no facts which really prove that acquired characters can be inherited, although their inheritance has, by most writers, been considered so probable as hardly to stand in need of direct proof.

We have already shown, in the earlier part of this chapter, that many instances of change, imputed to the inheritance of acquired variations, are really cases of selection; while the very fact that *use* implies *usefulness* renders it almost impossible to eliminate the action of selection in a state of nature. As regards mutilations, it is generally admitted that they are not hereditary, and there is ample evidence on this point. When it was the fashion to dock horses' tails, it was not found that horses were born with short tails; nor are Chinese women born with distorted feet; nor are any of the numerous forms of racial mutilation in man, which have in some cases been carried on for hundreds of generations, inherited. Nevertheless, a few cases of apparent inheritance of mutilations have been recorded,[1] and these, if trustworthy, are difficulties in the way of the theory. The undoubted inheritance of disease is hardly a difficulty, because the predisposition to disease is a congenital, not an acquired character, and as such would be the subject of inheritance. The often-quoted case of a disease induced by mutilation being inherited (Brown-Sequard's epileptic guinea-pigs) has been discussed by Professor Weismann, and shown to be not conclusive. The mutilation itself —a section of certain nerves—was never inherited, but

new forms. Again, in parthenogenetic females the complete apparatus for fertilisation remains unreduced; but if these varied as do sexually produced animals, the organs referred to, being unused, would become rudimentary.

Even more important is the significance of the "polar bodies," as explained by Weismann in one of his *Essays;* since, if his interpretation of them be correct, variability is a necessary consequence of sexual generation.

[1] Darwin's *Animals and Plants,* vol. ii. pp. 23, 24.

the resulting epilepsy, or a general state of weakness, deformity, or sores, was sometimes inherited. It is, however, possible that the mere injury introduced and encouraged the growth of certain microbes, which, spreading through the organism, sometimes reached the germ-cells, and thus transmitted a diseased condition to the offspring. Such a transference of microbes is believed to occur in syphilis and tuberculosis, and has been ascertained to occur in the case of the muscardine silkworm disease.[1]

The Theory of Instinct.

The theory now briefly outlined cannot be said to be proved, but it commends itself to many physiologists as being inherently probable, and as furnishing a good working hypothesis till displaced by a better. We cannot, therefore, accept any arguments against the agency of natural selection which are based upon the opposite and equally unproved theory that acquired characters are inherited; and as this applies to the whole school of what may be termed Neo-Lamarckians, their speculations cease to have any weight.

The same remark applies to the popular theory of instincts as being inherited habits; though Darwin gave very little weight to this, but derived almost all instincts from spontaneous useful variations which, like other spontaneous variations, are of course inherited. At first sight it appears as if the acquired habits of our trained dogs—pointers, retrievers, etc.—are certainly inherited; but this need not be the case, because there must be some structural or psychical peculiarities, such as modifications in the attachments of muscles, increased delicacy of smell or sight, or peculiar likes and dislikes, which are inherited; and from these, peculiar habits follow as a natural consequence, or are easily acquired. Now, as selection has been constantly at work in improving all our domestic animals, we have unconsciously modified the structure, while preserving only those animals which best served our purpose in their peculiar faculties, instincts, or habits.

[1] In his essay on "Heredity," Dr. Weismann discusses many other cases of supposed inheritance of acquired characters, and shows that they can all be explained in other ways. Shortsightedness among civilised nations, for example, is due partly to the absence of selection and consequent regression towards a mean, and partly to its individual production by constant reading.

Much of the mystery of instinct arises from the persistent refusal to recognise the agency of imitation, memory, observation, and reason as often forming part of it. Yet there is ample evidence that such agency must be taken into account. Both Wilson and Leroy state that young birds build inferior nests to old ones, and the latter author observes that the best nests are made by birds whose young remain longest in the nest. So, migration is now well ascertained to be effected by means of vision, long flights being made on bright moonlight nights when the birds fly very high, while on cloudy nights they fly low, and then often lose their way. Thousands annually fly out to sea and perish, showing that the instinct to migrate is imperfect, and is not a good substitute for reason and observation.

Again, much of the perfection of instinct is due to the extreme severity of the selection during its development, any failure involving destruction. The chick which cannot break the eggshell, the caterpillar that fails to suspend itself properly or to spin a safe cocoon, the bees that lose their way or that fail to store honey, inevitably perish. So the birds that fail to feed and protect their young, or the butterflies that lay their eggs on the wrong food-plant, leave no offspring, and the race with imperfect instincts perishes. Now, during the long and very slow course of development of each organism, this rigid selection at every step of progress has led to the preservation of every detail of structure, faculty, or habit that has been necessary for the preservation of the race, and has thus gradually built up the various instincts which seem so marvellous to us, but which can yet be shown to be in many cases still imperfect. Here, as everywhere else in nature, we find comparative, not absolute perfection, with every gradation from what is clearly due to imitation or reason up to what seems to us perfect instinct—that in which a complex action is performed without any previous experience or instruction.[1]

[1] Weismann explains instinct on similar lines, and gives many interesting illustrations (see *Essays on Heredity*). He holds "that all instinct is entirely due to the operation of natural selection, and has its foundation, not upon inherited experiences, but upon variations of the germ." Many interesting and difficult cases of instinct are discussed by Darwin in Chapter VIII of the *Origin of Species*, which should be read in connection with the above remarks.
Since this chapter was written my attention has been directed to Mr.

Concluding Remarks.

Having now passed in review the more important of the recent objections to, or criticisms of, the theory of natural selection, we have arrived at the conclusion that in no one case have the writers in question been able materially to diminish its importance, or to show that any of the laws or forces to which they appeal can act otherwise than in strict subordination to it. The direct action of the environment as set forth by Mr. Herbert Spencer, Dr. Cope, and Dr. Karl Semper, even if we admit that its effects on the individual are transmitted by inheritance, are so small in comparison with the amount of spontaneous variation of every part of the organism that they must be quite overshadowed by the latter. And if such direct action may, in some cases, have initiated certain organs or outgrowths, these must from their very first beginnings have been subject to variation and natural selection, and their further development have been almost wholly due to these ever-present and powerful causes.

Francis Galton's *Theory of Heredity* (already referred to at p. 417) which was published thirteen years ago as an alternative for Darwin's theory of pangenesis.

Mr. Galton's theory, although it attracted little attention, appears to me to be substantially the same as that of Professor Weismann. Galton's "stirp" is Weismann's "germ-plasm." Galton supposes the sexual elements in the offspring to be directly formed from the residue of the *stirp* not used up in the development of the body of the parent—Weismann's "continuity of the germ-plasm." Galton also draws many of the same conclusions from his theory. He maintains that characters acquired by the individual as the result of external influences cannot be inherited, unless such influences act directly on the reproductive elements—instancing the possible heredity of alcoholism, because the alcohol permeates the tissues and may reach the sexual elements. He discusses the supposed heredity of effects produced by use or disuse, and explains them much in the same manner as does Weismann. Galton is an anthropologist, and applies the theory, mainly, to explain the peculiarities of hereditary transmission in man, many of which peculiarities he discusses and elucidates. Weismann is a biologist, and is mostly concerned with the application of the theory to explain variation and instinct, and to the further development of the theory of evolution. He has worked it out more thoroughly, and has adduced embryological evidence in its support; but the views of both writers are substantially the same, and their theories were arrived at quite independently. The names of Galton and Weismann should therefore be associated as discoverers of what may be considered (if finally established) the most important contribution to the evolution theory since the appearance of the *Origin of Species.*

The same remark applies to the views of Professor Geddes on the laws of growth which have determined certain essential features in the morphology of plants and animals. The attempt to substitute these laws for those of variation and natural selection has failed in cases where we can apply a definite test, as in that of the origin of spines on trees and shrubs; while the extreme diversity of vegetable structure and form among the plants of the same country and of the same natural order, of itself affords a proof of the preponderating influence of variation and natural selection in keeping the many diverse forms in harmony with the highly complex and ever-changing environment.

Lastly, we have seen that Professor Weismann's theory of the continuity of the germ-plasm and the consequent non-heredity of acquired characters, while in perfect harmony with all the well-ascertained facts of heredity and development, adds greatly to the importance of natural selection as the one invariable and ever-present factor in all organic change, and that which can alone have produced the temporary fixity combined with the secular modification of species. While admitting, as Darwin always admitted, the co-operation of the fundamental laws of growth and variation, of correlation and heredity, in determining the direction of lines of variation or in the initiation of peculiar organs, we find that variation and natural selection are ever-present agencies, which take possession, as it were, of every minute change originated by these fundamental causes, check or favour their further development, or modify them in countless varied ways according to the varying needs of the organism. Whatever other causes have been at work, Natural Selection is supreme, to an extent which even Darwin himself hesitated to claim for it. The more we study it the more we are convinced of its overpowering importance, and the more confidently we claim, in Darwin's own words, that it "has been the most important, but not the exclusive, means of modification."

CHAPTER XV

DARWINISM APPLIED TO MAN

General identity of human and animal structure—Rudiments and variations showing relation of man to other mammals—The embryonic development of man and other mammalia—Diseases common to man and the lower animals—The animals most nearly allied to man—The brains of man and apes—External differences of man and apes—Summary of the animal characteristics of man—The geological antiquity of man—The probable birthplace of man—The origin of the moral and intellectual nature of man—The argument from continuity—The origin of the mathematical faculty—The origin of the musical and artistic faculties—Independent proof that these faculties have not been developed by natural selection—The interpretation of the facts—Concluding remarks.

OUR review of modern Darwinism might fitly have terminated with the preceding chapter; but the immense interest that attaches to the origin of the human race, and the amount of misconception which prevails regarding the essential teachings of Darwin's theory on this question, as well as regarding my own special views upon it, induce me to devote a final chapter to its discussion.

To any one who considers the structure of man's body, even in the most superficial manner, it must be evident that it is the body of an animal, differing greatly, it is true, from the bodies of all other animals, but agreeing with them in all essential features. The bony structure of man classes him as a vertebrate; the mode of suckling his young classes him as a mammal; his blood, his muscles, and his nerves, the structure of his heart with its veins and arteries, his lungs and his whole respiratory and circulatory systems, all closely correspond to those of other mammals, and are often almost identical with them. He possesses the same number of limbs terminating in the same number of digits as belong fundamentally to the mammalian class. His senses are identical with theirs, and his organs of sense are the same in number and occupy the same relative position. Every detail of structure which is common to the mammalia as a class is found also in man, while he only differs from them in such ways and degrees as the various species or groups of mammals differ from each other. If, then, we have good reason to believe that every existing group of mammalia has descended from some common ancestral form—as we saw to be so completely demonstrated in the case of the horse tribe,—and that each family, each order, and even the whole class must similarly have descended from some much more ancient and more generalised type, it would be in the highest degree improbable—so improbable as to be almost inconceivable—that man, agreeing with them so closely in every detail of his structure, should have had some quite distinct mode of origin. Let us, then, see what other evidence bears upon the question, and whether it is sufficient to convert the probability of his animal origin into a practical certainty.

Rudiments and Variations as Indicating the Relation of Man to other Mammals.

All the higher animals present rudiments of organs which, though useless to them, are useful in some allied group, and are believed to have descended from a common ancestor in which they were useful. Thus there are in ruminants rudiments of incisor teeth which, in some species, never cut through the gums; many lizards have external rudimentary legs; while many birds, as the Apteryx, have quite rudimentary wings. Now man possesses similar rudiments, sometimes constantly, sometimes only occasionally present, which serve intimately to connect his bodily structure with that of the lower animals. Many animals, for example, have a special muscle for moving or twitching the skin. In man there are remnants of this in certain parts of the body, especially in the forehead, enabling us to raise our eyebrows; but some persons have it in other parts. A few persons are able to move the whole scalp so as to throw off any object placed on the head, and this property has been proved, in one case, to be inherited. In the outer fold of the ear there is sometimes a projecting point, corresponding in position to the pointed ear of many animals, and believed to be a rudiment of it. In the alimentary canal there is a rudiment—the vermiform appendage of the cæcum—which is not only useless, but is sometimes a cause of disease and death in man; yet in many vegetable feeding animals it is very long, and even in the orang-utan it is of considerable length and convoluted. So, man possesses rudimentary bones of a tail concealed beneath the skin, and, in some rare cases, this forms a minute external tail.

The variability of every part of man's structure is very great, and many of these variations tend to approximate towards the structure of other animals. The courses of the arteries are eminently variable, so that for surgical purposes it has been necessary to determine the probable proportion of each variation. The muscles are so variable that in fifty cases the muscles of the foot were found to be not strictly alike in any two, and in some the deviations were considerable; while in thirty-six subjects Mr. J. Wood observed no fewer than 558 muscular variations. The same author states that in a single male subject there were no fewer than seven muscular variations, all of which plainly represented muscles proper to various kinds of apes. The muscles of the hands and arms—parts which are so eminently characteristic of man—are extremely liable to vary, so as to resemble the corresponding muscles of the lower animals. That such variations are due to reversion to a former state of existence Mr. Darwin thinks highly probable, and he adds: "It is quite incredible that a man should, through mere accident, abnormally resemble certain apes in no less than seven of his muscles, if there had been no genetic connection between them. On the other hand, if man is descended from some ape-like creature, no valid reason can be assigned why certain muscles should not suddenly reappear after an interval of many thousand generations, in the same manner as, with horses, asses, and mules, dark coloured stripes suddenly reappear on the legs and shoulders, after an interval of hundreds, or more probably of thousands of generations."[1]

[1] *Descent of Man*, pp. 41-43; also pp. 13-15.

The Embryonic Development of Man and other Mammalia.

The progressive development of any vertebrate from the ovum or minute embryonic egg affords one of the most marvellous chapters in Natural History. We see the contents of the ovum undergoing numerous definite changes, its interior dividing and subdividing till it consists of a mass of cells, then a groove appears marking out the median line or vertebral column of the future animal, and thereafter are slowly developed the various essential organs of the body. After describing in some detail what takes place in the case of the ovum of the dog, Professor Huxley continues: "The history of the development of any other vertebrate animal, lizard, snake, frog, or fish tells the same story. There is always to begin with, an egg having the same essential structure as that of the dog; the yelk of that egg undergoes division or segmentation, as it is called, the ultimate products of that segmentation constitute the building materials for the body of the young animal; and this is built up round a primitive groove, in the floor of which a notochord is developed. Furthermore, there is a period in which the young of all these animals resemble one another, not merely in outward form, but in all essentials of structure, so closely, that the differences between them are inconsiderable, while in their subsequent course they diverge more and more widely from one another. And it is a general law that the more closely any animals resemble one another in adult structure, the larger and the more intimately do their embryos resemble one another; so that, for example, the embryos of a snake and of a lizard remain like one another longer than do those of a snake and a bird; and the embryos of a dog and of a cat remain like one another for a far longer period than do those of a dog and a bird, or of a dog and an opossum, or even than those of a dog and a monkey."[1]

We thus see that the study of development affords a test of affinity in animals that are externally very much unlike each other; and we naturally ask how this applies to man. Is he developed in a different way from other mammals, as we should certainly expect if he has had a distinct and

[1] *Man's Place in Nature*, p. 64.

altogether different origin? "The reply," says Professor Huxley, "is not doubtful for a moment. Without question, the mode of origin and the early stages of the development of man are identical with those of the animals immediately below him in the scale." And again he tells us: "It is very long before the body of the young human being can be readily discriminated from that of the young puppy; but at a tolerably early period the two become distinguishable by the different forms of their adjuncts, the yelk-sac and the allantois;" and after describing these differences he continues: "But exactly in those respects in which the developing man differs from the dog, he resembles the ape. . . . So that it is only quite in the latter stages of development that the young human being presents marked differences from the young ape, while the latter departs as much from the dog in its development as the man does. Startling as this last assertion may appear to be, it is demonstrably true, and it alone appears to me sufficient to place beyond all doubt the structural unity of man with the rest of the animal world, and more particularly and closely with the apes."[1]

A few of the curious details in which man passes through stages common to the lower animals may be mentioned. At one stage the os coccyx projects like a true tail, extending considerably beyond the rudimentary legs. In the seventh month the convolutions of the brain resemble those of an adult baboon. The great toe, so characteristic of man, forming the fulcrum which most assists him in standing erect, in an early stage of the embryo is much shorter than the other toes, and instead of being parallel with them, projects at an angle from the side of the foot, thus corresponding with its permanent condition in the quadrumana. Numerous other examples might be quoted, all illustrating the same general law.

Diseases Common to Man and the Lower Animals.

Though the fact is so well known, it is certainly one of profound significance that many animal diseases can be communicated to man, since it shows similarity, if not identity, in

[1] *Man's Place in Nature*, p. 67. See Figs. of Embryos of Man and Dog in Darwin's *Descent of Man*, p. 10.

the minute structure of the tissues, the nature of the blood, the nerves, and the brain. Such diseases as hydrophobia, variola, the glanders, cholera, herpes, etc., can be transmitted from animals to man or the reverse; while monkeys are liable to many of the same non-contagious diseases as we are. Rengger, who carefully observed the common monkey (Cebus Azaræ) in Paraguay, found it liable to catarrh, with the usual symptoms, terminating sometimes in consumption. These monkeys also suffered from apoplexy, inflammation of the bowels, and cataract in the eye. Medicines produced the same effect upon them as upon us. Many kinds of monkeys have a strong taste for tea, coffee, spirits, and even tobacco. These facts show the similarity of the nerves of taste in monkeys and in ourselves, and that their whole nervous system is affected in a similar way. Even the parasites, both external and internal, that affect man are not altogether peculiar to him, but belong to the same families or genera as those which infest animals, and in one case, scabies, even the same species.[1] These curious facts seem quite inconsistent with the idea that man's bodily structure and nature are altogether distinct from those of animals, and have had a different origin; while the facts are just what we should expect if he has been produced by descent with modification from some common ancestor.

The Animals most nearly Allied to Man.

By universal consent we see in the monkey tribe a caricature of humanity. Their faces, their hands, their actions and expressions present ludicrous resemblances to our own. But there is one group of this great tribe in which this resemblance is greatest, and they have hence been called the anthropoid or man-like apes. These are few in number, and inhabit only the equatorial regions of Africa and Asia, countries where the climate is most uniform, the forests densest, and the supply of fruit abundant throughout the year. These animals are now comparatively well known, consisting of the orang-utan of Borneo and Sumatra, the chimpanzee and the gorilla of West Africa, and the group of gibbons or long-armed apes, consisting of many species and inhabiting South-Eastern

[1] *The Descent of Man*, pp. 7, 8.

Asia and the larger Malay Islands. These last are far less like man than the other three, one or other of which has at various times been claimed to be the most man-like of the apes and our nearest relations in the animal kingdom. The question of the degree of resemblance of these animals to ourselves is one of great interest, leading, as it does, to some important conclusions as to our origin and geological antiquity, and we will therefore briefly consider it.

If we compare the skeletons of the orang or chimpanzee with that of man, we find them to be a kind of distorted copy, every bone corresponding (with very few exceptions), but altered somewhat in size, proportions, and position. So great is this resemblance that it led Professor Owen to remark: "I cannot shut my eyes to the significance of that all-pervading similitude of structure—every tooth, every bone, strictly homologous—which makes the determination of the difference between *Homo* and *Pithecus* the anatomist's difficulty."

The actual differences in the skeletons of these apes and that of man—that is, differences dependent on the presence or absence of certain bones, and not on their form or position—have been enumerated by Mr. Mivart as follows:—(1) In the breast-bone consisting of but two bones, man agrees with the gibbons; the chimpanzee and gorilla having this part consisting of seven bones in a single series, while in the orang they are arranged in a double series of ten bones. (2) The normal number of the ribs in the orang and some gibbons is twelve pairs, as in man, while in the chimpanzee and gorilla there are thirteen pairs. (3) The orang and the gibbons also agree with man in having five lumbar vertebræ, while in the gorilla and the chimpanzee there are but four, and sometimes only three. (4) The gorilla and chimpanzee agree with man in having eight small bones in the wrist, while the orang and the gibbons, as well as all other monkeys, have nine.[1]

The differences in the form, size, and attachments of the various bones, muscles, and other organs of these apes and

[1] *Man and Apes.* By St. George Mivart, F.R.S., 1873. It is an interesting fact (for which I am indebted to Mr. E. B. Poulton) that the human embryo possesses the extra rib and wrist-bone referred to above in (2) and (4) as occurring in some of the apes.

man are very numerous and exceedingly complex, sometimes one species, sometimes another agreeing most nearly with ourselves, thus presenting a tangled web of affinities which it is very difficult to unravel. Estimated by the skeleton alone, the chimpanzee and gorilla seem nearer to man than the orang, which last is also inferior as presenting certain aberrations in the muscles. In the form of the ear the gorilla is more human than any other ape, while in the tongue the orang is the more man-like. In the stomach and liver the gibbons approach nearest to man, then come the orang and chimpanzee, while the gorilla has a degraded liver more resembling that of the lower monkeys and baboons.

The Brains of Man and Apes.

We come now to that part of his organisation in which man is so much higher than all the lower animals—the brain; and here, Mr. Mivart informs us, the orang stands highest in rank. The height of the orang's cerebrum in front is greater in proportion than in either the chimpanzee or the gorilla. "On comparing the brain of man with the brains of the orang, chimpanzee, and baboon, we find a successive decrease in the frontal lobe, and a successive and very great increase in the relative size of the occipital lobe. Concomitantly with this increase and decrease, certain folds of brain substance, called 'bridging convolutions,' which in man are conspicuously interposed between the parietal and occipital lobes, seem as utterly to disappear in the chimpanzee, as they do in the baboon. In the orang, however, though much reduced, they are still to be distinguished. . . . The actual and absolute mass of the brain is, however, slightly greater in the chimpanzee than in the orang, as is the relative vertical extent of the middle part of the cerebrum, although, as already stated, the frontal portion is higher in the orang; while, according to M. Gratiolet, the gorilla is not only inferior to the orang in cerebral development, but even to his smaller African congener, the chimpanzee."[1]

On the whole, then, we find that no one of the great apes can be positively asserted to be nearest to man in structure. Each of them approaches him in certain characteristics, while

[1] *Man and Apes*, pp. 138, 144.

in others it is widely removed, giving the idea, so consonant with the theory of evolution as developed by Darwin, that all are derived from a common ancestor, from which the existing anthropoid apes as well as man have diverged. When, however, we turn from the details of anatomy to peculiarities of external form and motions, we find that, in a variety of characters, all these apes resemble each other and differ from man, so that we may fairly say that, while they have diverged somewhat from each other, they have diverged much more widely from ourselves. Let us briefly enumerate some of these differences.

External Differences of Man and Apes.

All apes have large canine teeth, while in man these are no longer than the adjacent incisors or premolars, the whole forming a perfectly even series. In apes the arms are proportionately much longer than in man, while the thighs are much shorter. No ape stands really erect, a posture which is natural in man. The thumb is proportionately larger in man, and more perfectly opposable than in that of any ape. The foot of man differs largely from that of all apes, in the horizontal sole, the projecting heel, the short toes, and the powerful great toe firmly attached parallel to the other toes; all perfectly adapted for maintaining the erect posture, and for free motion without any aid from the arms or hands. In apes the foot is formed almost exactly like our hand, with a large thumb-like great toe quite free from the other toes, and so articulated as to be opposable to them; forming with the long finger-like toes a perfect grasping hand. The sole cannot be placed horizontally on the ground; but when standing on a level surface the animal rests on the outer edge of the foot with the finger and thumb-like toes partly closed, while the hands are placed on the ground resting on the knuckles. The illustration on the next page (Fig. 37) shows, fairly well, the peculiarities of the hands and feet of the chimpanzee, and their marked differences, both in form and use, from those of man.

The four limbs, with the peculiarly formed feet and hands, are those of arboreal animals which only occasionally and awkwardly move on level ground. The arms are used in progression equally with the feet, and the hands are only adapted for uses similar to those of our hands when the animal is at rest, and then but clumsily. Lastly, the apes are all hairy animals, like the majority of other mammals, man alone having a smooth and almost naked skin. These numerous and striking differences, even more than those of the skeleton and internal

FIG. 37.—Chimpanzee (Troglodytes niger).

anatomy, point to an enormously remote epoch when the race that was ultimately to develop into man diverged from that other stock which continued the animal type and ultimately produced the existing varieties of anthropoid apes.

Summary of the Animal Characteristics of Man.

The facts now very briefly summarised amount almost to a demonstration that man, in his bodily structure, has been derived from the lower animals, of which he is the culminating development. In his possession of rudimentary structures which are functional in some of the mammalia; in the numerous variations of his muscles and other organs agreeing with characters which are constant in some apes; in his embryonic development, absolutely identical in character with that of mammalia in general, and closely resembling in its details that of the higher quadrumana; in the diseases which he has in common with other mammalia; and in the wonderful approximation of his skeleton to those of one or other of the anthropoid apes, we have an amount of evidence in this direction which it seems impossible to explain away. And this evidence will appear more forcible if we consider for a moment what the rejection of it implies. For the only alternative supposition is, that man has been specially created—that is to say, has been produced in some quite different way from other animals and altogether independently of them. But in that case the rudimentary structures, the animal-like variations, the identical course of development, and all the other animal characteristics he possesses are deceptive, and inevitably lead us, as thinking beings making use of the reason which is our noblest and most distinctive feature, into gross error.

We cannot believe, however, that a careful study of the facts of nature leads to conclusions directly opposed to the truth; and, as we seek in vain, in our physical structure and the course of its development, for any indication of an origin independent of the rest of the animal world, we are compelled to reject the idea of "special creation" for man, as being entirely unsupported by facts as well as in the highest degree improbable.

The Geological Antiquity of Man.

The evidence we now possess of the exact nature of the resemblance of man to the various species of anthropoid apes, shows us that he has little special affinity for any one rather than another species, while he differs from them all in several important characters in which they agree with each other. The conclusion to be drawn from these facts is, that his points of affinity connect him with the whole group, while his special peculiarities equally separate him from the whole group, and that he must, therefore, have diverged from the common ancestral form before the existing types of anthropoid apes had diverged from each other. Now, this divergence almost certainly took place as early as the Miocene period, because in the Upper Miocene deposits of Western Europe remains of two species of ape have been found allied to the gibbons, one of them, Dryopithecus, nearly as large as a man, and believed by M. Lartet to have approached man in its dentition more than the existing apes. We seem hardly, therefore, to have reached, in the Upper Miocene, the epoch of the common ancestor of man and the anthropoids.

The evidence of the antiquity of man himself is also scanty, and takes us but very little way back into the past. We have clear proof of his existence in Europe in the latter stages of the glacial epoch, with many indications of his presence in interglacial or even pre-glacial times; while both the actual remains and the works of man found in the auriferous gravels of California deep under lava-flows of Pliocene age, show that he existed in the New World at least as early as in the Old.[1] These earliest remains of man have been received with doubt, and even with ridicule, as if there were some extreme improbability in them. But, in point of fact, the wonder is that human remains have not been found more frequently in pre-glacial deposits. Referring to the most ancient fossil remains found in Europe—the Engis and Neanderthal crania,—Professor Huxley makes the following weighty remark: "In conclusion, I may say, that the fossil remains of Man hitherto discovered do not seem to me to take us appreciably nearer to that lower pithecoid form, by the modification of which he has, probably, become what he is." The Californian remains and works of art, above referred to, give no indication of a specially low form of man; and it remains an unsolved problem why no traces of the long line of man's ancestors, back to the remote period when he first branched off from the pithecoid type, have yet been discovered.

It has been objected by some writers—notably by Professor Boyd Dawkins—that man did not probably exist in Pliocene times, because almost all the known mammalia of that epoch are distinct species from those now living on the earth, and that the same changes of the environment which led to

[1] For a sketch of the evidence of Man's Antiquity in America, see *The Nineteenth Century* for November 1887.

the modification of other mammalian species would also have led to a change in man. But this argument overlooks the fact that man differs essentially from all other mammals in this respect, that whereas any important adaptation to new conditions can be effected in them only by a change in bodily structure, man is able to adapt himself to much greater changes of conditions by a mental development leading him to the use of fire, of tools, of clothing, of improved dwellings, of nets and snares, and of agriculture. By the help of these, without any change whatever in his bodily structure, he has been able to spread over and occupy the whole earth; to dwell securely in forest, plain, or mountain; to inhabit alike the burning desert or the arctic wastes; to cope with every kind of wild beast, and to provide himself with food in districts where, as an animal trusting to nature's unaided productions, he would have starved.[1]

It follows, therefore, that from the time when the ancestral man first walked erect, with hands freed from any active part in locomotion, and when his brain-power became sufficient to cause him to use his hands in making weapons and tools, houses and clothing, to use fire for cooking, and to plant seeds or roots to supply himself with stores of food, the power of natural selection would cease to act in producing modifications of his body, but would continuously advance his mind through the development of its organ, the brain. Hence man may have become truly man—the species, Homo sapiens—even in the Miocene period; and while all other mammals were becoming modified from age to age under the influence of ever-changing physical and biological conditions, he would be advancing mainly in intelligence, but perhaps also in stature, and by that advance alone would be able to maintain himself as the master of all other animals and as the most widespread occupier of the earth. It is quite in accordance with this view that we find the most pronounced distinction between man and the anthropoid apes in the size and complexity of his brain. Thus, Professor Huxley tells us that "it may be doubted whether a healthy human adult brain ever weighed

[1] This subject was first discussed in an article in the *Anthropological Review*, May 1864, and republished in my *Contributions to Natural Selection*, chap. ix, in 1870.

less than 31 or 32 ounces, or that the heaviest gorilla brain has exceeded 20 ounces," although "a full-grown gorilla is probably pretty nearly twice as heavy as a Bosjes man, or as many an European woman."[1] The average human brain, however, weighs 48 or 49 ounces, and if we take the average ape brain at only 2 ounces less than the very largest gorilla's brain, or 18 ounces, we shall see better the enormous increase which has taken place in the brain of man since the time when he branched off from the apes; and this increase will be still greater if we consider that the brains of apes, like those of all other mammals, have also increased from earlier to later geological times.

If these various considerations are taken into account, we must conclude that the essential features of man's structure as compared with that of apes—his erect posture and free hands—were acquired at a comparatively early period, and were, in fact, the characteristics which gave him his superiority over other mammals, and started him on the line of development which has led to his conquest of the world. But during this long and steady development of brain and intellect, mankind must have continuously increased in numbers and in the area which they occupied—they must have formed what Darwin terms a "dominant race." For had they been few in numbers and confined to a limited area, they could hardly have successfully struggled against the numerous fierce carnivora of that period, and against those adverse influences which led to the extinction of so many more powerful animals. A large population spread over an extensive area is also needed to supply an adequate number of brain variations for man's progressive improvement. But this large population and long-continued development in a single line of advance renders it the more difficult to account for the complete absence of human or pre-human remains in all those deposits which have furnished, in such rich abundance, the remains of other land animals. It is true that the remains of apes are also very rare, and we may well suppose that the superior intelligence of man led him to avoid that extensive destruction by flood or in morass which seems to have often overwhelmed other animals. Yet, when

[1] *Man's Place in Nature*, p. 102.

we consider that, even in our own day, men are not unfrequently overwhelmed by volcanic eruptions, as in Java and Japan, or carried away in vast numbers by floods, as in Bengal and China, it seems impossible but that ample remains of Miocene and Pliocene man do exist buried in the most recent layers of the earth's crust, and that more extended research or some fortunate discovery will some day bring them to light.

The Probable Birthplace of Man.

It has usually been considered that the ancestral form of man originated in the tropics, where vegetation is most abundant and the climate most equable. But there are some important objections to this view. The anthropoid apes, as well as most of the monkey tribe, are essentially arboreal in their structure, whereas the great distinctive character of man is his special adaptation to terrestrial locomotion. We can hardly suppose, therefore, that he originated in a forest region, where fruits to be obtained by climbing are the chief vegetable food. It is more probable that he began his existence on the open plains or high plateaux of the temperate or sub-tropical zone, where the seeds of indigenous cereals and numerous herbivora, rodents, and game-birds, with fishes and molluscs in the lakes, rivers, and seas supplied him with an abundance of varied food. In such a region he would develop skill as a hunter, trapper, or fisherman, and later as a herdsman and cultivator,—a succession of which we find indications in the palæolithic and neolithic races of Europe.

In seeking to determine the particular areas in which his earliest traces are likely to be found, we are restricted to some portion of the Eastern hemisphere, where alone the anthropoid apes exist, or have apparently ever existed.

There is good reason to believe, also, that Africa must be excluded, because it is known to have been separated from the northern continent in early tertiary times, and to have acquired its existing fauna of the higher mammalia by a later union with that continent after the separation from it of Madagascar, an island which has preserved for us a sample, as it were, of the early African mammalian fauna, from which not only the anthropoid apes, but all the higher quadrumana are absent.[1] There remains only the great Euro-Asiatic continent; and its enormous plateaux, extending from Persia right across Tibet and Siberia to Manchuria, afford an area, some part or other of which probably offered suitable conditions, in late Miocene or early Pliocene times, for the development of ancestral man.

It is in this area that we still find that type of mankind—the Mongolian—which retains a colour of the skin midway between the black or brown-black of the negro, and the ruddy or olive-white of the Caucasian types, a colour which still prevails over all Northern Asia, over the American continents, and over much of Polynesia. From this primary tint arose, under the influence of varied conditions, and probably in correlation with constitutional changes adapted to peculiar climates, the varied tints which still exist among mankind. If the reasoning by which this conclusion is reached be sound, and all the earlier stages of man's development from an animal form occurred in the area now indicated, we can better understand how it is that we have as yet met with no traces of the missing links, or even of man's existence during late tertiary times, because no part of the world is so entirely unexplored by the geologist as this very region. The area in question is sufficiently extensive and varied to admit of primeval man having attained to a considerable population, and having developed his full human characteristics, both physical and mental, before there was any need for him to migrate beyond its limits. One of his earliest important migrations was probably into Africa, where, spreading westward, he became modified in colour and hair in correlation with physiological changes adapting him to the climate of the equatorial lowlands. Spreading north-westward into Europe the moist and cool climate led to a modification of an opposite character, and thus may have arisen the three great human types which still exist. Somewhat later, probably, he spread eastward into North-West America and soon scattered himself over the whole continent; and all this may well have occurred in early or middle Pliocene times. Thereafter, at very long intervals, successive waves of migration carried him into every

[1] For a full discussion of this question, see the author's *Geographical Distribution of Animals*, vol. i. p. 285.

part of the habitable world, and by conquest and intermixture led ultimately to that puzzling gradation of types which the ethnologist in vain seeks to unravel.

The Origin of the Moral and Intellectual Nature of Man.

From the foregoing discussion it will be seen that I fully accept Mr. Darwin's conclusion as to the essential identity of man's bodily structure with that of the higher mammalia, and his descent from some ancestral form common to man and the anthropoid apes. The evidence of such descent appears to me to be overwhelming and conclusive. Again, as to the cause and method of such descent and modification, we may admit, at all events provisionally, that the laws of variation and natural selection, acting through the struggle for existence and the continual need of more perfect adaptation to the physical and biological environments, may have brought about, first that perfection of bodily structure in which he is so far above all other animals, and in co-ordination with it the larger and more developed brain, by means of which he has been able to utilise that structure in the more and more complete subjection of the whole animal and vegetable kingdoms to his service.

But this is only the beginning of Mr. Darwin's work, since he goes on to discuss the moral nature and mental faculties of man, and derives these too by gradual modification and development from the lower animals. Although, perhaps, nowhere distinctly formulated, his whole argument tends to the conclusion that man's entire nature and all his faculties, whether moral, intellectual, or spiritual, have been derived from their rudiments in the lower animals, in the same manner and by the action of the same general laws as his physical structure has been derived. As this conclusion appears to me not to be supported by adequate evidence, and to be directly opposed to many well-ascertained facts, I propose to devote a brief space to its discussion.

The Argument from Continuity.

Mr. Darwin's mode of argument consists in showing that the rudiments of most, if not of all, the mental and moral faculties of man can be detected in some animals. The manifestations of intelligence, amounting in some cases to distinct acts of reasoning, in many animals, are adduced as exhibiting in a much less degree the intelligence and reason of man. Instances of curiosity, imitation, attention, wonder, and memory are given; while examples are also adduced which may be interpreted as proving that animals exhibit kindness to their fellows, or manifest pride, contempt, and shame. Some are said to have the rudiments of language, because they utter several different sounds, each of which has a definite meaning to their fellows or to their young; others the rudiments of arithmetic, because they seem to count and remember up to three, four, or even five. A sense of beauty is imputed to them on account of their own bright colours or the use of coloured objects in their nests; while dogs, cats, and horses are said to have imagination, because they appear to be disturbed by dreams. Even some distant approach to the rudiments of religion is said to be found in the deep love and complete submission of a dog to his master.[1]

Turning from animals to man, it is shown that in the lowest savages many of these faculties are very little advanced from the condition in which they appear in the higher animals; while others, although fairly well exhibited, are yet greatly inferior to the point of development they have reached in civilised races. In particular, the moral sense is said to have been developed from the social instincts of savages, and to depend mainly on the enduring discomfort produced by any action which excites the general disapproval of the tribe. Thus, every act of an individual which is believed to be contrary to the interests of the tribe, excites its unvarying disapprobation and is held to be immoral; while every act, on the other hand, which is, as a rule, beneficial to the tribe, is warmly and constantly approved, and is thus considered to be right or moral. From the mental struggle, when an act that would benefit self is injurious to the tribe, there arises conscience; and thus the social instincts are the foundation of the moral sense and of the fundamental principles of morality.[2]

The question of the origin and nature of the moral sense and of conscience is far too vast and complex to be discussed

[1] For a full discussion of all these points, see *Descent of Man*, chap. iii.

[2] *Descent of Man*, chap. iv.

here, and a reference to it has been introduced only to complete the sketch of Mr. Darwin's view of the continuity and gradual development of all human faculties from the lower animals up to savages, and from savage up to civilised man. The point to which I wish specially to call attention is, that to prove continuity and the progressive development of the intellectual and moral faculties from animals to man, is not the same as proving that these faculties have been developed by natural selection; and this last is what Mr. Darwin has hardly attempted, although to support his theory it was absolutely essential to prove it. Because man's physical structure has been developed from an animal form by natural selection, it does not necessarily follow that his mental nature, even though developed *pari passu* with it, has been developed by the same causes only.

To illustrate by a physical analogy. Upheaval and depression of land, combined with sub-aerial denudation by wind and frost, rain and rivers, and marine denudation on coast-lines, were long thought to account for all the modelling of the earth's surface not directly due to volcanic action; and in the early editions of Lyell's *Principles of Geology* these are the sole causes appealed to. But when the action of glaciers was studied and the recent occurrence of a glacial epoch demonstrated as a fact, many phenomena—such as moraines and other gravel deposits, boulder clay, erratic boulders, grooved and rounded rocks, and Alpine lake basins—were seen to be due to this altogether distinct cause. There was no breach of continuity, no sudden catastrophe; the cold period came on and passed away in the most gradual manner, and its effects often passed insensibly into those produced by denudation or upheaval; yet none the less a new agency appeared at a definite time, and new effects were produced which, though continuous with preceding effects, were not due to the same causes. It is not, therefore, to be assumed, without proof or against independent evidence, that the later stages of an apparently continuous development are necessarily due to the same causes only as the earlier stages. Applying this argument to the case of man's intellectual and moral nature, I propose to show that certain definite portions of it could not have been developed by variation and natural selection alone, and that, therefore, some other influence, law, or agency is required to account for them. If this can be clearly shown for any one or more of the special faculties of intellectual man, we shall be justified in assuming that the same unknown cause or power may have had a much wider influence, and may have profoundly influenced the whole course of his development.

The Origin of the Mathematical Faculty.

We have ample evidence that, in all the lower races of man, what may be termed the mathematical faculty is, either absent, or, if present, quite unexercised. The Bushmen and the Brazilian Wood-Indians are said not to count beyond two. Many Australian tribes only have words for one and two, which are combined to make three, four, five, or six, beyond which they do not count. The Damaras of South Africa only count to three; and Mr. Galton gives a curious description of how one of them was hopelessly puzzled when he had sold two sheep for two sticks of tobacco each, and received four sticks in payment. He could only find out that he was correctly paid by taking two sticks and then giving one sheep, then receiving two sticks more and giving the other sheep. Even the comparatively intellectual Zulus can only count up to ten by using the hands and fingers. The Ahts of North-West America count in nearly the same manner, and most of the tribes of South America are no further advanced.[1] The Kaffirs have great herds of cattle, and if one is lost they miss it immediately, but this is not by counting, but by noticing the absence of one they know; just as in a large family or a school a boy is missed without going through the process of counting. Somewhat higher races, as the Esquimaux, can count up to twenty by using the hands and the feet; and other races get even further than this by saying "one man" for twenty, "two men" for forty, and so on, equivalent to our rural mode of reckoning by scores. From the fact that so many of the existing savage races can only count to four or five, Sir John Lubbock thinks it improbable that our earliest ancestors could have counted as high as ten.[2]

[1] Lubbock's *Origin of Civilisation*, fourth edition, pp. 434-440; Tylor's *Primitive Culture*, chap. vii.

[2] It has been recently stated that some of these facts are erroneous, and that some Australians can keep accurate reckoning up to 100, or more, when

When we turn to the more civilised races, we find the use of numbers and the art of counting greatly extended. Even the Tongas of the South Sea islands are said to have been able to count as high as 100,000. But mere counting does not imply either the possession or the use of anything that can be really called the mathematical faculty, the exercise of which in any broad sense has only been possible since the introduction of the decimal notation. The Greeks, the Romans, the Egyptians, the Jews, and the Chinese had all such cumbrous systems, that anything like a science of arithmetic, beyond very simple operations, was impossible; and the Roman system, by which the year 1888 would be written MDCCCLXXXVIII, was that in common use in Europe down to the fourteenth or fifteenth centuries, and even much later in some places. Algebra, which was invented by the Hindoos, from whom also came the decimal notation, was not introduced into Europe till the thirteenth century, although the Greeks had some acquaintance with it; and it reached Western Europe from Italy only in the sixteenth century.[1] It was, no doubt, owing to the absence of a sound system of numeration that the mathematical talent of the Greeks was directed chiefly to geometry, in which science Euclid, Archimedes, and others made such brilliant discoveries. It is, however, during the last three centuries only that the civilised world appears to have become conscious of the possession of a marvellous faculty which, when supplied with the necessary tools in the decimal notation, the elements of algebra and geometry, and the power of rapidly communicating discoveries and ideas by the art of printing, has developed to an extent, the full grandeur of which can be appreciated only by those who have devoted some time (even if unsuccessfully) to the study.

The facts now set forth as to the almost total absence of mathematical faculty in savages and its wonderful development in quite recent times, are exceedingly suggestive, and in regard

required. But this does not alter the general fact that many low races, including the Australians, have no words for high numbers and never require to use them. If they are now, with a little practice, able to count much higher, this indicates the possession of a faculty which could not have been developed under the law of utility only, since the absence of words for such high numbers shows that they were neither used nor required.

[1] Article Arithmetic in *Eng. Cyc. of Arts and Sciences.*

to them we are limited to two possible theories. Either prehistoric and savage man did not possess this faculty at all (or only in its merest rudiments); or they did possess it, but had neither the means nor the incitements for its exercise. In the former case we have to ask by what means has this faculty been so rapidly developed in all civilised races, many of which a few centuries back were, in this respect, almost savages themselves; while in the latter case the difficulty is still greater, for we have to assume the existence of a faculty which had never been used either by the supposed possessors of it or by their ancestors.

Let us take, then, the least difficult supposition—that savages possessed only the mere rudiments of the faculty, such as their ability to count, sometimes up to ten, but with an utter inability to perform the very simplest processes of arithmetic or of geometry—and inquire how this rudimentary faculty became rapidly developed into that of a Newton, a La Place, a Gauss, or a Cayley. We will admit that there is every possible gradation between these extremes, and that there has been perfect continuity in the development of the faculty; but we ask, What motive power caused its development?

It must be remembered we are here dealing solely with the capability of the Darwinian theory to account for the origin of the *mind*, as well as it accounts for the origin of the *body* of man, and we must, therefore, recall the essential features of that theory. These are, the preservation of useful variations in the struggle for life; that no creature can be improved beyond its necessities for the time being; that the law acts by life and death, and by the survival of the fittest. We have to ask, therefore, what relation the successive stages of improvement of the mathematical faculty had to the life or death of its possessors; to the struggles of tribe with tribe, or nation with nation; or to the ultimate survival of one race and the extinction of another. If it cannot possibly have had any such effects, then it cannot have been produced by natural selection.

It is evident that in the struggles of savage man with the elements and with wild beasts, or of tribe with tribe, this faculty can have had no influence. It had nothing to do with the early migrations of man, or with the conquest and extermination of weaker by more powerful peoples. The Greeks did not successfully resist the Persian invaders by any aid from their few mathematicians, but by military training, patriotism, and self-sacrifice. The barbarous conquerors of the East, Timurlane and Gengkhis Khan, did not owe their success to any superiority of intellect or of mathematical faculty in themselves or their followers. Even if the great conquests of the Romans were, in part, due to their systematic military organisation, and to their skill in making roads and encampments, which may, perhaps, be imputed to some exercise of the mathematical faculty, that did not prevent them from being conquered in turn by barbarians, in whom it was almost entirely absent. And if we take the most civilised peoples of the ancient world—the Hindoos, the Arabs, the Greeks, and the Romans, all of whom had some amount of mathematical talent—we find that it is not these, but the descendants of the barbarians of those days—the Celts, the Teutons, and the Slavs—who have proved themselves the fittest to survive in the great struggle of races, although we cannot trace their steadily growing success during past centuries either to the possession of any exceptional mathematical faculty or to its exercise. They have indeed proved themselves, to-day, to be possessed of a marvellous endowment of the mathematical faculty; but their success at home and abroad, as colonists or as conquerors, as individuals or as nations, can in no way be traced to this faculty, since they were almost the last who devoted themselves to its exercise. We conclude, then, that the present gigantic development of the mathematical faculty is wholly unexplained by the theory of natural selection, and must be due to some altogether distinct cause.

The Origin of the Musical and Artistic Faculties.

These distinctively human faculties follow very closely the lines of the mathematical faculty in their progressive development, and serve to enforce the same argument. Among the lower savages music, as we understand it, hardly exists, though they all delight in rude musical sounds, as of drums, tom-toms, or gongs; and they also sing in monotonous chants. Almost exactly as they advance in general intellect, and in the arts of social life, their appreciation of music appears to rise in proportion; and we find among them rude stringed instruments and whistles, till, in Java, we have regular bands of skilled performers probably the successors of Hindoo musicians of the age before the Mahometan conquest. The Egyptians are believed to have been the earliest musicians, and from them the Jews and the Greeks, no doubt, derived their knowledge of the art; but it seems to be admitted that neither the latter nor the Romans knew anything of harmony or of the essential features of modern music.[1] Till the fifteenth century little progress appears to have been made in the science or the practice of music; but since that era it has advanced with marvellous rapidity, its progress being curiously parallel with that of mathematics, inasmuch as great musical geniuses appeared suddenly among different nations, equal in their possession of this special faculty to any that have since arisen.

As with the mathematical, so with the musical faculty, it is impossible to trace any connection between its possession and survival in the struggle for existence. It seems to have arisen as a *result* of social and intellectual advancement, not as a *cause*; and there is some evidence that it is latent in the lower races, since under European training native military bands have been formed in many parts of the world, which have been able to perform creditably the best modern music.

The artistic faculty has run a somewhat different course, though analogous to that of the faculties already discussed. Most savages exhibit some rudiments of it, either in drawing or carving human or animal figures; but, almost without exception, these figures are rude and such as would be executed by the ordinary inartistic child. In fact, modern savages are, in this respect hardly equal to those prehistoric men who represented the mammoth and the reindeer on pieces of horn or bone. With any advance in the arts of social life, we have a corresponding advance in artistic skill and taste, rising very high in the art of Japan and India, but culminating in the marvellous sculpture of the best period of Grecian history. In the Middle Ages art was chiefly manifested in

[1] See "History of Music," in *Eng. Cyc.*, Science and Arts Division.

ecclesiastical architecture and the illumination of manuscripts, but from the thirteenth to the fifteenth centuries pictorial art revived in Italy and attained to a degree of perfection which has never been surpassed. This revival was followed closely by the schools of Germany, the Netherlands, Spain, France, and England, showing that the true artistic faculty belonged to no one nation, but was fairly distributed among the various European races.

These several developments of the artistic faculty, whether manifested in sculpture, painting, or architecture, are evidently outgrowths of the human intellect which have no immediate influence on the survival of individuals or of tribes, or on the success of nations in their struggles for supremacy or for existence. The glorious art of Greece did not prevent the nation from falling under the sway of the less advanced Roman; while we ourselves, among whom art was the latest to arise, have taken the lead in the colonisation of the world, thus proving our mixed race to be the fittest to survive.

Independent Proof that the Mathematical, Musical, and Artistic Faculties have not been Developed under the Law of Natural Selection.

The law of Natural Selection or the survival of the fittest is, as its name implies, a rigid law, which acts by the life or death of the individuals submitted to its action. From its very nature it can act only on useful or hurtful characteristics, eliminating the latter and keeping up the former to a fairly general level of efficiency. Hence it necessarily follows that the characters developed by its means will be present in all the individuals of a species, and, though varying, will not vary very widely from a common standard. The amount of variation we found, in our third chapter, to be about one-fifth or one-sixth of the mean value—that is, if the mean value were taken at 100, the variations would reach from 80 to 120, or somewhat more, if very large numbers were compared. In accordance with this law we find, that all those characters in man which were certainly essential to him during his early stages of development, exist in all savages with some approach to equality. In the speed of running, in bodily strength, in skill with weapons, in acuteness of vision, or in power of following a trail, all are fairly proficient, and the differences of endowment do not probably exceed the limits of variation in animals above referred to. So, in animal instinct or intelligence, we find the same general level of development. Every wren makes a fairly good nest like its fellows; every fox has an average amount of the sagacity of its race; while all the higher birds and mammals have the necessary affections and instincts needful for the protection and bringing-up of their offspring.

But in those specially developed faculties of civilised man which we have been considering, the case is very different. They exist only in a small proportion of individuals, while the difference of capacity between these favoured individuals and the average of mankind is enormous. Taking first the mathematical faculty, probably fewer than one in a hundred really possess it, the great bulk of the population having no natural ability for the study, or feeling the slightest interest in it.[1] And if we attempt to measure the amount of variation in the faculty itself between a first-class mathematician and the ordinany run of people who find any kind of calculation confusing and altogether devoid of interest, it is probable that the former could not be estimated at less than a hundred times the latter, and perhaps a thousand times would more nearly measure the difference between them.

The artistic faculty appears to agree pretty closely with the mathematical in its frequency. The boys and girls who, going beyond the mere conventional designs of children, draw what they *see*, not what they *know* to be the shape of things; who naturally sketch in perspective, because it is thus they see objects; who see, and represent in their sketches, the light and shade as well as the mere outlines of objects; and who can draw recognisable sketches of every one they know, are certainly very few compared with those who are totally incap-

[1] This is the estimate furnished me by two mathematical masters in one of our great public schools of the proportion of boys who have any special taste or capacity for mathematical studies. Many more, of course, can be drilled into a fair knowledge of elementary mathematics, but only this small proportion possess the natural faculty which renders it possible for them ever to rank high as mathematicians, to take any pleasure in it, or to do any original mathematical work.

able of anything of the kind. From some inquiries I have made in schools, and from my own observation, I believe that those who are endowed with this natural artistic talent do not exceed, even if they come up to, one per cent of the whole population.

The variations in the amount of artistic faculty are certainly very great, even if we do not take the extremes. The gradations of power between the ordinary man or woman "who does not draw," and whose attempts at representing any object, animate or inanimate, would be laughable, and the average good artist who, with a few bold strokes, can produce a recognisable and even effective sketch of a landscape, a street, or an animal, are very numerous; and we can hardly measure the difference between them at less than fifty or a hundred fold.

The musical faculty is undoubtedly, in its lower forms, less uncommon than either of the preceding, but it still differs essentially from the necessary or useful faculties in that it is almost entirely wanting in one-half even of civilised men. For every person who draws, as it were instinctively, there are probably five or ten who sing or play without having been taught and from mere innate love and perception of melody and harmony.[1] On the other hand, there are probably about as many who seem absolutely deficient in musical perception, who take little pleasure in it, who cannot perceive discords or remember tunes, and who could not learn to sing or play with any amount of study. The gradations, too, are here quite as great as in mathematics or pictorial art, and the special faculty of the great musical composer must be reckoned many hundreds or perhaps thousands of times greater than that of the ordinary "unmusical" person above referred to.

It appears then, that, both on account of the limited number of persons gifted with the mathematical, the artistic, or the musical faculty, as well as from the enormous variations in its development, these mental powers differ widely from those which are essential to man, and are, for the most part, common to him and the lower animals; and that they could

[1] I am informed, however, by a music master in a large school that only about one per cent have real or decided musical talent, corresponding curiously with the estimate of the mathematicians.

not, therefore, possibly have been developed in him by means of the law of natural selection.

We have thus shown, by two distinct lines of argument, that faculties are developed in civilised man which, both in their mode of origin, their function, and their variations, are altogether distinct from those other characters and faculties which are essential to him, and which have been brought to their actual state of efficiency by the necessities of his existence. And besides the three which have been specially referred to, there are others which evidently belong to the same class. Such is the metaphysical faculty, which enables us to form abstract conceptions of a kind the most remote from all practical applications, to discuss the ultimate causes of things, the nature and qualities of matter, motion, and force, of space and time, of cause and effect, of will and conscience. Speculations on these abstract and difficult questions are impossible to savages, who seem to have no mental faculty enabling them to grasp the essential ideas or conceptions; yet whenever any race attains to civilisation, and comprises a body of people who, whether as priests or philosophers, are relieved from the necessity of labour or of taking an active part in war or government, the metaphysical faculty appears to spring suddenly into existence, although, like the other faculties we have referred to, it is always confined to a very limited proportion of the population.

In the same class we may place the peculiar faculty of wit and humour, an altogether natural gift whose development appears to be parallel with that of the other exceptional faculties. Like them, it is almost unknown among savages, but appears more or less frequently as civilisation advances and the interests of life become more numerous and more complex. Like them, too, it is altogether removed from utility in the struggle for life, and appears sporadically in a very small percentage of the population; the majority being, as is well known, totally unable to say a witty thing or make a pun even to save their lives.[1]

[1] In the latter part of his essay on Heredity (pp. 91-93 of the volume of *Essays*), Dr. Weismann refers to this question of the origin of "talents" in man, and, like myself, comes to the conclusion that they could not be developed

The Interpretation of the Facts.

The facts now set forth prove the existence of a number of mental faculties which either do not exist at all or exist in a very rudimentary condition in savages, but appear almost suddenly and in perfect development in the higher civilised races. These same faculties are further characterised by their sporadic character, being well developed only in a very small proportion of the community; and by the enormous amount of variation in their development, the higher manifestations of them being many times—perhaps a hundred or a thousand times—stronger than the lower. Each of these characteristics is totally inconsistent with any action of the law of natural selection in the production of the faculties referred to; and the facts, taken in their entirety, compel us to recognise some origin for them wholly distinct from that which has served to account for the animal characteristics—whether bodily or mental—of man.

under the law of natural selection. He says: "It may be objected that, in man, in addition to the instincts inherent in every individual, special individual predispositions are also found, of such a nature that it is impossible they can have arisen by individual variations of the germ-plasm. On the other hand, these predispositions—which we call talents—cannot have arisen through natural selection, because life is in no way dependent on their presence, and there seems to be no way of explaining their origin except by an assumption of the summation of the skill attained by exercise in the course of each single life. In this case, therefore, we seem at first sight to be compelled to accept the transmission of acquired characters." Weismann then goes on to show that the facts do not support this view; that the mathematical, musical, or artistic faculties often appear suddenly in a family whose other members and ancestors were in no way distinguished; and that even when hereditary in families, the talent often appears at its maximum at the commencement or in the middle of the series, not increasing to the end, as it should do if it depended in any way on the transmission of acquired skill. Gauss was not the son of a mathematician, nor Handel of a musician, nor Titian of a painter, and there is no proof of any special talent in the ancestors of these men of genius, who at once developed the most marvellous pre-eminence in their respective talents. And after showing that such great men only appear at certain stages of human development, and that two or more of the special talents are not unfrequently combined in one individual, he concludes thus—

"Upon this subject I only wish to add that, in my opinion, talents do not appear to depend upon the improvement of any special mental quality by continued practice, but they are the expression, and to a certain extent the bye-product, of the human mind, which is so highly developed in all directions."

It will, I think, be admitted that this view hardly accounts for the existence of the highly peculiar human faculties in question.

The special faculties we have been discussing clearly point to the existence in man of something which he has not derived from his animal progenitors—something which we may best refer to as being of a spiritual essence or nature, capable of progressive development under favourable conditions. On the hypothesis of this spiritual nature, superadded to the animal nature of man, we are able to understand much that is otherwise mysterious or unintelligible in regard to him, especially the enormous influence of ideas, principles, and beliefs over his whole life and actions. Thus alone we can understand the constancy of the martyr, the unselfishness of the philanthropist, the devotion of the patriot, the enthusiasm of the artist, and the resolute and persevering search of the scientific worker after nature's secrets. Thus we may perceive that the love of truth, the delight in beauty, the passion for justice, and the thrill of exultation with which we hear of any act of courageous self-sacrifice, are the workings within us of a higher nature which has not been developed by means of the struggle for material existence.

It will, no doubt, be urged that the admitted continuity of man's progress from the brute does not admit of the introduction of new causes, and that we have no evidence of the sudden change of nature which such introduction would bring about. The fallacy as to new causes involving any breach of continuity, or any sudden or abrupt change, in the effects, has already been shown; but we will further point out that there are at least three stages in the development of the organic world when some new cause or power must necessarily have come into action.

The first stage is the change from inorganic to organic, when the earliest vegetable cell, or the living protoplasm out of which it arose, first appeared. This is often imputed to a mere increase of complexity of chemical compounds; but increase of complexity, with consequent instability, even if we admit that it may have produced protoplasm as a chemical compound, could certainly not have produced *living* protoplasm—protoplasm which has the power of growth and of reproduction, and of that continuous process of development which has resulted in the marvellous variety and complex organisation of the whole vegetable kingdom. There is in all this something quite beyond and apart from chemical changes, however complex; and it has been well said that the first vegetable cell was a new thing in the world, possessing altogether new powers—that of extracting and fixing carbon from the carbon-dioxide of the atmosphere, that of indefinite reproduction, and, still more marvellous, the power of variation and of reproducing those variations till endless complications of structure and varieties of form have been the result. Here, then, we have indications of a new power at work, which we may term *vitality*, since it gives to certain forms of matter all those characters and properties which constitute Life.

The next stage is still more marvellous, still more completely beyond all possibility of explanation by matter, its laws and forces. It is the introduction of sensation or consciousness, constituting the fundamental distinction between the animal and vegetable kingdoms. Here all idea of mere complication of structure producing the result is out of the question. We feel it to be altogether preposterous to assume that at a certain stage of complexity of atomic constitution, and as a necessary result of that complexity alone, an *ego* should start into existence, a thing that *feels*, that is *conscious* of its own existence. Here we have the certainty that something new has arisen, a being whose nascent consciousness has gone on increasing in power and definiteness till it has culminated in the higher animals. No verbal explanation or attempt at explanation—such as the statement that life is the result of the molecular forces of the protoplasm, or that the whole existing organic universe from the amœba up to man was latent in the fire-mist from which the solar system was developed—can afford any mental satisfaction, or help us in any way to a solution of the mystery.

The third stage is, as we have seen, the existence in man of a number of his most characteristic and noblest faculties, those which raise him furthest above the brutes and open up possibilities of almost indefinite advancement. These faculties could not possibly have been developed by means of the same laws which have determined the progressive development of the organic world in general, and also of man's physical organism.[1]

[1] For an earlier discussion of this subject, with some wider applications, see the author's *Contributions to the Theory of Natural Selection*, chap. x.

These three distinct stages of progress from the inorganic world of matter and motion up to man, point clearly to an unseen universe—to a world of spirit, to which the world of matter is altogether subordinate. To this spiritual world we may refer the marvellously complex forces which we know as gravitation, cohesion, chemical force, radiant force, and electricity, without which the material universe could not exist for a moment in its present form, and perhaps not at all, since without these forces, and perhaps others which may be termed atomic, it is doubtful whether matter itself could have any existence. And still more surely can we refer to it those progressive manifestations of Life in the vegetable, the animal, and man—which we may classify as unconscious, conscious, and intellectual life,—and which probably depend upon different degrees of spiritual influx. I have already shown that this involves no necessary infraction of the law of continuity in physical or mental evolution; whence it follows that any difficulty we may find in discriminating the inorganic from the organic, the lower vegetable from the lower animal organisms, or the higher animals from the lowest types of man, has no bearing at all upon the question. This is to be decided by showing that a change in essential nature (due, probably, to causes of a higher order than those of the material universe) took place at the several stages of progress which I have indicated; a change which may be none the less real because absolutely imperceptible at its point of origin, as is the change that takes place in the curve in which a body is moving when the application of some new force causes the curve to be slightly altered.

Concluding Remarks.

Those who admit my interpretation of the evidence now adduced—strictly scientific evidence in its appeal to facts which are clearly what ought *not* to be on the materialistic theory—will be able to accept the spiritual nature of man, as not in any way inconsistent with the theory of evolution, but as dependent on those fundamental laws and causes which furnish the very materials for evolution to work with. They will also be relieved from the crushing mental burthen imposed upon those who—maintaining that we, in common with the rest of nature, are but products of the blind eternal forces of the universe, and believing also that the time must come when the sun will lose his heat and all life on the earth necessarily cease—have to contemplate a not very distant future in which all this glorious earth—which for untold millions of years has been slowly developing forms of life and beauty to culminate at last in man—shall be as if it had never existed; who are compelled to suppose that all the slow growths of our race struggling towards a higher life, all the agony of martyrs, all the groans of victims, all the evil and misery and undeserved suffering of the ages, all the struggles for freedom, all the efforts towards justice, all the aspirations for virtue and the wellbeing of humanity, shall absolutely vanish, and, "like the baseless fabric of a vision, leave not a wrack behind."

As contrasted with this hopeless and soul-deadening belief, we, who accept the existence of a spiritual world, can look upon the universe as a grand consistent whole adapted in all its parts to the development of spiritual beings capable of indefinite life and perfectibility. To us, the whole purpose, the only *raison d'être* of the world—with all its complexities of physical structure, with its grand geological progress, the slow evolution of the vegetable and animal kingdoms, and the ultimate appearance of man—was the development of the human spirit in association with the human body. From the fact that the spirit of man—the man himself—*is* so developed, we may well believe that this is the only, or at least the best, way for its development; and we may even see in what is usually termed "evil" on the earth, one of the most efficient means of its growth. For we know that the noblest faculties of man are strengthened and perfected by struggle and effort; it is by unceasing warfare against physical evils and in the midst of difficulty and danger that energy, courage, self-reliance, and industry have become the common qualities of the northern races; it is by the battle with moral evil in all its hydra-headed forms, that the still nobler qualities of justice and mercy and humanity and self-sacrifice have been steadily increasing in the world. Beings thus trained and strengthened by their surroundings, and possessing latent faculties capable of such noble development, are surely destined for a higher and more permanent existence; and we may confidently believe with our greatest living poet—

That life is not as idle ore,
But iron dug from central gloom,
And heated hot with burning fears,
And dipt in baths of hissing tears,
And batter'd with the shocks of doom
To shape and use.

We thus find that the Darwinian theory, even when carried out to its extreme logical conclusion, not only does not oppose, but lends a decided support to, a belief in the spiritual nature of man. It shows us how man's body may have been developed from that of a lower animal form under the law of natural selection; but it also teaches us that we possess intellectual and moral faculties which could not have been so developed, but must have had another origin; and for this origin we can only find an adequate cause in the unseen universe of Spirit.

INDEX

A

B

C

D

E

F

G

H

I

J

K

N

O

P

R

T

U

Y

Z

THE END